MATHEMATICS RESEARCH DEVELOPMENTS

CURRENT RESEARCH TOPICS IN GALOIS GEOMETRY

Mathematics Research Developments

Additional books in this series can be found on Nova's website
under the Series tab.

Additional E-books in this series can be found on Nova's website
under the E-books tab.

Cryptography, Steganography and Data Security

Additional books in this series can be found on Nova's website
under the Series tab.

Additional E-books in this series can be found on Nova's website
under the E-books tab.

MATHEMATICS RESEARCH DEVELOPMENTS

CURRENT RESEARCH TOPICS IN GALOIS GEOMETRY

LEO STORME
AND
JAN DE BEULE
EDITORS

Nova Science Publishers, Inc.
New York

For permission to use material from this book please contact us:
Telephone 631-231-7269; Fax 631-231-8175
Web Site: http://www.novapublishers.com

Library of Congress Cataloging-in-Publication Data

Storme, Leo.
Current research topics on Galois geometrics / Leo Storme and Jan de Beule.
p. cm.
Includes index.
ISBN 978-1-61209-523-3 (harcover)
1. Galois theory. 2. Geometry, Algebraic. I. Beule, Jan de. II. Title.
QA214.S76 2011
516'.11--dc22
2011003567

Published by Nova Science Publishers, Inc. † New York

Contents

PREFACE

Galois geometry is in our mind a field of mathematics that deals with structures living in projective spaces over a finite field. This is a very rough description, but as with any field in mathematics, its borders and contents are not clearly defined.

Several facts have influenced the list of topics that are covered by the chapters in this volume. A wide list of topics are *fundamental* in the sense that many results in Galois geometry rely on them, not only in the field itself, but also in the wider field of finite (incidence) geometry. We especially think of those structures living in a projective space that are used to build models of interesting (non-classical) incidence structures. Topics that are related to Galois geometry and the study of some of its substructures, but that can also be seen as research topics in algebra, have been included. This brings us to the list of *related topics*. Two fields play a special role: *coding theory* and *cryptography*. The reason, for us, that they play a special role, is that research in these fields does not only use results from Galois geometry, but is also inspiring and influencing research in Galois geometry. Therefore, the list of covered topics has a rather large intersection with these two fields, but we took care that the links with Galois geometry were always undoubtedly present.

With these ideas in mind, we can survey the list of topics present in the different chapters. The first two chapters each discuss a variety of substructures in Galois geometry. They are specifically intended for readers wishing to obtain a broad overview of Galois geometry. In particular, Chapter 1 presents results on *classical objects* in the projective space $\mathrm{PG}(n,q)$, such as arcs, ovals, ovoids and unitals, and Chapter 2 presents results on *substructures of classical polar spaces*. Polar spaces are incidence structures described by different axioms than projective spaces, but the classical ones are represented completely by symplectic, quadratic and sesquilinear forms on a projective space, and as such, their study fits completely in the field of Galois geometry. Results on (partial) ovoids, (partial) spreads, covers and blocking sets are presented. The next two chapters discuss specific substructures in projective spaces. Chapter 3 discusses results on *blocking sets in projective spaces*. Blocking sets occur within many problems in Galois geometry, thus giving them a central place within Galois geometry. Chapter 4 presents results on *large caps in projective spaces*; here a topic is discussed which has a well-known link to coding theory, i.e., to the cap-codes. The next two chapters diverge a bit from the four preceding chapters to introduce two important related topics. Chapter 5 is entitled *The polynomial method in Galois geometries*, discusses a powerful technique within Galois geometry, and presents results that have a strong algebraic nature, but that have immediate consequences for some of the mentioned structures in projective spaces, especially blocking sets of projective spaces. Chapter 6 presents results on finite semifields; these are algebraic structures that are related to e.g. spreads of projective spaces. Also this chapter has a more algebraic nature, but its connections with Galois geometry are clearly described. These first 6 chapters can be thought of as the part of the collection that deals with fundamentals of Galois geometry.

Chapter 7 presents results on *codes over ring geometries*. Chapter 8 presents results on *linear codes*, and in particular on those geometric objects related to linear codes. Similarly, Chapter 9 presents *applications of Galois Geometry to Cryptology*. Finally, Chapter 10 presents results on *LDPC codes* and on their links to Galois geometry. These last four chapters can be thought of as the part of the collection that deals with applications, but as we explained, many of the topics presented here have influenced and inspired research in Galois geometry, and this fact can be found throughout these chapters.

Undoubtedly, (many) more topics could have been included in this volume.

During the editorial process, cross references between chapters originated, but each chapter of this volume is a self contained paper, and it can be read independently of the others.

We assume that the reader is familiar with the basic concepts of Galois Geometry. A thorough introduction to Galois geometry can be found in the three fundamental volumes of Hirschfeld on Galois geometry, of which Thas is the co-author for the third volume; [1–3].

The aim and hope of this collected volume on Galois geometry is to give the readers a survey of current important research topics in Galois geometry, describing to them the main results, main techniques and ideas that led to these results, and to present them open problems for future research. We hope that the chapters of this collected work inspire and motivate the readers to contribute to Galois geometry, and encourage them to continue or initiate research on Galois geometry. There is something for everybody's taste in Galois geometry!

Jan De Beule and Leo Storme
April 10, 2010

References

[1] J. W. P. Hirschfeld, *Finite projective spaces of three dimensions*, Oxford Mathematical Monographs, The Clarendon Press Oxford University Press, New York, 1985. Oxford Science Publications.

[2] ——, *Projective geometries over finite fields*, Oxford Mathematical Monographs, The Clarendon Press Oxford University Press, New York, second ed., 1998.

[3] J. W. P. Hirschfeld and J. A. Thas, *General Galois geometries*, Oxford Mathematical Monographs, The Clarendon Press Oxford University Press, New York, 1991. Oxford Science Publications.

In: Current Research Topics in Galois Geometry
Editors: J. De Beule and L. Storme

ISBN: 978-1-61209-523-3

Chapter 1

CONSTRUCTIONS AND CHARACTERIZATIONS OF CLASSICAL SETS IN PG(n,q)

***Frank De Clerck*[1,*] *and Nicola Durante*[2,†]**
[1] Ghent University, Department of Mathematics, Krijgslaan
Gent, Belgium
[2] Dipartimento di Matematica e Applicazioni Caccioppoli,
Università di Napoli "Federico II", Complesso di Monte S. Angelo–Edificio T.,
via Cintia, Napoli

Abstract

In this article we are interested in characterization theorems of the point sets of classical objects such as conics, quadrics, Hermitian varieties and (Baer) subgeometries in terms of their intersection with respect to subspaces. We will give some constructions of sets that have the same type of intersection with subspaces as the classical example.

Key Words: Quadrics, Conics, Maximal arcs, Hermitian Varieties, Subgeometries, Unitals, Ovoids.

AMS Subject Classification: 51E, 05B.

1. Introduction

A set $\mathcal{K}$ of points of PG(n,q) is said to be a *set of class* $[m_1,\ldots,m_k]_r$, $1 \leq r \leq n-1$, if for every r-dimensional subspace π, $|\pi \cap \mathcal{K}| = m_i$, $1 \leq i \leq k$. It is said to be a *set of type* $(m_1,\ldots,m_k)_r$ if every m_i actually occurs for some r-dimensional space π.

In this article, our main interest will go to sets in PG(n,q) of class $[m_1,\ldots,m_k]_1$ or of class $[m_1,\ldots,m_k]_{n-1}$ for which the number of *intersection numbers* m_i is small and such that there is a "classical" example known. Actually, if the dimension r of the intersecting subspace is clear, we often will omit the index r in this notation.

*E-mail address: fdc@cage.ugent.be
†E-mail address: ndurante@unina.it

The classical examples we have in mind are mostly sets of absolute points of an orthogonal or Hermitian polarity. We will assume that the reader is familiar with the concept of classical polar spaces, but see e.g. [41] for more details. For this article we can restrict to the following definitions and notations.

A point P of a subset $\mathcal{K}$ of the point set of $\mathrm{PG}(n,q)$ is a *singular* point of $\mathcal{K}$ if all lines on P intersect $\mathcal{K}$ either in 1 or in $q+1$ points. The point set $\mathcal{K}$ is *singular* if it has a singular point.

A *quadric* (sometimes called a *hyperquadric* or *quadratic variety*) $\mathcal{Q}$ in $\mathrm{PG}(n,q)$ is a variety that can be described by a quadratic form

$$Q(x_0,x_1,\ldots,x_n)=\sum_{i,j=0}^{n} a_{i,j}x_ix_j.$$

If q is odd and the quadric is non-singular, the points on the quadric can be regarded as the set of absolute points of an orthogonal polarity. Quadrics in $\mathrm{PG}(2,q)$ are called *conics*.

A *Hermitian variety* of $\mathrm{PG}(n,q^2)$ is the set of absolute points of a non-degenerate unitary polarity. If $n=2$, it is called a *Hermitian curve*, if $n=3$, it is called a *Hermitian surface*.

A *subgeometry of order* p^t of $\mathrm{PG}(n,q)$, $q=p^h$, $t|h$, is the projective subgeometry induced by a subset of points of $\mathrm{PG}(n,q)$ whose coordinates, with respect to a suitable frame, are in $\mathrm{GF}(p^t)$. If h is even, a subgeometry of order $p^{h/2}$ of $\mathrm{PG}(n,q)$ is called a *Baer subgeometry*; in particular a *Baer subline* if $n=1$ and a *Baer subplane* if $n=2$. We remark that Hermitian varieties of $\mathrm{PG}(1,q^2)$ and Baer sublines are the same objects.

Let $\mathcal{K}$ be any set of points in $\mathrm{PG}(n,q)$, and embed $\mathrm{PG}(n,q)$ as a hyperplane Σ_∞ in $\mathrm{PG}(n+1,q)$. The point-line geometry $T_n^*(\mathcal{K})$, called the *linear representation* of $\mathcal{K}$, is constructed as follows: the points of the geometry $T_n^*(\mathcal{K})$ are the points of the affine space $\mathrm{AG}(n+1,q)=\mathrm{PG}(n+1,q)\setminus\Sigma_\infty$ and the lines of the geometry $T_n^*(\mathcal{K})$ are the lines of $\mathrm{PG}(n+1,q)$ not in Σ_∞ meeting Σ_∞ in a point of $\mathcal{K}$. The point graph of the geometry $T_n^*(\mathcal{K})$ is denoted by $\Gamma(\mathcal{K})$.

Let $\mathcal{K}$ be a set of points in $\mathrm{PG}(n,q)$ which is of type $(m_1,m_2)_{n-1}$, also called a *two-character set* with *characters* m_1,m_2; then the following nice theorem by Delsarte is commonly known.

Theorem 1.1 ([47]). *Let* $\mathcal{K}=\{P_i : i=1,2,\ldots,|\mathcal{K}|\}$, *where each* P_i *is an element of* $\mathrm{GF}(q)^{n+1}$, *be a two-character set in* $\mathrm{PG}(n,q)$, *with characters* m_1,m_2. *If* $\mathcal{K}$ *generates* $\mathrm{PG}(n,q)$, *then*

1. *the graph* $\Gamma(\mathcal{K})$ *is a strongly regular graph;*
2. *the code* $\{(x\cdot P_1,x\cdot P_2,\ldots,x\cdot P_{|\mathcal{K}|}) : x\in\mathrm{GF}(q)^{n+1}\}$ *is a linear two-weight* $[|\mathcal{K}|,n+1]$-*code with weights* $|\mathcal{K}|-m_1,|\mathcal{K}|-m_2$;
3. *the set* $\mathcal{D}=\{v\in\mathrm{GF}(q)^{n+1} : \langle v\rangle\in\mathcal{K}\}$ *is a* $\{\lambda_1,\lambda_2\}$-*difference set over* $\mathrm{GF}(q)$, *for some* $\{\lambda_1,\lambda_2\}$.

See the overview by Calderbank and Kantor [36] for a comprehensive survey of two-character sets, two-weight codes, $\{\lambda_1,\lambda_2\}$-difference sets and the related strongly regular graphs. See also [103].

2. Classical Sets with Few Intersection Numbers in PG$(2,q)$

2.1. Conics, Ovals and Hyperovals

A *k-arc* $\mathcal{K}$ in PG$(2,q)$ is a set of k points which is of class $[0,1,2]$. It is immediately clear that $|\mathcal{K}| \leq q+2$. A $(q+2)$-arc is called a *hyperoval* and can only exist if q is even, an example being a conic $\mathcal{C}$ together with its nucleus N, also called a *regular hyperoval* (or *hyperconic*). A $(q+1)$-arc in PG$(2,q)$ is called an *oval*. Assume q is even; take a regular hyperoval $\mathcal{C} \cup \{N\}$ and delete any point P different from the nucleus N, then this yields an oval $(\mathcal{C} \cup \{N\}) \setminus \{P\}$ also called a *pointed conic*. It has canonical form $\{(1,t,\sqrt{t}) : t \in \mathrm{GF}(q)\} \cup \{(0,0,1)\}$ and cannot be a conic if $q \geq 8$ as two different conics have at most 4 points in common.

In a very well known theorem, Segre [104] proves however that every $(q+1)$-arc in PG$(2,q)$, q odd, is a conic. The method of proof of Segre's theorem is ingenious. We may take three points of the oval to be $P_1(1,0,0)$, $P_2(0,1,0)$, and $P_3(0,0,1)$ and if $P(a_0,a_1,a_2)$ is a further point on the oval and $x_1 = \lambda_0 x_2$, $x_2 = \lambda_1 x_0$, $x_0 = \lambda_2 x_1$ are the three secants PP_1, PP_2, PP_3, then immediately $\lambda_0\lambda_1\lambda_2 = 1$. Since the product of all non-zero elements in the field is -1, it will follow (known as *The Lemma of Tangents*) that for the tangents at P_1, P_2, P_3 being $x_1 = k_0 x_2$, $x_2 = k_1 x_0$, $x_0 = k_2 x_1$, it holds that $k_0 k_1 k_2 = -1$. From this follows that the inscribed triangle and its circumscribed triangle are perspective with respect to the center $(1, k_0 k_1, -k_1)$. It follows generally that every inscribed triangle and its circumscribed triangle are perspective. Using this relation Segre proves that the coordinates of P satisfy a quadratic equation, hence describe a conic $\mathcal{C}$.

2.1.1. Known Hyperovals

A hyperoval $\mathcal{O}$ in PG$(2,q)$ $(q = 2^h, h > 1)$ contains at least 4 points, no three of which are collinear. Without any restriction we may assume that $\mathcal{O}$ passes through the four points $(1,0,0), (0,1,0), (0,0,1)$ and $(1,1,1)$, which implies that it is completely determined by its affine points $(x,y,1)$. We define $y = f(x)$ if and only if $(x,y,1)$ is a point of $\mathcal{O}$. It is easily seen that $f(x)$ is a permutation polynomial which is called an *o-polynomial*.

A lot of the known examples can be described by an o-polynomial of the form $f(x) = x^k$, also called a *monomial o-polynomial*. Define $\mathcal{D}(h) = \{k \| x^k \text{ is an o-polynomial over } \mathrm{GF}(2^h)\}$.

Theorem 2.1. *If $k \in \mathcal{D}(h)$ then $1/k, 1-k, 1/(1-k), k/(k-1)$ and $(k-1)/k$ (all taken modulo $2^h - 1$) are also elements of $\mathcal{D}(h)$ and yield projectively equivalent hyperovals.*

We give a short description of the known elements in $\mathcal{D}(h)$; the related hyperovals are called *monomial hyperovals*.

1. It is clear that $2 \in \mathcal{D}(h)$, for all h and gives the regular hyperoval. Actually it is known that if $h \leq 3$, every hyperoval in PG$(2,2^h)$ is a regular hyperoval.

2. It was proved by Segre [105] that $2^i \in \mathcal{D}(h)$ if and only if $\gcd(i,h) = 1$. These hyperovals are called *translation hyperovals* since they admit as an automorphism group, a group of translations acting transitively on the affine points of the hyperoval.

When $i \neq 1, h-1$, these hyperovals are not equivalent to regular hyperovals and examples exist for $h \geq 5$, but $h \neq 6$.

3. Another class of monomial hyperovals is given by $f(x) = x^6$, in case h is odd. These hyperovals were also discovered by Segre [108] in 1962, see also [109] for more details.

4. Let σ and γ be automorphisms of $\mathrm{GF}(2^h)$, h odd, such that $\gamma^4 \equiv \sigma^2 \equiv 2 \pmod{2^h - 1}$ then Glynn [70] proved that $\gamma + \sigma$ and $3\sigma + 4$ are elements of $\mathcal{D}(h)$.

Remarks

1. Glynn [71] has checked by computer the possible values for monomial o-polynomials, given h, and from his search follows that no other monomial o-polynomials exist for $h \leq 28$.

2. There are several hyperovals known which are not of the monomial type, but it should take us too far to go into this. We refer to the nice electronic overview of Cherowitzo [38] for all updated information on hyperovals.

3. The smallest plane that can contain an irregular hyperoval, is the plane $\mathrm{PG}(2,16)$, which contains up to isomorphism exactly one irregular hyperoval, the Lunelli-Sce hyperoval [89]. Its automorphism group has order 144 and is acting transitively on the points of the hyperoval. For a more detailed discussion on this hyperoval, we refer again to Bill Cherowitzo's hyperoval webpage [38], where also a full description of the 6 non-equivalent hyperovals in $\mathrm{PG}(2,32)$ is given.

4. For more details on the codes related to hyperovals we refer to e.g. [85].

2.1.2. Characterization Theorems of Conics and Related Sets

The theorem by Segre stipulates that a set of size $q+1$ and of type $(0,1,2)$ in a Desarguesian projective plane $\mathrm{PG}(2,q)$, q odd, is a conic $\mathcal{C}$. A point P in the plane not on the conic $\mathcal{C}$ is called an *interior point* of $\mathcal{C}$ if it is incident with no tangent to $\mathcal{C}$ while it is called an *external point* of $\mathcal{C}$ if it is incident with exactly two tangents to $\mathcal{C}$. A conic $\mathcal{C}$ has $\frac{1}{2}q(q+1)$ external points and $\frac{1}{2}q(q-1)$ interior points. The polar line of an external point with respect to the conic $\mathcal{C}$ is a secant to $\mathcal{C}$, while the polar line of an interior point with respect to $\mathcal{C}$ is an exterior line to $\mathcal{C}$.

The set of external points of a conic in $\mathrm{PG}(2,q)$, q odd, is clearly a set of type $\left(\frac{1}{2}(q-1), \frac{1}{2}(q+1), q\right)$. One can wonder whether this characterizes this set indeed. The following theorem from 1983 gives the status of knowledge until 2007.

Theorem 2.2 ([62]). *If $\mathcal{K}$ is a set in $\mathrm{PG}(2,q)$, q odd, which is of type $\left(\frac{1}{2}(q-1), \frac{1}{2}(q+1), q\right)$ and $|\mathcal{K}| < q(q+1)/2 + q/5$, then $|\mathcal{K}| = q(q+1)/2$ and $\mathcal{K}$ is the set of external points of a conic.*

In 2007, the theorem has been improved as follows.

Theorem 2.3 ([42]). *If $\mathcal{K}$ is a set in $\mathrm{PG}(2,q)$, q odd, which is of type $\left(\frac{1}{2}(q-1),\frac{1}{2}(q+1),q\right)$, then $|\mathcal{K}|=q(q+1)/2$ and $\mathcal{K}$ is the set of external points of a conic.*

The set of internal points with respect to a conic in $\mathrm{PG}(2,q)$, q odd, is a set of type $\left(0,\frac{1}{2}(q-1),\frac{1}{2}(q+1)\right)$. Several theorems, trying to characterize the set of internal points of a conic have been proved in the 80's. See for instance [1] and the following theorem.

Theorem 2.4 ([61]). *If $\mathcal{K}$ is a set in $\mathrm{PG}(2,q)$, q odd, which is of type $\left(0,\frac{1}{2}(q-1),\frac{1}{2}(q+1)\right)$, then $(q^2-2q-1)/2<|\mathcal{K}|<(q^2+1)/2$.*

The following theorem gives however the best characterization.

Theorem 2.5 ([42]). *If $\mathcal{K}$ is a set in $\mathrm{PG}(2,q)$, q odd, which is of type $\left(0,\frac{1}{2}(q-1),\frac{1}{2}(q+1)\right)$, then $|\mathcal{K}|=q(q-1)/2$ and $\mathcal{K}$ is the set of interior points of a conic.*

Remarks

It is not difficult to prove the following results as a corollary of the above theorems.

1. A set of type $\left(1,\frac{1}{2}(q+1),\frac{1}{2}(q+3)\right)$ in $\mathrm{PG}(2,q)$, q odd, is the union of a non-degenerate conic $\mathcal{C}$ and its internal points.
2. A set of type $\left(\frac{1}{2}(q+1),\frac{1}{2}(q+3),q+1\right)$ in $\mathrm{PG}(2,q)$, q odd, is the union of a non-degenerate conic $\mathcal{C}$ and its external points.
3. Actually, in [42] all sets of class $\left[0,\frac{1}{2}(q-1),\frac{1}{2}(q+1),q\right]$ in $\mathrm{PG}(2,q)$, q odd, are classified, so also their complements being sets of class $\left[1,\frac{1}{2}(q+1),\frac{1}{2}(q+3),q+1\right]$.

2.2. Maximal Arcs

2.2.1. Introduction

A $\{k;d\}$*-arc* $\mathcal{K}$, in a finite projective plane of order q, is a non-empty proper subset of k points such that some line meets $\mathcal{K}$ in d points, but no line meets $\mathcal{K}$ in more than d points. For given q and d, an easy counting argument shows that $k\le q(d-1)+d$. If equality holds, $\mathcal{K}$ is called a *maximal arc of degree d*. A maximal arc $\mathcal{K}$ can be defined as a non-empty, proper subset of points of the projective plane such that every line meets the set in 0 or d points, for some d. Trivial examples are the following ones.

- Any point of a projective plane of order q is a maximal $\{1;1\}$-arc in that plane.
- An affine plane of order q in a projective plane of order q is a $\{q^2;q\}$-arc.

For the remainder we will discard these trivial examples. A point of the plane not on the maximal arc $\mathcal{K}$ is incident with $q+1-q/d$ lines each intersecting $\mathcal{K}$ in d points. Hence, if $\mathcal{K}$ is a maximal arc of degree d, then d should divide q. Moreover, from this observation follows that the set of external lines to the maximal arc (i.e. the lines not intersecting $\mathcal{K}$) constitutes a maximal arc $\mathcal{K}'$ of degree q/d in the dual projective plane. Hence, any maximal arc $\mathcal{K}$ of degree d in $\mathrm{PG}(2,q)$ yields another maximal arc $\mathcal{K}'$, also called the *dual* *maximal arc of degree q/d*, in $\mathrm{PG}(2,q)$. Examples of non-trivial maximal arcs in even characteristic planes are known since 1969 (the Denniston maximal arcs, see further for the construction). It has been conjectured by several authors that non-trivial maximal arcs could not exist in $\mathrm{PG}(2,q)$, q odd. It has taken more than 25 years to prove this.

Theorem 2.6 ([11]). *No non-trivial maximal arcs exist in* $\mathrm{PG}(2,q)$ *when q is odd.*

For the polynomial technique used to prove this result and for a more recent and more general theorem we refer to [10]. Hence, from now on we may assume that q and d are powers of 2. In the next section we will describe the known examples and give some characterization theorems.

2.2.2. The Known Constructions of Maximal Arcs

We will describe the constructions of maximal arcs, known so far. The oldest construction is due to Denniston [48], while the most recent construction is due to Mathon [90]. Both constructions are algebraic while the two constructions of Thas [116, 117] are more geometric. The Denniston maximal arcs are a special case of those of Mathon type, and hence we will start with this more general class of maximal arcs of Mathon type.

The Construction by R. Mathon

Let Tr be the usual absolute trace map from $\mathrm{GF}(q)$ onto $\mathrm{GF}(2)$. Represent the points of $\mathrm{PG}(2,q)$ via homogeneous coordinates (a,b,c), the lines as triples $[u,v,w]$ over $\mathrm{GF}(q)$ and the incidence by the usual inner product $au+bv+cw=0$.

For $\alpha,\beta \in \mathrm{GF}(q)$, with $\mathrm{Tr}(\alpha\beta)=1$, and $\lambda \in \mathrm{GF}(q)$, define $F_{\alpha,\beta,\lambda}$ to be the conic

$$F_{\alpha,\beta,\lambda}=\{(x,y,z):\alpha x^2+xy+\beta y^2+\lambda z^2=0\}$$

and let $\mathcal{F}$ be the union of all such conics. All conics in $\mathcal{F}$ have the point $F_0=(0,0,1)$ as their nucleus.

For given $\lambda\neq\lambda'$, define a composition $F_{\alpha,\beta,\lambda}\oplus F_{\alpha',\beta',\lambda'}=F_{\alpha\oplus\alpha',\beta\oplus\beta',\lambda\oplus\lambda'}$, where the operator $\oplus$ is defined on $\mathrm{GF}(q)\times\mathrm{GF}(q)$ by

$$\alpha\oplus\alpha'=\frac{\alpha\lambda+\alpha'\lambda'}{\lambda+\lambda'},\qquad \beta\oplus\beta'=\frac{\beta\lambda+\beta'\lambda'}{\lambda+\lambda'},\qquad \lambda\oplus\lambda'=\lambda+\lambda'.$$

Lemma 2.7 ([90]). *Two non-degenerate conics $F_{\alpha,\beta,\lambda}$, $F_{\alpha',\beta',\lambda'}$, $\lambda\neq\lambda'$, and their composition $F_{\alpha,\beta,\lambda}\oplus F_{\alpha',\beta',\lambda'}$ are mutually disjoint if* $\mathrm{Tr}((\alpha\oplus\alpha')(\beta\oplus\beta'))=1$.

Given some subset $\mathcal{C}$ of $\mathcal{F}$, we say $\mathcal{C}$ is *closed* if for every $F_{\alpha,\beta,\lambda}\neq F_{\alpha',\beta',\lambda'}\in\mathcal{C}$, $F_{\alpha\oplus\alpha',\beta\oplus\beta',\lambda+\lambda'}\in\mathcal{C}$.

Theorem 2.8 ([90]). *Let $\mathcal{C}$ be a closed set of conics with a common nucleus F_0 in* $\mathrm{PG}(2,q)$, *q even. Then the union of the points of the conics of $\mathcal{C}$ together with F_0 form a maximal arc of degree $|\mathcal{C}|+1$ in* $\mathrm{PG}(2,q)$.

For examples of maximal arcs of Mathon type and more information, we refer to [64], [65], [73], [74], [75], [90].

The Construction by R. Denniston

The maximal arcs of Denniston type are a special case of those of Mathon type. Choose $\alpha \in GF(q)$ such that $Tr(\alpha) = 1$. Let A be a subset of $GF(q)^\star$ such that $H = A \cup \{0\}$ is closed under addition. Then the set of conics $\{F_{\alpha,1,\lambda} : \lambda \in A\}$ together with the nucleus F_0 is the set of points of a maximal arc of degree $|H|$ in $PG(2,q)$, which yields exactly the construction of Denniston [48]. Actually, it is known that the dual of a Denniston maximal arc is again of Denniston type (see for instance [76] for a proof).

The Constructions by J. A. Thas

In 1974, Thas [116] gave the following construction of maximal arcs of degree q in translation planes of order q^2. We first quickly describe the so-called André-Bruck-Bose representation ([8], [29], [30]) of these planes. Let $PG(3,q)$ be embedded as a hyperplane Σ_∞ in $PG(4,q)$. Let $\mathcal{S}$ be a spread of Σ_∞. Then the following incidence structure π of points and lines is an affine plane of order q^2, known as a *translation plane* (with kernel containing $GF(q)$). The points of π are the points of $PG(4,q) \setminus \Sigma_\infty$, the lines of π are the planes of $PG(4,q)$ meeting Σ_∞ in a line of $\mathcal{S}$; incidence is the natural inclusion. The affine plane π can be completed to a projective plane by adding the points at infinity represented by the elements of $\mathcal{S}$. The projective plane is Desarguesian if and only if the spread $\mathcal{S}$ is regular (i.e. the regulus defined by any 3 lines of $\mathcal{S}$ is completely contained in $\mathcal{S}$).

The construction of Thas goes as follows. Let $\mathcal{O}$ be an ovoid and let $\mathcal{S}$ be a spread of Σ_∞ such that each line of $\mathcal{S}$ has exactly one point in common with $\mathcal{O}$. If $X \in PG(4,q) \setminus \Sigma_\infty$, then the union $\mathcal{K}$ of points on the lines joining X and $\mathcal{O}$ forms a maximal $\{q^3 - q^2 + q; q\}$-arc in the translation plane π of order q^2 defined by $\mathcal{S}$. The known ovoids of $PG(3,q)$, q even, are the elliptic quadrics and the Tits ovoids defined for $q = 2^{2e+1}$, $e \geq 1$ (see Section 3.1.3.). If $\mathcal{S}$ is the regular spread and the ovoid $\mathcal{O}$ is the elliptic quadric, then the maximal arc is of Denniston type (see also [76]). If $\mathcal{O}$ is the Tits ovoid and $\mathcal{S}$ is the regular spread, then it is not of Denniston type neither of Mathon type. Hamilton and Penttila [76] found that there are exactly two orbits on Tits ovoids in the stabilizer of a regular spread, yielding two families of Thas maximal arcs of degree q in $PG(2,q^2)$, $q = 2^{2e+1}$, $e \geq 1$, associated with Tits ovoids.

In 1980, Thas [117] employed quadrics and spreads in projective spaces to construct degree q^{t-1} maximal arcs in symplectic translation planes of order q^t. Let $\mathcal{Q}^- = \mathcal{Q}^-(2t-1,q)$ be a non-singular elliptic quadric and let $\mathcal{S}$ be a $(t-1)$-spread in $PG(2t-1,q)$ of which the restriction to $\mathcal{Q}^-$ is a $(t-2)$-spread. Embed $PG(2t-1,q)$ as a hyperplane Σ_∞ in $PG(2t,q)$. If $X \in PG(2t,q) \setminus \Sigma_\infty$ then the union $\mathcal{K}$ of points on the lines joining X and $\mathcal{Q}^-$ form a maximal $\{q^{2t-1} - q^t + q^{t-1}; q^{t-1}\}$-arc in the translation plane π of order $q^{2(t-1)}$ defined by $\mathcal{S}$. We note that for $t = 2$ we obtain the former construction of Thas in which the ovoid is an elliptic quadric. If $\mathcal{S}$ is a spread such that the plane π is Desarguesian then $\mathcal{K}$ is a Denniston maximal arc, i.e. all the Thas maximal arcs of this type in Desarguesian planes are of Denniston type (see also [76]).

Remark

In [21] it has been proved that a $(t-1)$-spread $\mathcal{S}$ in $\mathrm{PG}(2t-1,q)$ of which the restriction to $\mathcal{Q}^-$ is a $(t-2)$-spread cannot exist if q is odd, and hence the construction of Thas does not work indeed for q odd.

2.2.3. Some Characterization Theorems for Maximal Arcs

In this section we will give some characterization theorems that we think are important, but this is of course not the complete list of all characterization theorems.

The following theorem due to V. Abatangelo and B. Larato gives up to our knowledge a first characterization of the maximal arcs of Denniston type that can be seen indeed as the pencil $\mathcal{K} = \bigcup_{\lambda \in H} F_\lambda$ of conics $F_\lambda = F_{\alpha,1,\lambda}$, H an additive subgroup of order d of $\mathrm{GF}(q),+$; $q = 2^h$, $\mathrm{Tr}(\alpha) = 1$.

Theorem 2.9 ([3]). 1. *If $\mathcal{K} = \bigcup_{\lambda \in H} F_\lambda$ is a maximal arc for some subset H of $\mathrm{GF}(q)$, then H must be a subgroup of the additive group of $\mathrm{GF}(q)$.*

2. *If a maximal arc of $\mathrm{PG}(2,q)$, q even, is invariant under a cyclic linear collineation group of order $q+1$, then it is a Denniston arc.*

Actually, the full stabilizer of a Denniston maximal arc has been completely described by Hamilton and Penttila [76].

Theorem 2.10 ([76]). *Let $\mathcal{K}$ be a maximal arc of Denniston type in $\mathrm{PG}(2,2^h)$, $h > 2$, which is of degree d, $2 < d < q/2$. Let H be the additive subgroup of $\mathrm{GF}(q),+$ of order d defining the maximal arc. Define the group G acting on $\mathrm{GF}(2^h)$ by $G = \{x \mapsto ax^\sigma : a \in \mathrm{GF}(2^h)^*, \sigma \in \mathrm{Aut}(\mathrm{GF}(2^{2h}))\}$. Then the collineation stabilizer of $\mathcal{K}$ is isomorphic to $C_{2^h+1} \rtimes G_H$, the semidirect product of a cyclic group of order 2^h+1 with the stabilizer of H in G.*

As far as the Thas maximal arcs of degree q in $\mathrm{PG}(2,q^2)$ are concerned, as already mentioned they are isomorphic to a Denniston maximal arc if the ovoid $\mathcal{O}$ is an elliptic quadric $\mathcal{Q}^-(3,q)$. When the ovoid is the Tits ovoid, then it yields two non-isomorphic maximal arcs which are not of Denniston type. The following theorem gives all information on the stabilizer of these maximal arcs.

Theorem 2.11 ([76]). *There are, up to equivalence under $\mathrm{P\Gamma L}(3,q^2)$, $q = 2^{2e+1}$, $e \geq 1$, two maximal arcs of Thas type in $\mathrm{PG}(2,q^2)$ arising from Tits ovoids. They have stabilizers in $\mathrm{P\Gamma L}(3,q^2)$ given by the semidirect product of a dihedral group of order $4(q \pm (2q)^{\frac{1}{2}} + 1)(q-1)$ by a cyclic group of order $2e+1$.*

Finally, the next theorem is also proved in [76].

Theorem 2.12 ([76]). *Let $\mathcal{K}$ be a non-trivial maximal arc in $\mathrm{PG}(2,q)$, $q > 2$, such that the collineation stabilizer of $\mathcal{K}$ acts transitively on the points of $\mathcal{K}$, then $\mathcal{K}$ is isomorphic to one of the following.*

1. *A regular hyperoval in $\mathrm{PG}(2,2)$ or $\mathrm{PG}(2,4)$, or a Lunelli-Sce hyperoval in $\mathrm{PG}(2,16)$.*

2. *The dual of a translation hyperoval in $\mathrm{PG}(2,q)$ for any even $q = 2^h$.*

We close this section with the following nice theorem

Theorem 2.13 ([74]). *Let $\mathcal{K}$ be a degree d maximal arc in* PG$(2,q)$ *of Mathon type, then $\mathcal{K}$ is of Denniston type if and only if its dual contains a regular hyperoval.*

From this theorem follows that the dual of a *proper Mathon maximal arc* (i.e a Mathon arc that is not of Denniston type) is not of Mathon type.

2.2.4. Maximal Arcs in Small Desarguesian Planes

1. The plane PG$(2,8)$ has, up to isomorphism, only one maximal arc of degree 4; it is of Denniston type and is the dual of the regular hyperoval.

2. The plane PG$(2,16)$ has, up to isomorphism, two maximal arcs of degree 8: the dual of the regular hyperoval which is of Denniston type, and the dual of the Lunelli-Sce hyperoval which is of Mathon type. It has two non-isomorphic maximal arcs of degree 4, both of Denniston type and both self-dual.

3. The plane PG$(2,32)$ has six non-isomorphic hyperovals and hence the same number of maximal arcs of degree 16. The dual of the regular hyperoval is a Denniston arc, while the dual of the Cherowitzo hyperoval is a proper Mathon maximal arc. There is one Denniston arc of degree 4, its dual being a Denniston arc of degree 8. Mathon gives in his original paper [90] a construction of three maximal arcs of degree 8 (and hence of three maximal arcs of degree 4), which are not of Denniston type. It has been proved in [43] that there are no other maximal arcs of Mathon type of degree 8 and moreover a geometric construction of these three maximal arcs of degree 8 of Mathon type has been given.

4. Mathon mentions in his paper [90] that, neglecting the hyperovals and their duals, the known maximal arcs in the plane PG$(2,64)$ are as follows.

   - There are 94 non-isomorphic maximal arcs of degree 16 of Mathon type known, four of them are of Denniston type. Hence, there are also 94 non-isomorphic maximal arcs known of degree 4.
   - There are 71 non-isomorphic maximal arcs of degree 8 known, two of them are of Thas type and are related to the Tits ovoid, and are self-dual, seven of them are of Denniston type and are also self-dual, the others are all of proper Mathon type. Mathon found 31 of them by computer, none of them self-dual, yielding in total 62 maximal arcs of degree 8 of proper Mathon type.

2.3. Hermitian Curves and Unitals

2.3.1. Definitions and Constructions

A *unital* or *Hermitian arc* in any projective plane π of order q^2 is a set $\mathcal{U}$ of q^3+1 points such that every line of the plane contains 1 or $q+1$ points of $\mathcal{U}$. Given a unital $\mathcal{U}$ and a point P off $\mathcal{U}$, there are $q+1$ tangent lines to $\mathcal{U}$ from P giving, as intersection with $\mathcal{U}$, $q+1$ points called the *feet* of P.

Although many examples of unitals in non-Desarguesian planes do exist (and even unitals as $2-(q^3+1,q+1,1)$ designs, non-embeddable in projective planes) we will investigate only unitals in Desarguesian planes. The classical example is the *Hermitian curve*, also called the *classical unital* in $\mathrm{PG}(2,q^2)$ that has as canonical equation $x_0^{q+1}+x_1^{q+1}+x_2^{q+1}=0$.

Let $\mathcal{H}$ be a Hermitian curve of $\mathrm{PG}(2,q^2)$, then on every point of $\mathcal{H}$ there is a unique tangent line. Every line which is not a tangent line meets $\mathcal{H}$ in a Baer subline. The $q+1$ points of the Baer subline $\ell\cap\mathcal{H}$ are the feet of the polar point $\ell^{\perp}$.

Every unital in $\mathrm{PG}(2,4)$ is a Hermitian curve. However, for every $q>2$, there are unitals in $\mathrm{PG}(2,q^2)$ that are not Hermitian curves.

Buekenhout [35] has constructed unitals in translation planes π of order q^2 using the André-Bruck-Bose representation (see Section 2.2.2.). He proved that if $\mathcal{H}$ is a classical unital, then the corresponding set $\mathcal{H}^*$ in $\mathrm{PG}(4,q)$ is either a quadric $Q(4,q)$ intersecting the space Σ_∞ in a regulus of the regular spread $\mathcal{S}$ (if ℓ_∞ is a secant of $\mathcal{H}$) or it is a quadratic cone with vertex a point V on a line t of the regular spread at infinity and base an elliptic quadric meeting Σ_∞ at a point of $t\setminus\{V\}$ (if ℓ_∞ is a tangent line to $\mathcal{H}$). Conversely, if $\mathcal{U}^*$ is an ovoidal cone in $\mathrm{PG}(4,q)$ with base an ovoid $\mathcal{O}$ of a $\mathrm{PG}(3,q)$ meeting Σ_∞ in a tangent plane to $\mathcal{O}$, containing a line t of $\mathcal{S}$ and with vertex a point V on the line t such that $\mathcal{U}^*\cap\Sigma_\infty=t$, then the corresponding set $\mathcal{U}$ of points in $\mathrm{PG}(2,q^2)$ forms a unital which has the line at infinity, say ℓ_∞, as a tangent line. Hence, the construction by Buekenhout gives new unitals for $q=2^{2e+1}, h\geqslant 1$, by choosing $\mathcal{O}$ a Tits ovoid (see Section 3.1.3.).

R. Metz [92] proved that in $\mathrm{PG}(4,q)\setminus\Sigma_\infty$ a conic ℓ^* can be chosen such that ℓ is a set of $q+1$ collinear points of $\mathrm{PG}(2,q^2)\setminus\ell_\infty$ not corresponding to a Baer subline. Such a conic can always be completed to an ovoid (which has to be an elliptic quadric, see Theorem 3.10) giving by Buekenhout's construction a non-classical unital in the plane $\mathrm{PG}(2,q^2)$.

Remarks

1. Note that if the spread $\mathcal{S}$ is not a regular spread, then Buekenhout's construction yields a unital in the translation plane π constructed from $\mathcal{S}$.

2. In the next sections we will use the following standard terminology.

   - A unital coming from Buekenhout's construction using a quadric $Q(4,q)$ will be called a *Buekenhout unital.*
   - A unital coming from Buekenhout's construction using a cone with base an ovoid is called an *ovoidal Buekenhout-Metz unital*; it is called an *orthogonal Buekenhout-Metz unital* if the base is an elliptic quadric. Hence, the classical unital in $\mathrm{PG}(2,q^2)$ is an orthogonal Buekenhout-Metz unital, but from R. Metz [92] it follows that there exist in this plane orthogonal Buekenhout-Metz unitals that are not classical. An ovoidal Buekenhout-Metz unital with base a Tits ovoid is also called a *Buekenhout-Tits unital.*

3. There are no other unitals embedded in $\mathrm{PG}(2,q^2)$ known at this moment.

2.3.2. Characterization Theorems

It is known that every Buekenhout unital in $\mathrm{PG}(2,q^2)$ is classical (see e.g. [14]).

For $q \leq 3$ it is known that every unital embedded in $\mathrm{PG}(2,q^2)$ is an orthogonal Buekenhout-Metz unital. (See [97] for $\mathrm{PG}(2,9)$).

As all ovoids of $\mathrm{PG}(3,q)$, q odd, are elliptic quadrics (see Section 3.1.3.), the orthogonal Buekenhout-Metz unitals are the only possible ovoidal Buekenhout-Metz unitals in $\mathrm{PG}(2,q^2)$, q odd.

One of the first results on unitals in projective planes is due to Tallini-Scafati [114], who proved the following theorem.

Theorem 2.14 ([114]). *Let $\mathcal{U}$ be a unital in a projective plane π of order q^2. The set of tangents to $\mathcal{U}$ forms a unital $\mathcal{U}^d$ in the dual plane π^d; $\mathcal{U}^d$ is called* the dual unital *of $\mathcal{U}$.*

It is well known that the dual of a classical unital is classical. Also the dual of both Buekenhout-Metz and Buekenhout-Tits unitals in $\mathrm{PG}(2,q^2)$ are of the same type [9], [56], [37]. Hence every known unital embedded in $\mathrm{PG}(2,q^2)$ is either an orthogonal Buekenhout-Metz unital or a Buekenhout-Tits unital.

A special class of orthogonal Buekenhout-Metz unitals has been constructed by Hirschfeld and Szőnyi [80] as the union of a partial pencil of conics in $\mathrm{PG}(2,q^2)$, q odd. These are the only known unitals containing conics. They use coordinates to describe these unitals. Later, a similar description in coordinates has been given that we summarize in the following theorem.

Theorem 2.15 ([9], [56]). *Let*

$$U_{\alpha,\beta} = \{(x, \alpha x^2 + \beta x^{q+1} + r, 1) : x \in \mathrm{GF}(q^2), r \in \mathrm{GF}(q)\} \cup \{(0,1,0)\}$$

for some $\alpha, \beta \in \mathrm{GF}(q^2)$ such that $d = (\beta^q - \beta)^2 - 4\alpha^{q+1}$ is a non-square in $\mathrm{GF}(q)$ if q is odd, while $\mathrm{Tr}(\alpha^{q+1}/(\beta^q+\beta)^2) = 0$, $\beta \notin \mathrm{GF}(q)$, if $q > 2$ is even. Then $U_{\alpha,\beta}$ is an orthogonal Buekenhout-Metz unital in $\mathrm{PG}(2,q^2)$.

Coordinates have also been given by Ebert [57] for Buekenhout-Tits unitals obtaining the following theorem.

Theorem 2.16 ([57]). *Let $q = 2^{2e+1}$, $e \geqslant 1$, let $\{1,\delta\}$ be a basis of $\mathrm{GF}(q^2)$ over $\mathrm{GF}(q)$ and let σ be the automorphism of $\mathrm{GF}(q)$ defined by $\sigma : x \mapsto x^{2^{e+1}}$. Let*

$$\mathcal{U} = \{(x_0 + x_1\delta, y_0 + (x_0^{\sigma+2} + x_1^{\sigma} + x_0x_1)\delta, 1) : x_0, x_1, y_0 \in \mathrm{GF}(q)\} \cup \{(0,1,0)\}.$$

Then $\mathcal{U}$ is a Buekenhout-Tits unital in $\mathrm{PG}(2,q^2)$.

Recall that a *blocking set* in a projective plane π_n of order n is a set of points meeting every line and containing no line. It is *minimal* if no proper subset is again a blocking set.

Theorem 2.17 ([31], [34]). *Let $\mathcal{B}$ be a minimal blocking set in π_n. Then*

$$n + \sqrt{n} + 1 \leqslant |\mathcal{B}| \leqslant n\sqrt{n} + 1.$$

Moreover, if $n = q^2$ and $|\mathcal{B}| = n + \sqrt{n} + 1$, then $\mathcal{B}$ is a Baer subplane; if $n = q^2$ and $|\mathcal{B}| = n\sqrt{n} + 1$, then $\mathcal{B}$ is a unital.

Many characterization theorems for the known unitals embedded in $\mathrm{PG}(2,q^2)$ are known. We will just recall a few ones.

Theorem 2.18 ([114]). *Let $\mathcal{U}$ be a unital in π_{q^2}. For a point P_i on $\mathcal{U}$, denote by ℓ_i the tangent line to $\mathcal{U}$ at P_i. Suppose that $\mathcal{U}$ is* reciprocal, *i.e., for every $P_1,P_2,P_3,P_4 \in \mathcal{U}$ no three collinear, $\ell_1 \cap \ell_2 \in P_3P_4$ implies $\ell_3 \cap \ell_4 \in P_1P_2$. Then $\mathcal{U}$ is the set of absolute points of a polarity. Hence, if $\pi_{q^2} = \mathrm{PG}(2,q^2)$, then $\mathcal{U}$ is classical.*

Theorem 2.19 ([88], [59]). *A unital $\mathcal{U}$ in $\mathrm{PG}(2,q^2)$ with $q > 2$ such that every secant line meets $\mathcal{U}$ in a Baer subline is classical.*

This result has been improved by Ball, Blokhuis and O'Keefe in case q is a prime p.

Theorem 2.20 ([12]). *In $\mathrm{PG}(2,p^2)$, p prime, a unital $\mathcal{U}$ such that $p(p^2-2)$ secant lines meet $\mathcal{U}$ in a Baer subline is classical.*

The following theorem is a strong characterization theorem of the Hermitian curve.

Theorem 2.21 ([119]). *A unital $\mathcal{U}$ of $\mathrm{PG}(2,q^2)$ such that the tangents of $\mathcal{U}$ at collinear points of $\mathcal{U}$ are concurrent, is classical.*

The proof of this theorem is based on Segre's "Lemma of tangents", and on Theorem 2.19. The hypothesis of the previous theorem has been weakened by the following theorem.

Theorem 2.22 ([7]). *Let $\mathcal{U}$ be a unital in $\mathrm{PG}(2,q^2)$, $q > 2$. If there are two points $P_1,P_2 \in \mathcal{U}$ with tangent lines ℓ_1,ℓ_2, respectively, such that for all points $Q \in \ell_1 \setminus \{P_1\}$ and $R \in \ell_2 \setminus \{P_2\}$, the corresponding feet are collinear, then $\mathcal{U}$ is classical.*

The following theorems are nice characterization theorems of a classical unital as an ovoidal Buekenhout-Metz unital.

Theorem 2.23 ([17]). *Let $\mathcal{U}$ be an ovoidal Buekenhout-Metz unital with the line l_∞ tangent at P_∞ in $\mathrm{PG}(2,q^2)$. If there is a secant line not through P_∞ that intersects $\mathcal{U}$ in a Baer subline, then $\mathcal{U}$ is classical.*

The next theorem by K. Metsch [91] embeds $\mathrm{PG}(4,q)$ in $\mathrm{PG}(4,q^2)$. It is well known that the form of an elliptic quadric in $\mathrm{PG}(3,q)$ yields a hyperbolic quadric in $\mathrm{PG}(3,q^2)$, and hence the elliptic cone of $\mathrm{PG}(4,q)$, used to construct the orthogonal Buekenhout-Metz unital, becomes a hyperbolic cone. As far as the spread $\mathcal{S}$ in $\mathrm{PG}(3,q)$ is concerned, there exist exactly two disjoint lines L and L' of $\mathrm{PG}(3,q^2)$, missing $\mathrm{PG}(3,q)$, and conjugate under the Baer involution of $\mathrm{PG}(3,q^2)$ fixing $\mathrm{PG}(3,q)$, such that $\mathcal{S}$ is the set of lines (regarded as lines of $\mathrm{PG}(3,q^2)$) intersecting L and L'. These two lines are commonly called the *generator lines* of the regular spread $\mathcal{S}$.

Theorem 2.24 ([91]). *Let $\mathcal{U}$ be an orthogonal Buekenhout-Metz unital. If L is a generator line of the regular spread S, then $\mathcal{U}$ is classical if and only if L lies on the hyperbolic cone (defined by $\mathcal{U}$) in $\mathrm{PG}(4,q^2)$.*

More characterization theorems of orthogonal Buekenhout-Metz unitals are known, see for instance [98] and [118] where they use the method of the so-called field reduction, but it would bring us too far to give these theorems in detail.

In terms of algebraic curves, the following nice characterization theorem for classical unitals is known.

Theorem 2.25 ([79]). *If $\mathcal{U}$ is an algebraic curve of degree $q+1$ with $|\mathcal{U}| \geqslant q^3+1$ and without a linear component in* PG$(2,q^2)$, *then $\mathcal{U}$ is a classical unital.*

For characterization theorems of ovoidal Buekenhout-Metz unitals in terms of intersections with lines, the following theorems are worthwhile to mention.

Theorem 2.26 ([87]). *Let $\mathcal{U}$ be a unital in* PG$(2,q^2)$, *$q>2$, and let ℓ be some tangent line to $\mathcal{U}$. If all Baer sublines having a point on ℓ, intersect $\mathcal{U}$ in $0,1,2$, or $q+1$ points, then $\mathcal{U}$ is an ovoidal Buekenhout-Metz unital.*

Theorem 2.27 ([86]). *Let $\mathcal{U}$ be a unital in* PG$(2,q^2)$, *q odd, and let ℓ be a tangent line to $\mathcal{U}$ at P. Then $\mathcal{U}$ is an ovoidal Buekenhout-Metz unital if and only if for any two lines ℓ_1 and ℓ_2 such that $\ell_1 \cap \ell_2 = P$, there is a Baer subplane $\mathcal{B}$ having ℓ as a secant line and satisfying $\mathcal{B} \cap \mathcal{U} = (\ell_1 \cap \mathcal{U}) \cup (\ell_2 \cap \mathcal{U})$.*

These theorems have been improved as follows.

Theorem 2.28 ([37], [99]). *If $\mathcal{U}$ is a unital of* PG$(2,q^2)$, *$q>2$, containing a point P such that each of the q^2 secant lines through P meets $\mathcal{U}$ in a Baer subline, then $\mathcal{U}$ is an ovoidal Buekenhout-Metz unital.*

The proof ([37] covers the case q even and $q=3$, while [99] covers q odd, $q>3$) is a sequence of lemmas using the André-Bruck-Bose representation for PG$(2,q^2)$, carefully analyzing the set $\mathcal{U}^*$ of points in PG$(4,q)$ corresponding to the unital $\mathcal{U}$.

Regarding the intersection between a unital and a Baer subplane, the following result is known.

Theorem 2.29 ([15], Lemma 2.9). *Let $\mathcal{H}$ be a Hermitian curve and let $\mathcal{B}$ be a Baer subplane in* PG$(2,q^2)$. *Then $\mathcal{H} \cap \mathcal{B}$ is a (possibly degenerate) conic of $\mathcal{B}$. Hence $|\mathcal{H} \cap \mathcal{B}| \in \{1, q+1, 2q+1\}$.*

The converse of this result is also valid as the following theorem proves.

Theorem 2.30 ([16]). *Let $\mathcal{U}$ be a unital in* PG$(2,q^2)$. *If every Baer subplane meets $\mathcal{U}$ in 1, $q+1$ or $2q+1$ points, then $\mathcal{U}$ is classical.*

As a general result regarding the intersection between a unital and a Baer subplane, the following theorem is worthwhile to mention.

Theorem 2.31 ([33], [72]). *Let π be a projective plane of order q^2 containing a unital $\mathcal{U}$ and a Baer subplane $\mathcal{B}$ and let b be the number of lines secant to $\mathcal{B}$ and tangent to $\mathcal{U}$. Then $|\mathcal{B} \cap U| = 2(q+1) - b$.*

Regarding the intersection of an ovoidal Buekenhout-Metz unital and a Baer subplane, the following result is known.

Theorem 2.32 ([16]). *Let $\mathcal{U}$ be a unital tangent to ℓ_∞ in a translation plane π_{q^2} with kernel containing* GF(q). *If every Baer subplane secant to ℓ_∞ meets $\mathcal{U}$ in $1, q+1$ or $2q+1$ points, then $\mathcal{U}$ is an ovoidal Buekenhout-Metz unital.*

O'Nan showed in [95] that the classical unital does not contain a configuration of four distinct lines which meet in six points (called an *O'Nan configuration*).

Wilbrink proved the following theorem.

Theorem 2.33 ([126]). *Let $\mathcal{U}$ be a unital in* $\mathrm{PG}(2,q^2)$*, q even. If*

- *$\mathcal{U}$ contains no O'Nan configuration;*
- *for each secant ℓ of $\mathcal{U}$, point $X \in \mathcal{U} \setminus \ell$ and secant m on X meeting ℓ in a point of $\mathcal{U}$, it holds that if Y is a point of $(m \cap \mathcal{U}) \setminus \{X\}$, then there exists a secant ℓ' distinct from m on Y which meets each secant on X that meets ℓ in a point of $\mathcal{U}$.*

Then $\mathcal{U}$ is classical.

We conclude this section discussing the configurations arising as intersection of two Hermitian curves and with some group theoretical characterizations of the unitals embedded in $\mathrm{PG}(2,q^2)$. First a more general result.

Theorem 2.34 ([20]). *Let $\mathcal{H}$ be a Hermitian curve and let $\mathcal{U}$ be a unital of* $\mathrm{PG}(2,q^2)$*, $q = p^h$, p prime, $h \geq 1$. Then $|\mathcal{H} \cap \mathcal{U}| \equiv 1$ (mod p).*

Theorem 2.35 ([84]). *Let $\mathcal{H}$ and $\mathcal{H}'$ be two distinct Hermitian curves of* $\mathrm{PG}(2,q^2)$*. Then one of the following configurations occurs for $\mathcal{H} \cap \mathcal{H}'$.*

- *A point; a Baer subline; a Kestenband (q^2-q+1)-complete arc.*
- *A set of q^2+1 points on q Baer sublines with a point P in common. The q lines containing these Baer sublines together with the tangent line at P form a dual Baer subline.*
- *A set of q^2+1 points on $q-1$ Baer sublines on lines with a point in common plus two other points such that the full configuration is contained in a dual Baer subline.*
- *A set of q^2+q+1 points on $q+1$ Baer sublines with a point in common on a dual Baer subline.*
- *A set of $(q+1)^2$ points on $q+1$ Baer sublines on lines with a point in common forming a dual Baer subline.*

The point sets with q^2+q+1 and $(q+1)^2$ points together with the point set with q^2+1 points on $q-1$ Baer sublines plus two points have been studied in detail in [49]. By using the André-Bruck-Bose representation of $\mathrm{PG}(2,q^2)$, it is proved that these sets correspond to the elliptic or hyperbolic quadrics or to a quadratic cone in a hyperplane $\Pi \neq \Sigma_\infty$ of $\mathrm{PG}(4,q)$. The other point set with q^2+1 points has been studied in [51] and also corresponds to a quadratic cone in a hyperplane $\Pi \neq \Sigma_\infty$ of $\mathrm{PG}(4,q)$. For a study of the groups stabilizing $\mathcal{H} \cap \mathcal{H}'$, see Giuzzi [67].

On the groups of the known unitals embedded in $\mathrm{PG}(2,q^2)$, we just recall that the group of the classical unital is the unitary group $\mathrm{PGU}(3,q^2)$, it has order $q^3(q^3+1)(q^2-1)$ and acts as a 2-transitive group on the points of the Hermitian curve.

Some group theoretical characterizations of the classical unital are given in the following theorems.

Theorem 2.36 ([19]). *Let $\mathcal{U}$ be a unital in $\mathrm{PG}(2,q^2)$. If the group G of collineations fixing $\mathcal{U}$ is transitive on secant lines to $\mathcal{U}$ and is generated by involutions, then $\mathcal{U}$ is classical.*

Theorem 2.37 ([40]). *Let $\mathcal{U}$ be a unital in $\mathrm{PG}(2,q^2)$ fixed by a Singer subgroup of order q^2-q+1 of $\mathrm{PGL}(3,q^2)$. Then $\mathcal{U}$ is classical.*

The groups of the other orthogonal Buekenhout-Metz unitals and of the Buekenhout-Tits unitals can be found in [15]. Some group theoretical characterizations of both orthogonal Buekenhout-Metz and Buekenhout-Tits unitals in $\mathrm{PG}(2,q^2)$ are given in the following theorems.

Theorem 2.38 ([2]). *Let $\mathcal{U}$ be an ovoidal Buekenhout-Metz unital in $\mathrm{PG}(2,q^2)$. If there is a cyclic group of collineations of order q^2-1 fixing two points of $\mathcal{U}$ and stabilizing $\mathcal{U}$, then $\mathcal{U}$ is classical.*

Theorem 2.39 ([4], [2]). *If $\mathcal{U}$ is a unital in $\mathrm{PG}(2,q^2)$ fixed by a subgroup G of $\mathrm{PGL}(3,q^2)$ such that:*

- *there is a point P of $\mathcal{U}$ fixed by G;*
- *G has a normal subgroup acting transitively on $\mathcal{U}\setminus\{P\}$;*
- *the stabilizer in G of a point $Q\in\mathcal{U}\setminus\{P\}$ has a cyclic subgroup of order $q-1$.*

Then $\mathcal{U}$ is an orthogonal Buekenhout-Metz unital.

Theorem 2.40 ([5]). *Let $\mathcal{U}$ be an ovoidal Buekenhout-Metz unital in $\mathrm{PG}(2,q^2)$. If there is a point $P\in\mathcal{U}$ such that the stabilizer of $\mathcal{U}$ in $\mathrm{PGL}(3,q^2)$ has a subgroup that acts sharply transitive on $\mathcal{U}\setminus\{P\}$, then $\mathcal{U}$ is an orthogonal Buekenhout-Metz unital.*

Theorem 2.41 ([58]). *Let $\mathcal{U}$ be a unital in $\mathrm{PG}(2,q^2)$ which is fixed by a subgroup of $\mathrm{PGL}(3,q^2)$ which is a semidirect product of a group of order q^3 and a group of order $q-1$. Then $\mathcal{U}$ is an orthogonal Buekenhout-Metz unital.*

Very recently along these lines the following theorems have been proved.

Theorem 2.42 ([53]). *Let $\mathcal{U}$ be a unital in $\mathrm{PG}(2,q^2)$ fixed by a subgroup G of $\mathrm{PGL}(3,q^2)$ of elations of order q with center a point A and suppose there is a subgroup of $\mathrm{PGL}(3,q^2)$ of order a divisor of $q-1$ greater than $2(\sqrt{q}-1)$ fixing both A and another point B of $\mathcal{U}\setminus\{A\}$, then $\mathcal{U}$ is an ovoidal Buekenhout-Metz unital with respect to A.*

Theorem 2.43 ([53]). *Suppose that q is either odd or $q\in\{2,4\}$. A unital $\mathcal{U}$ in $\mathrm{PG}(2,q^2)$ is an orthogonal Buekenhout-Metz unital if and only if there exist two distinct points A and B on $\mathcal{U}$ such that there exists a subgroup G of $\mathrm{PGL}(3,q^2)$ of elations with center A of order q and a subgroup of $\mathrm{PGL}(3,q^2)$ fixing both A and B of order a divisor of $q-1$ greater than $2(\sqrt{q}-1)$ stabilizing $\mathcal{U}$.*

Theorem 2.44 ([53]). *Let $\mathcal{U}$ be a unital in $\mathrm{PG}(2,p^2)$, p a prime. The unital $\mathcal{U}$ is an orthogonal Buekenhout-Metz unital if and only if there exists a non-identity elation stabilizing $\mathcal{U}$.*

2.4. Characterizing Subplanes of $\mathrm{PG}(2,q)$

In this section we discuss characterization theorems regarding Baer subplanes and subplanes of smaller order of $\mathrm{PG}(2,q)$ and some related results.

Theorem 2.45 ([113]). *Let $\mathcal{K}$ be a set of type $(1,k)$ in a projective plane π_n of order n. Then $n=q^2, k=q+1$ and $\mathcal{K}$ is either a Baer subplane or a unital.*

As far as the embedding of Baer subplanes is concerned, a more general theorem is known.

Theorem 2.46 ([28]). *Let π_m be a subplane of order m of a projective plane π_n of order n. Then either $m^2=n$, that is π_m is a Baer subplane of π_n, or $m^2+m \leq n$.*

Note that in $\mathrm{PG}(2,q)$, $q=p^h$, there is a subplane $\mathrm{PG}(2,p^t)$ for any $t|h$. Regarding sets of class $[0,1,n]$ in $\mathrm{PG}(2,q)$ the following results are known.

Theorem 2.47 ([82]). *If $\mathcal{K}$ is a point set of type $(0,1,n)$, $n>4$, in $\mathrm{PG}(2,q)$, then $|\mathcal{K}| < q\sqrt{q}+1$.*

Theorem 2.48 ([124]). *If $\mathcal{K}$ is a proper point set of class $[0,1,n]$ in $\mathrm{PG}(2,q)$ with $n \geqslant \sqrt{q}+1$, then $\mathcal{K}$ is one of the following:*

- *either one or n collinear points;*
- *a Baer subplane;*
- *a unital;*
- *a maximal arc.*

Remark

The condition $n \geqslant \sqrt{q}+1$ in the above theorem is certainly necessary. Brouwer constructed in [24] a Steiner triple system on 19 points in $\mathrm{PG}(2,11)$.

What about the configuration arising from the intersection of two Baer subplanes or subplanes of smaller order? The study of the intersection of Baer subplanes in $\mathrm{PG}(2,q^2)$ started in [39] and was carried on in [23], [125]. In these papers many properties regarding the intersection of two Baer subplanes of $\mathrm{PG}(2,q^2)$ were found. The main result was that two Baer subplanes have as many points as lines in common. Later, all possible intersection configurations of two Baer subplanes in $\mathrm{PG}(2,q^2)$ have been determined.

Theorem 2.49 ([111], [112], [83]). *In $\mathrm{PG}(2,q^2)$ let π_q and π'_q be two distinct Baer subplanes, then $\pi_q \cap \pi'_q$ is one of the following configurations:*

- *the empty set;*
- *a point; two points; three points forming a triangle;*
- *a Baer subline ℓ_0 plus possibly a point not on ℓ_0.*

Moreover, all these configurations occur.

Recently, the previous theorem has been generalized for subplanes of any order. Indeed the following theorem has been achieved.

Theorem 2.50 ([52]). *In* $\mathrm{PG}(2,q)$, $q=p^h$, *let* π *be a subplane of order* p^t *and let* π' *be a subplane of order* $p^{t'}$ *with* $t|h$ *and* $t'|h$. *Then* $\pi\cap\pi'$ *is one of the following configurations:*

- *the empty set;*
- *a point; two points; three points forming a triangle;*
- *a subline* ℓ_0 *over the biggest common subfield of* $\mathrm{GF}(p^t)$ *and* $\mathrm{GF}(p^{t'})$ *plus possibly a point not on* ℓ_0.

Moreover, if $t|t'$, *then all these configurations occur.*

Observe that the configurations coming from the intersection of two subplanes of $\mathrm{PG}(2,q), q=p^h$, are of class $[0,1,2,p^s+1]$, for some $s|h$.

3. Classical Sets with Few Intersection Numbers in $\mathrm{PG}(n,q)$, $n\geq 3$

3.1. Quadrics and Quasi-quadrics

3.1.1. Definitions

In this section we are only interested in quadrics with an irreducible quadratic form and which can not be described in fewer variables (also known as *non-singular quadrics*). So we neglect the cones with vertex an m-dimensional space projecting a quadric in an $(n-m-1)$-dimensional space skew to the vertex, as well as the quadrics which degenerate in the union of subspaces. For more details on the general theory of quadrics we refer for instance to [78]. The following projective classification is part of standard knowledge.

Theorem 3.1. *If* $\mathcal{Q}$ *is a non-singular quadric, then it is of one of the following types.*

- $n=2m$, $\mathcal{Q}$ *is called* parabolic, *also denoted by* $Q(2m,q)$ *and the quadratic form is equivalent to the following canonical form*

$$Q(x_0,x_1,\ldots,x_{2m})=x_0^2+x_1x_2+\cdots+x_{2m-1}x_{2m}.$$

- $n=2m-1$, *in which case there are two non-equivalent quadrics.*
  - *The* hyperbolic quadric $Q^+(2m-1,q)$ *with quadratic form equivalent to the following canonical form*

$$Q(x_0,x_1,\ldots,x_{2m-1})=x_0x_1+x_2x_3+\cdots+x_{2m-2}x_{2m-1}.$$

– *The* elliptic quadric $Q^-(2m-1,q)$ *with quadratic form equivalent to the following canonical form*

$$Q(x_0,x_1,\ldots,x_{2m-1}) = f(x_0,x_1)+x_2x_3+\cdots+x_{2m-2}x_{2m-1},$$

with $f(x_0,x_1)$ *an irreducible quadratic form over* $\mathrm{GF}(q)$.

The subspaces on $\mathcal{Q}$ of maximal dimension (also called the *projective dimension of the quadric*) are called the *generators* of the quadric. It is commonly known that the projective dimension of the parabolic quadric $Q(2m,q)$ and of the hyperbolic quadric $Q^+(2m-1,q)$ is $m-1$, while the one of the elliptic quadric $Q^-(2m-1,q)$ is $m-2$.

It is a standard exercise to count the number of points $|\mathcal{Q}|$ of a quadric $\mathcal{Q}$.

$$\begin{aligned}
|Q(2m,q)| &= \frac{q^{2m}-1}{q-1},\\
|Q^+(2m-1,q)| &= \frac{(q^{m-1}+1)(q^m-1)}{q-1},\\
|Q^-(2m-1,q)| &= \frac{(q^{m-1}-1)(q^m+1)}{q-1}.
\end{aligned}$$

If q is odd, the set of points of the quadric $\mathcal{Q}$ in $\mathrm{PG}(n,q)$ can be regarded as the set of absolute points of an orthogonal polarity. If q is even and n is odd, the quadric defines a symplectic polarity, if q and n are both even then no polarity is defined and the tangent hyperplanes to $Q(2m,q)$ all have a unique point in common, the *nucleus* of the parabolic quadric (generalizing the same property for conics in $\mathrm{PG}(2,2^h)$).

3.1.2. Characterization Theorems

The set of points of an elliptic quadric as well as of a hyperbolic quadric in $\mathrm{PG}(2m-1,q)$ is a two-character set with respect to hyperplanes.

Indeed let $Q^-(2m-1,q)$ be an elliptic quadric, then its point set is a set $\mathcal{K}$ of $\frac{(q^{m-1}-1)(q^m+1)}{q-1}$ points such that a hyperplane either intersects the quadric in a parabolic quadric, hence in $\frac{q^{2m-2}-1}{q-1}$ points, or is a tangent hyperplane in which case it intersects $Q^-(2m-1,q)$ in a cone with vertex the tangent point X projecting an elliptic quadric $Q^-(2m-3,q)$, from which follows that such a tangent hyperplane intersects the elliptic quadric in $\frac{q(q^{m-1}+1)(q^{m-2}-1)}{q-1}+1$ points.

The same argument holds for the hyperbolic quadric $Q^+(2m-1,q)$ with point set $\mathcal{K}$ a set of $\frac{(q^{m-1}+1)(q^m-1)}{q-1}$ points such that each hyperplane meets it in either $\frac{q^{2m-2}-1}{q-1}$ or $\frac{q(q^{m-2}+1)(q^{m-1}-1)}{q-1}+1$ points.

Every set $\mathcal{K}$ being of the same type with respect to hyperplanes as the elliptic quadric (respectively hyperbolic quadric) is called in [45] an *elliptic quasi-quadric* (respectively *hyperbolic quasi-quadric*) and is denoted by $\mathcal{K}^-$ (respectively $\mathcal{K}^+$). In that paper the authors construct elliptic and hyperbolic quasi-quadrics. We will give here only one construction.

Let $Q^\varepsilon(2m-1,q)$ be a non-degenerate quadric in $\mathrm{PG}(2m-1,q)$ ($\varepsilon = -$ for the elliptic quadric and $\varepsilon = +$ for the hyperbolic quadric), $m > 2$. Let X be a point of $Q^\varepsilon(2m-1,q)$. The tangent (polar) space $X^\perp$ of X with respect to the quadratic form for $Q^\varepsilon(2m-1,q)$ is then of dimension $2m-2$, and $X^\perp \cap Q^\varepsilon(2m-1,q)$ is the cone $XQ^\varepsilon(2m-3,q)$ with vertex X and base a non-degenerate quadric $Q^\varepsilon(2m-3,q)$ in some subspace Σ_{2m-3} of $X^\perp$ of dimension $2m-3$ disjoint from X.

Let Q' be a (quasi-)quadric in Σ_{2m-3} with the same parameters as $Q^\varepsilon(2m-3,q)$. We then call the set $(Q^\varepsilon(2m-1,q) - \backslash XQ^\varepsilon(2m-3,q)) \cup XQ'$ a *pivoted set* of $Q^\varepsilon(2m-1,q)$ with respect to X. Note that the size of a pivoted set is the same as the size of $Q^\varepsilon(2m-1,q)$.

Theorem 3.2 ([45]). *Every pivoted set with respect to a point of $Q^-(2m-1,q)$ (respectively $Q^+(2m-1,q)$) is an elliptic (respectively hyperbolic) quasi-quadric.*

Remarks

1. If $q=2$, there are more possible constructions of elliptic and hyperbolic quasi-quadrics, see [44] and [45] for more details; in this case these quasi-quadrics give rise to other structures such as symmetric designs with the symmetric difference property, Reed-Muller codes and bent functions.

2. A quasi-quadric in PG$(3,2)$ is a quadric. In PG$(5,2)$ there are five projectively inequivalent quasi-quadrics of elliptic type and seven of hyperbolic type, see [123] for more details.

3. There also exist parabolic quasi-quadrics that are not quadrics; see [45] for more information.

Note that the pivoting construction is breaking up the lines and subspaces on the quadric, moreover one can repeat pivoting as much as one wants, implying that the family of quasi-quadrics is quite wild. However, the following theorems are worthwhile to mention in this context.

Theorem 3.3 ([55]). *Let $\mathcal{K}$ be a set of points in* PG$(3,q)$*, with $|\mathcal{K}| = q^2+q+1$, and suppose that $\mathcal{K}$ contains at least two lines and intersects every plane of* PG$(3,q)$ *in $1, q+1$ or $2q+1$ points. Then $\mathcal{K}$ is a cone projecting an oval in a plane π from a point v not in π.*

Theorem 3.4 ([63]). *Let $\mathcal{K}$ be a set of points in* PG(n,q)*, where $n \geqslant 4$ and $|\mathcal{K}| \geq q^3+q^2+q+1$, such that $\mathcal{K}$ intersects all planes in 1, a, or b points, $b \geq 2q+1$, $\mathcal{K}$ intersects all solids in c, $c+q$, or $c+2q$ points, $c \leq q^2+1$, and there exist solids intersecting $\mathcal{K}$ in c points and in $c+q$ points; then $\mathcal{K}$ is a non-singular quadric of* PG$(4,q)$.

De Winter and Schillewaert proved the following results in the same style.

Theorem 3.5 ([101]).

1. *If a set $\mathcal{K}$ of points in* PG$(4,q)$ *intersects all planes and all solids in the same number of points as quadrics do, then $\mathcal{K}$ is a parabolic quadric $Q(4,q)$.*

2. *If a set $\mathcal{K}$ of points in* PG(n,q)*, $n \geqslant 4$, intersects planes and solids in the same number of points as a quadric of* PG(n,q) *does, then $\mathcal{K}$ is one of the following:*

   (a) the space PG(n,q)*,*

(b) a hyperplane of $\mathrm{PG}(n,q)$*,*

(c) a quadric of $\mathrm{PG}(n,q)$*,*

(d) a cone with vertex an $(n-3)$*-dimensional space and base an oval,* q *even,*

(e) a cone with vertex an $(n-4)$*-dimensional space and base an ovoid,* q *even.*

Theorem 3.6 ([46]). *An elliptic quasi-quadric in* $\mathrm{PG}(n,q)$*,* $n \geq 4$*,* $q > 2$*, or a hyperbolic quasi-quadric in* $\mathrm{PG}(n,q)$*,* $n \geq 3$*,* $q > 2$*, such that it also has the same characters with respect to codimension 2 spaces is a quadric.*

Remark

In cryptography, one is studying for instance maximum non-linear functions. Geometrically, these functions correspond to quasi-quadrics, see for instance [77] for more details.

3.1.3. Ovoids and Generalizations

The elliptic quadric $Q^-(3,q)$ is a set of q^2+1 points, no three on a line. Every set $\mathcal{K}$ in $\mathrm{PG}(n,q)$ with $|\mathcal{K}| = k$ and the property that no three points are on a line is called a *k-cap*. If $n = 2$, a k-cap is a k-arc and this case has already been treated in the beginning of this article. Hence, from now on $n \geq 3$. A line of $\mathrm{PG}(n,q)$ will be called *external*, *tangent*, or *secant* to a cap according to whether it contains zero, one, or two points of the cap. A k-cap of maximal size has been called an *ovaloid* by Segre [106] and if $q > 2$ the maximal size is indeed q^2+1, which was first proved by Bose [22] for q odd, by Seiden [110] for $q = 4$, and by Qvist [100] for $q > 2$ and even. Note that if $q = 2$, the eight points of an affine subspace of $\mathrm{PG}(3,2)$ is a maximal set of points no three on a line. We will discard this case for the rest of the section. If $\mathcal{O}$ is an ovaloid of $\mathrm{PG}(3,q)$, $q > 2$, then for every point P on $\mathcal{O}$, the tangent lines through P are in a plane, the tangent plane, and hence there are q^2+1 tangent planes and the other q^3+q planes intersect $\mathcal{O}$ in an oval. Actually, Tits [120] defined an *ovoid* to be a set $\mathcal{O}$ of points in a projective geometry (not required to be finite nor Desarguesian) such that for any point $P \in \mathcal{O}$ the union of all lines ℓ with $\ell \cap \mathcal{O} = \{P\}$ is a hyperplane. In $\mathrm{PG}(n,q)$ an ovoid can only exist if $n \leq 3$. It is immediate from the definition of an ovoid that in $\mathrm{PG}(3,q)$ it has size q^2+1. Thus an ovoid of $\mathrm{PG}(3,q)$ is an ovaloid for $q > 2$, and we will use the term ovoid for the rest of this section.

The concept of an ovoid has been generalized in many ways; it would bring us too far to discuss also these generalizations. However, a set $\mathcal{O}$ of points (so not necessarily a cap) in $\mathrm{PG}(n,q)$, $n \geq 3$, such that the union of the tangents (1-secants) at each point is in a hyperplane, is called a *semi-ovoid*. It has been proved by Thas [116] that no semi-ovoids exist in $\mathrm{PG}(n,q)$, $n > 3$, and that it is an ovoid if $n = 3$.

One might wonder whether an ovoid of $\mathrm{PG}(3,q)$ is necessarily an elliptic quadric. The answer is affirmative if q is odd, as proved independently by Barlotti [13] and Panella [96]. It is however not the case if q is even. In this case, there is an ovoid known which is not the elliptic quadric; the so-called *Tits ovoid* that exists if $q = 2^{2e+1}$, $e \geq 1$, and it has the following canonical form

$$\{(1, zu + z^{\sigma+2} + u^{\sigma}, z, u) : z, u \in \mathrm{GF}(q)\} \cup \{(0,1,0,0)\},$$

with $\sigma : x \mapsto x^{2^{e+1}}$. One of the motivations of the study of Tits into ovoids of PG$(3,q)$ and his construction of this ovoid in [121], is the fact that the full stabilizer in PGL$(4,q)$ of the ovoid is the simple group of Suzuki, for this reason the ovoid is sometimes also called the *Tits-Suzuki ovoid.* Together with the elliptic quadric $Q^-(3,q)$ they are characterized by the fact that their full stabilizer acts doubly transitive on the ovoid [122].

While the non-tangent plane sections of an elliptic quadric all are conics, those of the Tits ovoid all are *translation ovals* (i.e., ovals invariant under a group E of elations of order q such that all the elations in E have a common axis) which are not conics.

No other ovoid is known and actually the ovoids in PG$(2,2^h)$ are classified for $h \leq 5$. One of the big research issues of the last years is to find whether the elliptic quadric and the Tits ovoid are the only ovoids in PG$(3,q)$ or not.

There are very nice characterization theorems known for the elliptic quadric and the Tits ovoid. We will give a few. The following theorem is a theorem by Brown.

Theorem 3.7 ([27]). *An ovoid of* PG$(3,q)$ *is stabilized by a central collineation if and only if it is an elliptic quadric.*

As already mentioned, Barlotti characterized the elliptic quadrics as the ovoids in PG$(3,q)$, q odd. Actually, he proved a more general theorem, not assuming q odd.

Theorem 3.8 ([13]). *If every non-tangent plane intersects the ovoid $\mathcal{O}$ of* PG$(3,q)$, $q > 2$, *in a conic, then $\mathcal{O}$ is the elliptic quadric.*

Segre improved this theorem in 1959.

Theorem 3.9 ([106]). *An ovoid of* PG$(3,q)$, $q \geq 8$, *which contains at least* $\frac{1}{2}(q^3 - q^2 + 2q)$ *conics must be an elliptic quadric.*

However, also this result has been improved by Brown in 2000 (using the theorem of Barlotti).

Theorem 3.10 ([26]). *An ovoid of* PG$(3,q)$, *q even, such that there is a plane intersecting the ovoid in a conic, is an elliptic quadric.*

So, the question arises what can be said if one of the non-tangent planes of PG$(3,q)$, q even, intersects the ovoid in an oval which is not a conic. One of the theorems we want to mention in this context is the following one.

Theorem 3.11 ([93], [94]). *Suppose that $\mathcal{O}$ is an ovoid in* PG$(3,q)$, *q even.*

1. *$\mathcal{O}$ has a pencil of translation ovals if and only if $\mathcal{O}$ is either an elliptic quadric or a Tits ovoid.*

2. *If each non-tangent plane section is an oval contained in a translation hyperoval, then $\mathcal{O}$ is an elliptic quadric or a Tits ovoid.*

Finally, here is a theorem which is in the same style as Theorem 3.10.

Theorem 3.12 ([25]). *Suppose that $\mathcal{O}$ is an ovoid of* PG$(3,q)$, $q = 2^h$, $h > 1$. *If there is a plane intersecting $\mathcal{O}$ in a pointed conic, then either $q = 4$ and $\mathcal{O}$ is an elliptic quadric, or $q = 8$ and $\mathcal{O}$ is the Tits ovoid.*

3.2. Hermitian Varieties

In this section we discuss characterization theorems regarding Hermitian varieties of $\mathrm{PG}(n,q^2)$. A Hermitian variety of $\mathrm{PG}(n,q^2)$ is a set of type $(0,1,q+1,q^2+1)_1$ and it is a two-character set with respect to hyperplanes. Hermitian varieties have been characterized by using their intersection numbers with lines. The following theorem is a combination of papers by Tallini-Scafati [115], Hirschfeld and Thas [81] and Glynn [69].

Theorem 3.13 ([115], [81], [69]). *Let $\mathcal{K}$ be a non-singular point set of type $(1,r,q^2+1)_1$ in $\mathrm{PG}(n,q^2)$, $n \geqslant 4, q > 2$, such that $3 \leq r \leq q^2-1$ and there is no plane π such that $\pi \cap \mathcal{K}$ is of type (r,q^2+1) in π. Then $\mathcal{K}$ is the point set of a Hermitian variety $H(n,q^2)$.*

Another characterization is using the intersection of a Hermitian variety with planes instead of lines. But this time not just the intersection numbers are required but also the intersection structure.

Theorem 3.14 ([60]). *Let $\mathcal{K}$ be a point set of $\mathrm{PG}(n,q)$ such that every plane section is a (possibly degenerate) Hermitian curve. Then $\mathcal{K}$ is a Hermitian variety.*

Recently, using the intersection numbers with more than one family of subspaces, the following characterizations have been obtained.

Theorem 3.15 ([102]). *Let $\mathcal{K}$ be a non-singular point set of $\mathrm{PG}(n,q^2), n \geqslant 4$, having the same intersection numbers with respect to planes and solids as $H(n,q^2)$. Then $\mathcal{K}$ is the point set of $H(n,q^2)$.*

It is however impossible to characterize Hermitian varieties using just their intersection numbers with respect to hyperplanes since quasi-Hermitian varieties can be constructed in the same way (using pivoting) as was done in an earlier section for quasi-quadrics. Asking however that the point set has also the same intersection numbers with respect to codimension two subspaces the following characterization has been obtained.

Theorem 3.16 ([46]). *Let $\mathcal{K}$ be a point set of $\mathrm{PG}(n,q^2), n \geqslant 3$, having the same intersection numbers with respect to hyperplanes and codimension two subspaces as $H(n,q^2)$. Then $\mathcal{K}$ is the point set of $H(n,q^2)$.*

About the intersection of two Hermitian surfaces $\mathcal{H}$ and $\mathcal{H}'$ of $\mathrm{PG}(3,q^2)$ we recall that in [68] Giuzzi describes all possible intersection configurations $\mathcal{H} \cap \mathcal{H}'$ under the hypothesis that the pencil generated by $\mathcal{H}$ and $\mathcal{H}'$ contains at least one degenerate Hermitian surface (obtaining several possible intersection configurations).

In [107] B. Segre defines two Hermitian surfaces in $\mathrm{PG}(3,q^2)$ to be *permutable* if and only if their associated polarities u, respectively u', commute and he proves the following theorem.

Theorem 3.17 ([107]). *If q is odd and $\mathcal{H}$, $\mathcal{H}'$ are permutable Hermitian surfaces of $\mathrm{PG}(3,q^2)$, then uu' is a projectivity with two skew lines of fixed points, called the fundamental lines of $\mathcal{H}$ and $\mathcal{H}'$.*

A point set of q^2+1 mutually skew lines in $\mathrm{PG}(3,q^2)$ with exactly two transversals is called a *pseudo-regulus*. This notion was introduced by J. Freeman in [66], where he proved that any pseudo-regulus can be extended to a spread of $\mathrm{PG}(3,q^2)$. The set of $(q^2+1)^2$ points covered by a pseudo-regulus is called a *hyperbolic* Q_F*-set* in [50]. It is one of the possible intersection configurations of two Hermitian surfaces. Indeed the following holds.

Theorem 3.18 ([6]). *Let $\mathcal{H}$ and $\mathcal{H}'$ be two permutable Hermitian surfaces of* $\mathrm{PG}(3,q^2)$*, q odd. If the fundamental lines are contained in $\mathcal{H}\cap\mathcal{H}'$, then $\mathcal{H}\cap\mathcal{H}'$ is the point set of a pseudo-regulus.*

The hypotheses in the previous theorem are weakened in [51].

Theorem 3.19 ([51]). *Let $\mathcal{H}$ and $\mathcal{H}'$ be two distinct Hermitian surfaces in* $\mathrm{PG}(3,q^2)$ *with associated polarities u and u', respectively. Suppose that L and M are two skew lines contained in $\mathcal{B}=\mathcal{H}\cap\mathcal{H}'$. Then $\mathcal{B}$ is a hyperbolic Q_F-set (point set of a pseudo-regulus) with transversals L and M if and only if u and u' agree on the points of $L\cup M$.*

Finally the next theorem yields a complete classification for $\mathcal{H}\cap\mathcal{H}'$.

Theorem 3.20 ([54]). *Let $\mathcal{H}$ and $\mathcal{H}'$ be two non-degenerate Hermitian surfaces in* $\mathrm{PG}(3,q^2)$ *and let $\mathcal{B}=\mathcal{H}\cap\mathcal{H}'$. If the Hermitian pencil they generate contains only non-degenerate Hermitian surfaces, then one of the following four cases must occur:*

- *$\mathcal{H}$ contains exactly two skew lines and q^4-1 other points;*
- *$\mathcal{H}$ contains exactly two skew lines L and M, a third line N intersecting both L and M, and q^4-q^2 other points;*
- *$\mathcal{H}$ contains exactly four lines forming a quadrangle and q^4-2q^2+1 other points;*
- *$\mathcal{H}$ is ruled by a pseudo-regulus.*

Moreover, all these cases occur.

Note that the intersection of two Hermitian varieties in $\mathrm{PG}(n,q^2)$ is always a set of class $[0,1,2,q+1,q^2+1]_1$, but very little is known regarding the intersection of two Hermitian varieties of $\mathrm{PG}(n,q^2)$ for $n\geqslant 4$.

3.3. Subgeometries

In Theorem 2.48 we have described the classification of sets of class $[0,1,r]_1$ in $\mathrm{PG}(2,q)$ under the condition that $r\geq\sqrt{q}+1$. Actually, in the same paper Ueberberg has given, under the same condition, the classification of sets of class $[0,1,r]_1$ in $\mathrm{PG}(n,q)$. He proved the following theorem.

Theorem 3.21 ([124]). *Let $\mathcal{K}$ be a proper point set of class $[0,1,r]_1$ in* $\mathrm{PG}(n,q)$ *with $r\geq\sqrt{q}+1$ and such that $\mathcal{K}$ spans the full space. Then it is a Baer subgeometry of* $\mathrm{PG}(n,q)$ *or an affine subspace of* $\mathrm{PG}(n,q)$*.*

The study of the intersection of two Baer subgeometries of $\mathrm{PG}(n,q^2)$ has been carried out in [32] and [111]. In these papers the authors prove that the number of common points of two Baer subgeometries of $\mathrm{PG}(n,q^2)$ equals the number of common hyperplanes (see [18] for a generalization of this result to other subspaces different from hyperplanes). In [111] also all possible intersection configurations in $\mathrm{PG}(3,q^2)$ have been conjectured. Finally in [83] a complete determination of the structure of these intersections has been determined, solving Sved's conjecture in the positive.

Theorem 3.22 ([83]). *Let $\mathcal{B}_1,\dots,\mathcal{B}_k$ be Baer subgeometries of subspaces of $\mathrm{PG}(n,q^2)$. The following statements are equivalent.*

1. *The Baer subgeometries $\mathcal{B}_1,\dots,\mathcal{B}_k$ satisfy the following two conditions:*
   - $k \le q+1$
   - $\langle \mathcal{B}_1,\dots,\mathcal{B}_{i-1},\mathcal{B}_{i+1},\dots,\mathcal{B}_k\rangle \cap \langle \mathcal{B}_i\rangle = \emptyset$, *for every* $i = 1,\dots,k$.
2. *There exist two Baer subgeometries $\mathcal{B}$ and $\mathcal{B}'$ of $\mathrm{PG}(n,q^2)$ such that $\mathcal{B}\cap\mathcal{B}' = \mathcal{B}_1 \cup \dots \cup \mathcal{B}_k$.*

The previous theorem has been recently generalized by determining all possible intersection configurations of any two subgeometries of $\mathrm{PG}(n,q)$.

Theorem 3.23 ([52]). *Let $\mathcal{G}$ and $\mathcal{G}'$ be two subgeometries of $\mathrm{PG}(n,q)$, $q = p^h$, of order p^t and $p^{t'}$ respectively, with $t \le t'$, and let $m = \gcd(t,t')$.*

If $\mathcal{G}\cap\mathcal{G}'$ is non-empty, then $\mathcal{G}\cap\mathcal{G}' = \mathcal{G}_1\cup\dots\cup\mathcal{G}_k$, with $k \le \frac{q-1}{p^{t'}-1}$ and with $\mathcal{G}_1,\dots,\mathcal{G}_k$ subgeometries of order p^m of independent subspaces of $\mathrm{PG}(n,q)$.

In the same paper the authors prove also the vice versa of the last theorem under the assumption $t|t'$.

Theorem 3.24 ([52]). *Let t and t' be two positive divisors of h with $t|t'$. Let $k \le \min\{n+1, \frac{q-1}{p^{t'}-1}\}$ and let $\mathcal{G}_1,\dots,\mathcal{G}_k$ be subgeometries of order p^t of independent subspaces of $\mathrm{PG}(n,q)$. Then there exist two subgeometries $\mathcal{G}$ and $\mathcal{G}'$ of order p^t and $p^{t'}$, respectively, of $\mathrm{PG}(n,q)$ such that $\mathcal{G}\cap\mathcal{G}' = \mathcal{G}_1\cup\dots\cup\mathcal{G}_k$.*

Open Problems

1. Is it possible to find all maximal arcs in $\mathrm{PG}(2,32)$?
2. The dual of a proper Mathon maximal arc in $\mathrm{PG}(2,q)$, q even, is not of Mathon type. Geometrically one can describe this dual arc as an intersection of Denniston arcs, but does there exist an algebraic description of the dual of a proper Mathon arc?
3. The Lunelli-Sce hyperoval in $\mathrm{PG}(2,16)$ as well as the Cherowitzo hyperoval in $\mathrm{PG}(2,32)$ are both duals of proper Mathon arcs. Are there any other proper Mathon arcs of degree $q/2$ in $\mathrm{PG}(2,q)$?
4. What is the minimum number of secant lines being Baer sublines one needs to conclude that a unital $\mathcal{U}$ is a Buekenhout-Metz unital ?

5. Determine all subsets of class $[0,1,r]$ in $\mathrm{PG}(2,q)$ with $r < \sqrt{q}+1$.

6. Remove the hypothesis $t|t'$ in Theorem 2.50.

7. Let $\mathcal{H}$ be a point set of $\mathrm{PG}(n,q^2)$ with $|\mathcal{H}|$ equals to the number of points of a Hermitian variety and such that all hyperplane sections are (possibly degenerate) Hermitian varieties. Is it true that $\mathcal{H}$ is a Hermitian variety?

8. Determine all the possible intersections of two Hermitian varieties of $\mathrm{PG}(n,q^2)$, $n \geqslant 4$.

9. Determine all subsets of class $[0,1,r]_1$ in $\mathrm{PG}(n,q)$, with $r < \sqrt{q}+1$, generating $\mathrm{PG}(n,q)$.

10. Remove the hypothesis $t|t'$ in Theorem 3.24.

11. Let A be a point of a unital $\mathcal{U}$ in $\mathrm{PG}(2,q^2)$. Suppose there is a group of elations of $\mathrm{PG}(2,q^2)$ with center A (and axis the tangent line t_A at A to $\mathcal{U}$) stabilizing $\mathcal{U}$. Is it true that $\mathcal{U}$ is a Buekenhout-Metz unital?

12. Are there other unitals, non isomorphic to Buekenhout-Metz unitals, in $\mathrm{PG}(2,q^2)$?

13. Suppose there is a point P of a unital $\mathcal{U}$ such that for every point of $t_P \setminus \{P\}$ the feet are collinear. Is it true that $\mathcal{U}$ has to be an ovoidal Buekenhout-Metz unital?

References

[1] L. M. Abatangelo and M. Pertichino, *A characterization of sets of class* $[0,1,n-1,n,2n-1]$, Riv. Mat. Univ. Parma (4), 12 (1986), pp. 293–302 (1987).

[2] V. Abatangelo, *On Buekenhout-Metz unitals in* $\mathrm{PG}(2,q^2)$, *q even*, Arch. Math. (Basel), 59 (1992), pp. 197–203.

[3] V. Abatangelo and B. Larato, *A characterization of Denniston's maximal arcs*, Geom. Dedicata, 30 (1989), pp. 197–203.

[4] ——, *A group-theoretical characterization of parabolic Buekenhout-Metz unitals*, Boll. Un. Mat. Ital. A (7), 5 (1991), pp. 195–206.

[5] ——, *A characterization of Buekenhout-Metz unitals in* $\mathrm{PG}(2,q^2)$, *q even*, Geom. Dedicata, 59 (1996), pp. 137–145.

[6] A. Aguglia, A. Cossidente, and G. L. Ebert, *On pairs of permutable Hermitian surfaces*, Discrete Math., 301 (2005), pp. 28–33.

[7] A. Aguglia and G. L. Ebert, *A combinatorial characterization of classical unitals*, Arch. Math. (Basel), 78 (2002), pp. 166–172.

[8] J. André, *Über nicht-Desarguessche Ebenen mit transitiver Translationsgruppe*, Math. Z., 60 (1954), pp. 156–186.

[9] R. D. Baker and G. L. Ebert, *On Buekenhout-Metz unitals of odd order*, J. Combin. Theory Ser. A, 60 (1992), pp. 67–84.

[10] S. Ball, *The polynomial method in Galois geometries*, in Current research topics in Galois geometry, Nova Sci. Publ., New York, 2011, ch. 5, pp. 103–128.

[11] S. Ball, A. Blokhuis, and F. Mazzocca, *Maximal arcs in Desarguesian planes of odd order do not exist*, Combinatorica, 17 (1997), pp. 31–41.

[12] S. Ball, A. Blokhuis, and C. M. O'Keefe, *On unitals with many Baer sublines*, Des. Codes Cryptogr., 17 (1999), pp. 237–252.

[13] A. Barlotti, *Un'estensione del teorema di Segre-Kustaanheimo*, Boll. Un. Mat. Ital. (3), 10 (1955), pp. 498–506.

[14] S. G. Barwick, *A characterization of the classical unital*, Geom. Dedicata, 52 (1994), pp. 175–180.

[15] S. G. Barwick and G. L. Ebert, *Unitals in projective planes*, Springer Monographs in Mathematics, Springer, New York, 2008.

[16] S. G. Barwick, C. M. O'Keefe, and L. Storme, *Unitals which meet Baer subplanes in 1 modulo q points*, J. Geom., 68 (2000), pp. 16–22.

[17] S. G. Barwick and C. T. Quinn, *Generalising a characterisation of Hermitian curves*, J. Geom., 70 (2001), pp. 1–7.

[18] A. Beutelspacher and J. Ueberberg, *On the intersection of Baer subspaces*, Arch. Math. (Basel), 56 (1991), pp. 203–208.

[19] P. Biscarini, *Hermitian arcs of* $\mathrm{PG}(2,q^2)$ *with a transitive collineation group on the set of* $(q+1)$*-secants*, in Proceedings of the conference on combinatorial and incidence geometry: principles and applications (La Mendola, 1982), vol. 7 of Rend. Sem. Mat. Brescia, Milan, 1984, Vita e Pensiero, pp. 111–124.

[20] A. Blokhuis, A. E. Brouwer, and H. A. Wilbrink, *Hermitian unitals are code words*, Discrete Math., 97 (1991), pp. 63–68.

[21] A. Blokhuis, N. Hamilton, and H. A. Wilbrink, *On the non-existence of Thas maximal arcs in odd order projective planes*, European J. Combin., 19 (1998), pp. 413–417.

[22] R. C. Bose, *Mathematical theory of the symmetrical factorial design*, Sankhyā, 8 (1947), pp. 107–166.

[23] R. C. Bose, J. W. Freeman, and D. G. Glynn, *On the intersection of two Baer subplanes in a finite projective plane*, Utilitas Math., 17 (1980), pp. 65–77.

[24] A. E. Brouwer, *A series of separable designs with application to pairwise orthogonal Latin squares*, European J. Combin., 1 (1980), pp. 39–41.

[25] M. R. Brown, *The determination of ovoids of* PG$(3,q)$ *containing a pointed conic*, J. Geom., 67 (2000), pp. 61–72. Second Pythagorean Conference (Pythagoreion, 1999).

[26] ——, *Ovoids of* PG$(3,q), q$ *even, with a conic section*, J. London Math. Soc. (2), 62 (2000), pp. 569–582.

[27] ——, *Ovoids of* PG$(3,q)$ *stabilized by a central collineation*, European J. Combin., 24 (2003), pp. 409–412.

[28] R. H. Bruck, *Difference sets in a finite group*, Trans. Amer. Math. Soc., 78 (1955), pp. 464–481.

[29] R. H. Bruck and R. C. Bose, *The construction of translation planes from projective spaces*, J. Algebra, 1 (1964), pp. 85–102.

[30] ——, *Linear representations of projective planes in projective spaces*, J. Algebra, 4 (1966), pp. 117–172.

[31] A. A. Bruen, *Baer subplanes and blocking sets*, Bull. Amer. Math. Soc., 76 (1970), pp. 342–344.

[32] ——, *Intersection of Baer subgeometries*, Arch. Math. (Basel), 39 (1982), pp. 285–288.

[33] A. A. Bruen and J. W. P. Hirschfeld, *Intersections in projective space. I. Combinatorics*, Math. Z., 193 (1986), pp. 215–225.

[34] A. A. Bruen and J. A. Thas, *Blocking sets*, Geom. Dedicata, 6 (1977), pp. 193–203.

[35] F. Buekenhout, *Existence of unitals in finite translation planes of order* q^2 *with a kernel of order* q, Geom. Dedicata, 5 (1976), pp. 189–194.

[36] R. Calderbank and W. M. Kantor, *The geometry of two-weight codes*, Bull. London Math. Soc., 18 (1986), pp. 97–122.

[37] L. R. A. Casse, C. M. O'Keefe, and T. Penttila, *Characterizations of Buekenhout-Metz unitals*, Geom. Dedicata, 59 (1996), pp. 29–42.

[38] B. Cherowitzo, *Bill Cherowtizo's Hyperoval Page*. `http://www-math.cudenver.edu/~wcherowi/research/hyperoval/hypero.html`, 1999.

[39] J. Cofman, *Baer subplanes in finite projective and affine planes*, Canad. J. Math., 24 (1972), pp. 90–97.

[40] A. Cossidente, G. L. Ebert, and G. Korchmáros, *A group-theoretic characterization of classical unitals*, Arch. Math. (Basel), 74 (2000), pp. 1–5.

[41] J. De Beule, A. Klein, and K. Metsch, *Substructures of finite classical polar spaces*, in Current research topics in Galois geometry, Nova Sci. Publ., New York, 2011, ch. 2, pp. 33–59.

[42] F. De Clerck and N. De Feyter, *A characterization of the sets of internal and external points of a conic*, European J. Combin., 28 (2007), pp. 1910–1921.

[43] F. De Clerck, S. De Winter, and T. Maes, *A geometric approach to Mathon maximal arcs*, J. Combin. Theory Ser. A, 118 (2011), pp. 1196–1211.

[44] F. De Clerck and M. Delanote, *Two-weight codes, partial geometries and Steiner systems*, Des. Codes Cryptogr., 21 (2000), pp. 87–98. Special issue dedicated to Dr. Jaap Seidel on the occasion of his 80th birthday (Oisterwijk, 1999).

[45] F. De Clerck, N. Hamilton, C. M. O'Keefe, and T. Penttila, *Quasi-quadrics and related structures*, Australas. J. Combin., 22 (2000), pp. 151–166.

[46] S. De Winter and J. Schillewaert, *Characterizations of finite classical polar spaces by intersection numbers with hyperplanes and spaces of codimension 2*, Combinatorica, 30 (2010), pp. 25–45.

[47] P. Delsarte, *Weights of linear codes and strongly regular normed spaces*, Discrete Math., 3 (1972), pp. 47–64.

[48] R. H. F. Denniston, *Some maximal arcs in finite projective planes*, J. Combinatorial Theory, 6 (1969), pp. 317–319.

[49] G. Donati and N. Durante, *Some subsets of the Hermitian curve*, European J. Combin., 24 (2003), pp. 211–218.

[50] ——, *A subset of the Hermitian surface*, Innov. Incidence Geom., 3 (2006), pp. 13–23.

[51] ——, *On the intersection of Hermitian curves and of Hermitian surfaces*, Discrete Math., 308 (2008), pp. 5196–5203.

[52] ——, *On the intersection of two subgeometries of* $\mathrm{PG}(n,q)$, Des. Codes Cryptogr., 46 (2008), pp. 261–267.

[53] ——, *Group theoretic characterizations of Buekenhout-Metz unitals in* $PG(2,q^2)$, J. Algebraic Combin., 33 (2011), pp. 401–407.

[54] N. Durante and G. L. Ebert, *On the intersection of Hermitian surfaces*, Innov. Incidence Geom., 6/7 (2007/08), pp. 153–167.

[55] N. Durante, V. Napolitano, and D. Olanda, *On quadrics of* $PG(3,q)$, in Trends in Incidence and Galois Geometries: a Tribute to Giuseppe Tallini, F. Mazzocca, N. Melone, and D. Olanda, eds., vol. 19 of Quad. Mat., Aracne Editrice, Roma, 2010, pp. 67–76.

[56] G. L. Ebert, *On Buekenhout-Metz unitals of even order*, European J. Combin., 13 (1992), pp. 109–117.

[57] ——, *Buekenhout-Tits unitals*, J. Algebraic Combin., 6 (1997), pp. 133–140.

[58] G. L. Ebert and K. Wantz, *A group-theoretic characterization of Buekenhout-Metz unitals*, J. Combin. Des., 4 (1996), pp. 143–152.

[59] G. Faina and G. Korchmáros, *A graphic characterization of Hermitian curves*, in Combinatorics '81 (Rome, 1981), vol. 18 of Ann. Discrete Math., North-Holland, Amsterdam, 1983, pp. 335–342.

[60] K. B. Farmer, *Hermitian geometries in projective space*, Linear Algebra Appl., 35 (1981), pp. 37–50.

[61] O. Ferri, *A graphical characterization of the set of external points of an oval in a plane Π_q (q odd)*, Rend. Mat. (7), 1 (1981), pp. 31–38.

[62] ——, *k-sets of class $[0, (q-1)/2, (q+1)/2, q]$ in a projective plane of odd order q*, Rend. Mat. (7), 3 (1983), pp. 33–41.

[63] O. Ferri and G. Tallini, *A characterization of nonsingular quadrics in* PG$(4,q)$, Rend. Mat. Appl. (7), 11 (1991), pp. 15–21.

[64] F. Fiedler, K. H. Leung, and Q. Xiang, *On Mathon's construction of maximal arcs in Desarguesian planes*, Adv. Geom., (2003), pp. S119–S139. Special issue dedicated to Adriano Barlotti.

[65] ——, *On Mathon's construction of maximal arcs in Desarguesian planes. II*, J. Combin. Theory Ser. A, 108 (2004), pp. 99–122.

[66] J. W. Freeman, *Reguli and pseudoreguli in* PG$(3,s^2)$, Geom. Dedicata, 9 (1980), pp. 267–280.

[67] L. Giuzzi, *Collineation groups of the intersection of two classical unitals*, J. Combin. Des., 9 (2001), pp. 445–459.

[68] ——, *On the intersection of Hermitian surfaces*, J. Geom., 85 (2006), pp. 49–60.

[69] D. G. Glynn, *On the characterization of certain sets of points in finite projective geometry of dimension three*, Bull. London Math. Soc., 15 (1983), pp. 31–34.

[70] D. G. Glynn, *Two new sequences of ovals in finite Desarguesian planes of even order*, in Combinatorial mathematics, X (Adelaide, 1982), vol. 1036 of Lecture Notes in Math., Springer, Berlin, 1983, pp. 217–229.

[71] D. G. Glynn, *A condition for the existence of ovals in* PG$(2,q)$, *q even*, Geom. Dedicata, 32 (1989), pp. 247–252.

[72] K. Grüning, *A class of unitals of order q which can be embedded in two different planes of order q^2*, J. Geom., 29 (1987), pp. 61–77.

[73] N. Hamilton, *Degree 8 maximal arcs in* PG$(2,2^h)$, *h odd*, J. Combin. Theory Ser. A, 100 (2002), pp. 265–276.

[74] N. Hamilton and R. Mathon, *More maximal arcs in Desarguesian projective planes and their geometric structure*, Adv. Geom., 3 (2003), pp. 251–261.

[75] ——, *On the spectrum of non-Denniston maximal arcs in* $\mathrm{PG}(2,2^h)$, European J. Combin., 25 (2004), pp. 415–421.

[76] N. Hamilton and T. Penttila, *Groups of maximal arcs*, J. Combin. Theory Ser. A, 94 (2001), pp. 63–86.

[77] D. Hertel and A. Pott, *Two results on maximum nonlinear functions*, Des. Codes Cryptogr., 47 (2008), pp. 225–235.

[78] J. W. P. Hirschfeld, *Projective geometries over finite fields*, Oxford Mathematical Monographs, The Clarendon Press Oxford University Press, New York, second ed., 1998.

[79] J. W. P. Hirschfeld, L. Storme, J. A. Thas, and J. F. Voloch, *A characterization of Hermitian curves*, J. Geom., 41 (1991), pp. 72–78.

[80] J. W. P. Hirschfeld and T. Szőnyi, *Sets in a finite plane with few intersection numbers and a distinguished point*, Discrete Math., 97 (1991), pp. 229–242.

[81] J. W. P. Hirschfeld and J. A. Thas, *Sets of type* $(1,n,q+1)$ *in* $\mathrm{PG}(d,q)$, Proc. London Math. Soc. (3), 41 (1980), pp. 254–278.

[82] X. Hubaut, *Limitation du nombre de points d'un* (k,n)*-arc régulier d'un plan projectif fini*, Atti Accad. Naz. Lincei Rend. Cl. Sci. Fis. Mat. Natur. (8), 48 (1970), pp. 490–493.

[83] I. Jagos, G. Kiss, and A. Pór, *On the intersection of Baer subgeometries of* $\mathrm{PG}(n,q^2)$, Acta Sci. Math. (Szeged), 69 (2003), pp. 419–429.

[84] B. C. Kestenband, *Unital intersections in finite projective planes*, Geom. Dedicata, 11 (1981), pp. 107–117.

[85] I. Landjev and L. Storme, *Galois geometries and coding theory*, in Current research topics in Galois geometry, Nova Sci. Publ., New York, 2011, ch. 8, pp. 185–212.

[86] B. Larato, *A characterization of the parabolic unitals of Buekenhout-Metz*, Le Matematiche (Catania), 38 (1983), pp. 95–98 (1987).

[87] C. Lefèvre-Percsy, *Characterization of Buekenhout-Metz unitals*, Arch. Math. (Basel), 36 (1981), pp. 565–568.

[88] ——, *Characterization of Hermitian curves*, Arch. Math. (Basel), 39 (1982), pp. 476–480.

[89] L. Lunelli and M. Sce, *k-archi completi nei piani proiettivi desarguesiani di rango* 8 *e* 16, Centro di Calcoli Numerici, Politecnico di Milano, Milan, 1958.

[90] R. Mathon, *New maximal arcs in Desarguesian planes*, J. Combin. Theory Ser. A, 97 (2002), pp. 353–368.

[91] K. Metsch, *A note on Buekenhout-Metz unitals*, in Geometry, combinatorial designs and related structures (Spetses, 1996), vol. 245 of London Math. Soc. Lecture Note Ser., Cambridge Univ. Press, Cambridge, 1997, pp. 177–180.

[92] R. Metz, *On a class of unitals*, Geom. Dedicata, 8 (1979), pp. 125–126.

[93] C. M. O'Keefe and T. Penttila, *Ovoids with a pencil of translation ovals*, Geom. Dedicata, 62 (1996), pp. 19–34.

[94] ——, *Ovals in translation hyperovals and ovoids*, European J. Combin., 18 (1997), pp. 667–683.

[95] M. E. O'Nan, *A characterization of* $U_3(q)$, J. Algebra, 22 (1972), pp. 254–296.

[96] G. Panella, *Caratterizzazione delle quadriche di uno spazio (tridimensionale) lineare sopra un corpo finito*, Boll. Un. Mat. Ital. (3), 10 (1955), pp. 507–513.

[97] T. Penttila and G. F. Royle, *Sets of type* (m,n) *in the affine and projective planes of order nine*, Des. Codes Cryptogr., 6 (1995), pp. 229–245.

[98] O. Polverino, *Linear representation of Buekenhout-Metz unitals*, Discrete Math., 267 (2003), pp. 247–252. Combinatorics 2000 (Gaeta).

[99] C. T. Quinn and L. R. A. Casse, *Concerning a characterisation of Buekenhout-Metz unitals*, J. Geom., 52 (1995), pp. 159–167.

[100] B. Qvist, *Some remarks concerning curves of the second degree in a finite plane*, Ann. Acad. Sci. Fennicae. Ser. A. I. Math.-Phys., 1952 (1952), p. 27.

[101] J. Schillewaert, *A characterization of quadrics by intersection numbers*, Des. Codes Cryptogr., 47 (2008), pp. 165–175.

[102] J. Schillewaert and J. A. Thas, *Characterizations of Hermitian varieties by intersection numbers*, Des. Codes Cryptogr., 50 (2009), pp. 41–60.

[103] B. Schmidt and C. White, *All two-weight irreducible cyclic codes?*, Finite Fields Appl., 8 (2002), pp. 1–17.

[104] B. Segre, *Ovals in a finite projective plane*, Canad. J. Math., 7 (1955), pp. 414–416.

[105] ——, *Sui k-archi nei piani finiti di caratteristica due*, Rev. Math. Pures Appl., 2 (1957), pp. 289–300.

[106] ——, *Le geometrie di Galois. Archi ed ovali; calotte ed ovaloidi*, Confer. Sem. Mat. Univ. Bari, 43-44 (1959), p. 31 pp. (1959).

[107] ——, *Lectures on modern geometry*, Edizioni Cremonese, Rome, 1961.

[108] ——, *Ovali e curve σ nei piani di Galois di caratteristica due.*, Atti Accad. Naz. Lincei Rend. Cl. Sci. Fis. Mat. Nat. (8), 32 (1962), pp. 785–790.

[109] B. Segre and U. Bartocci, *Ovali ed altre curve nei piani di Galois di caratteristica due*, Acta Arith., 18 (1971), pp. 423–449.

[110] E. Seiden, *A theorem in finite projective geometry and an application to statistics*, Proc. Amer. Math. Soc., 1 (1950), pp. 282–286.

[111] M. Sved, *On configurations of Baer subplanes of the projective plane over a finite field of square order*, in Combinatorial mathematics, IX (Brisbane, 1981), vol. 952 of Lecture Notes in Math., Springer, Berlin, 1982, pp. 423–443.

[112] ——, *Baer subspaces in the n-dimensional projective space*, in Combinatorial mathematics, X (Adelaide, 1982), vol. 1036 of Lecture Notes in Math., Springer, Berlin, 1983, pp. 375–391.

[113] M. Tallini Scafati, *$\{k, n\}$-archi di un piano grafico finito, con particolare riguardo a quelli con due caratteri. I, II*, Atti Accad. Naz. Lincei Rend. Cl. Sci. Fis. Mat. Natur. (8), 40 (1966), pp. 812–818; 1020–1025.

[114] ——, *Una proprietà grafica caratteristica delle forme hermitiane in uno spazio di Galois*, Univ. e Politec. Torino Rend. Sem. Mat., 26 (1966/1967), pp. 33–41.

[115] ——, *Caratterizzazione grafica delle forme hermitiane di un $S_{r,q}$*, Rend. Mat. e Appl. (5), 26 (1967), pp. 273–303.

[116] J. A. Thas, *Construction of maximal arcs and partial geometries*, Geometriae Dedicata, 3 (1974), pp. 61–64.

[117] ——, *Construction of maximal arcs and dual ovals in translation planes*, European J. Combin., 1 (1980), pp. 189–192.

[118] ——, *Semipartial geometries and spreads of classical polar spaces*, J. Combin. Theory Ser. A, 35 (1983), pp. 58–66.

[119] ——, *A combinatorial characterization of Hermitian curves*, J. Algebraic Combin., 1 (1992), pp. 97–102.

[120] J. Tits, *Ovoïdes à translations*, Rend. Mat. e Appl. (5), 21 (1962), pp. 37–59.

[121] ——, *Ovoïdes et groupes de Suzuki*, Arch. Math., 13 (1962), pp. 187–198.

[122] ——, *Une propriété caractéristique des ovoïdes associés aux groupes de Suzuki*, Arch. Math. (Basel), 17 (1966), pp. 136–153.

[123] V. D. Tonchev, *Quasi-symmetric designs, codes, quadrics, and hyperplane sections*, Geom. Dedicata, 48 (1993), pp. 295–308.

[124] J. Ueberberg, *On regular $\{v,n\}$-arcs in finite projective spaces*, J. Combin. Des., 1 (1993), pp. 395–409.

[125] K. Vedder, *A note on the intersection of two Baer subplanes*, Arch. Math. (Basel), 37 (1981), pp. 287–288.

[126] H. A. Wilbrink, *A characterization of the classical unitals*, in Finite geometries (Pullman, Wash., 1981), vol. 82 of Lecture Notes in Pure and Appl. Math., Dekker, New York, 1983, pp. 445–454.

In: Current Research Topics in Galois Geometry
Editors: J. De Beule and L. Storme
ISBN: 978-1-61209-523-3

Chapter 2

Substructures of Finite Classical Polar Spaces

Jan De Beule[1,*]***, Andreas Klein***[1,†] ***and Klaus Metsch***[2,‡]
[1] Ghent University, Department of Mathematics,
Gent, Belgium
[2] Universität Gießen, Mathematisches Institut, Arndtstraße 2,
Gießen, Germany

Abstract

We survey results and particular facts about (partial) ovoids, (partial) spreads, m-systems, m-ovoids, covers and blocking sets in finite classical polar spaces.

Key Words: finite classical polar space, quadric, hermitian variety, (partial) ovoid, (partial) spread, cover, blocking set, m-ovoid, m-system.

AMS Subject Classification: 51A50, 05B25.

1. Finite Classical Polar Spaces

The *finite classical polar spaces* are the geometries that are associated with non-degenerate reflexive sesquilinear and non-singular quadratic forms on vector spaces over a finite field. Given a projective space $\mathrm{PG}(d,q)$, then a polar space $\mathcal{P}$ in $\mathrm{PG}(d,q)$ consists of the projective subspaces of $\mathrm{PG}(d,q)$ that are totally isotropic with relation to a given non-degenerate reflexive sesquilinear form or that are totally singular with relation to a given non-singular quadratic form. The projective space $\mathrm{PG}(d,q)$ is called the *ambient projective space* of $\mathcal{P}$. In this article, with "polar space" we always refer to "finite classical polar space".

A projective subspace of maximal dimension in a polar space $\mathcal{P}$ is called a *generator*. One can prove (see [45], Theorem 26.1.2) that all generators have the same dimension

*E-mail address: jdebeule@cage.ugent.be; This author is a Postdoctoral Research Fellow of the Research Foundation – Flanders (Belgium) (FWO).
†E-mail address: klein@cage.ugent.be
‡E-mail address: Klaus.Metsch@math.uni-giessen.de

$r-1$. We call r the *rank* of the polar space. A polar space of rank 1 only contains projective points. There exist five different types of finite classical polar spaces, which are, up to transformation of the coordinate system, described as follows:

- The elliptic quadric $\mathrm{Q}^-(2n+1,q)$, $n \geq 1$, formed by all points of $\mathrm{PG}(2n+1,q)$ which satisfy the standard equation $x_0x_1 + \cdots + x_{2n-2}x_{2n-1} + f(x_{2n},x_{2n+1}) = 0$, where f is a homogeneous irreducible polynomial of degree 2 over $\mathbb{F}_q$.
- The parabolic quadric $\mathrm{Q}(2n,q)$, $n \geq 1$, formed by all points of $\mathrm{PG}(2n,q)$ which satisfy the standard equation $x_0x_1 + \cdots + x_{2n-2}x_{2n-1} + x_{2n}^2 = 0$.
- The hyperbolic quadric $\mathrm{Q}^+(2n+1,q)$, $n \geq 0$, formed by all points of $\mathrm{PG}(2n+1,q)$ which satisfy the standard equation $x_0x_1 + \cdots + x_{2n}x_{2n+1} = 0$.
- The symplectic polar space $\mathrm{W}(2n+1,q)$, $n \geq 0$, which consists of all points of $\mathrm{PG}(2n+1,q)$ together with the totally isotropic subspaces with respect to the standard symplectic form $\theta(x,y) = x_0y_1 - x_1y_0 + \cdots + x_{2n}y_{2n+1} - x_{2n+1}y_{2n}$.
- The hermitian variety $\mathrm{H}(n,q^2)$, $n \geq 1$, formed by all points of $\mathrm{PG}(n,q^2)$ which satisfy the standard equation $x_0^{q+1} + \cdots + x_n^{q+1} = 0$.

In the above list, the polar space of a given type has rank 1 for the smallest n that is allowed. Remark also that a quadric (also called an *orthogonal polar space*), and a hermitian variety, is determined completely by its point set, and can be described as above as a set of points whose coordinates satisfy an equation, which is of course derived from the sesquilinear or quadratic form.

Let P be a point of a polar space $\mathcal{P}$. Then $P^\perp$ is the set of points whose coordinates are orthogonal to P with respect to the underlying sesquilinear or quadratic form[1], so $P^\perp$ is the set of points of a hyperplane $T_P(\mathcal{P})$, called the *tangent hyperplane at P to $\mathcal{P}$*, and $P^\perp \cap \mathcal{P}$ is necessarily the set of points of $\mathcal{P}$ that lie on a line through P contained in $\mathcal{P}$. For any set A of points, $A^\perp := \cap_{P\in A} P^\perp$. The following result is fundamental in the theory of finite classical polar spaces.

Result 1.1. *Suppose that $\mathcal{P}_r$ is a finite classical polar space of rank $r \geq 2$. Then for any point P of $\mathcal{P}_r$, the set $P^\perp \cap \mathcal{P}_r$ is a cone with base $\mathcal{P}_{r-1}$ and vertex P, with $\mathcal{P}_{r-1}$ a finite classical polar space of rank $r-1$ of the same type as $\mathcal{P}_r$.*

From this theorem, it follows that the quotient space of a point P of $\mathcal{P}_r$, i.e. the set of all subspaces of $\mathcal{P}_r$ through P, is a polar space of rank $r-1$ of the same type as $\mathcal{P}_r$.

We define $\theta_i(q) := \frac{q^{i+1}-1}{q-1}$ for all integers $i \geq 0$, i.e. the number of points in $\mathrm{PG}(i,q)$.

Theorem 1.2. *The rank, the number of points, and the number of generators of all finite classical polar spaces are given in Table 1.*

Proof. We demonstrate the proof for $\mathrm{Q}^+(2n+1,q)$, the proofs for the other polar spaces are, mutatis mutandis, the same.

We prove the results by induction on n. For $n=0$, the hyperbolic quadric $x_0x_1 = 0$ contains two points on a line and $2 = \frac{(q^0+1)(q^1-1)}{q^1-1}$. Formally, we set $|\mathrm{Q}^+(-1,q)| = 0$. Note that this definition fits with the general formula. Now suppose that $n \geq 1$. Take a line l of

[1] When q is even, the quadratic form f determining $\mathcal{P}$, determines a possibly singular symplectic form σ. Two points P and Q are orthogonal with respect to f if, by definition, they are orthogonal with respect to σ.

$Q^+(2n+1,q)$. Then $l^\perp$ intersects $Q^+(2n+1,q)$ in a cone over a $Q^+(2n-3,q)$ which by induction has $a = q+1+q^2\frac{(q^{n-2}+1)(q^{n-1}-1)}{q-1}$ points.

For each point $P \notin l^\perp$ there exists exactly one point $R \in l$ with $R \in P^\perp$ or $P \in R^\perp$. Now $R^\perp$ intersects $Q^+(2n+1,q)$ in a cone over a $Q^+(2n-1,q)$ which by induction has $b = 1+q\frac{(q^{n-1}+1)(q^n-1)}{q-1}$ points. Thus $|Q^+(2n+1,q)| = (q+1)(b-a)+a = \frac{(q^n+1)(q^{n+1}-1)}{q-1} = (q^n+1)\theta_n(q)$.

Now we count the number of generators, again using induction on n. For $n = 0$ the 2 points of the hyperbolic quadric are its generators, hence it is a polar space of rank 1.

Now assume that $n \geq 1$, and that $Q^+(2n-1,q)$ is a polar space of rank n. Let P be a point of $Q^+(2n+1,q)$. Then $P^\perp$ intersects $Q^+(2n+1,q)$ in cone over a $Q^+(2n-1,q)$. Hence, by induction P lies on $2(q+1)(q^2+1)\cdots(q^{n-1}+1)$ generators. On the other hand, a generator of $Q^+(2n-1,q)$ contains $\frac{q^{n+1}-1}{q-1}$ points. Double counting gives for the number g of generators in $Q^+(2n+1,q)$ the equation

$$g\frac{q^{n+1}-1}{q-1} = |Q^+(2n-1,q)|2(q+1)(q^2+1)\cdots(q^{n-1}+1).$$

Solving the equation yields the number of generators. Finally, the dimension of the generators of $Q^+(2n+1,q)$ is one more than the dimension of the generators of $Q^+(2n-1,q)$. □

Table 1. Rank, number of points and number of generators of finite classical polar spaces

polar space	rank	number of points	number of generators
$Q^-(2n+1,q)$	n	$(q^{n+1}+1)\theta_{n-1}(q)$	$(q^2+1)(q^3+1)\cdots(q^{n+1}+1)$
$Q(2n,q)$	n	$(q^n+1)\theta_{n-1}(q)$	$(q+1)(q^2+1)(q^3+1)\cdots(q^n+1)$
$Q^+(2n+1,q)$	$n+1$	$(q^n+1)\theta_n(q)$	$2(q+1)(q^2+1)\cdots(q^n+1)$
$W(2n+1,q)$	$n+1$	$(q^{n+1}+1)\theta_n(q)$	$(q+1)(q^2+1)\cdots(q^{n+1}+1)$
$H(2n,q^2)$	n	$(q^{2n+1}+1)\theta_{n-1}(q^2)$	$(q^3+1)(q^5+1)\cdots(q^{2n+1}+1)$
$H(2n+1,q^2)$	$n+1$	$(q^{2n+1}+1)\theta_n(q^2)$	$(q+1)(q^3+1)\cdots(q^{2n+1}+1)$

It is well known (see e.g. [44]) that the generators of $Q^+(2n+1,q)$ fall into two equivalence classes, denoted by the sets $\mathcal{G}_1$ and $\mathcal{G}_2$. Recall that the rank of $Q^+(2n+1,q)$ is $n+1$. The following result is well known and can be found in [45] (Theorem 22.4.12 and its Corollary).

Result 1.3. *Let g_1 and g_2 be distinct generators of $Q^+(2n+1,q)$. If $n = 2s$, then*

$$\dim(g_1 \cap g_2) = \begin{cases} 0,2,4,\ldots,2s-2 & \text{if } g_1 \text{ and } g_2 \text{ belong to the same class} \\ -1,1,3,\ldots,2s-1 & \text{if } g_1 \text{ and } g_2 \text{ belong to a different class;} \end{cases}$$

and if $n = 2s+1$, then

$$\dim(g_1 \cap g_2) = \begin{cases} -1,1,3,\ldots,2s-1 & \text{if } g_1 \text{ and } g_2 \text{ belong to the same class} \\ 0,2,4,\ldots,2s & \text{if } g_1 \text{ and } g_2 \text{ belong to a different class.} \end{cases}$$

2. Isomorphisms of Finite Classical Polar Spaces

For q even, $Q(2n,q) \subseteq PG(2n,q)$ has a *nucleus*, i.e. a point $N \in PG(2n,q) \setminus Q(2n,q)$ contained in all tangent hyperplanes to $Q(2n,q)$. Projecting the elements of $Q(2n,q)$ from N yields a polar space isomorphic to $W(2n-1,q)$ (see e.g. [45]), so $Q(2n,q)$ and $W(2n-1,q)$ are isomorphic when q is even. The existence of this isomorphism implies that any result proved in one of these spaces, is also valid in the other one.

A *duality* δ between two rank 2 geometries $\mathcal{S} = (\mathcal{P},\mathcal{L},\mathrm{I})$ and $\mathcal{S}' = (\mathcal{P}',\mathcal{L}',\mathrm{I}')$ is an incidence preserving map from $\mathcal{P}$ to $\mathcal{L}'$, $\mathcal{L}$ to $\mathcal{P}'$, and from $\mathcal{L}'$ to $\mathcal{P}$, $\mathcal{P}'$ to $\mathcal{L}$, such that δ^2 is the identity mapping. There exist dualities between different types of finite classical polar spaces of rank 2 [68].

- $Q(4,q)$ is isomorphic to the dual of $W(3,q)$. This means that interchanging the role of the points and generators of $Q(4,q)$ yields an incidence geometry isomorphic to $W(3,q)$, and vice versa. As a consequence, for q even, $Q(4,q)$ and $W(3,q)$ are self-dual.
- $Q^-(5,q)$ is isomorphic to the dual of $H(3,q^2)$.

Consider now $Q^+(7,q)$ and define a rank 4 incidence geometry Ω as follows. $\Omega = (\mathcal{P},\mathcal{L},\mathcal{G}_1,\mathcal{G}_2)$, where $\mathcal{P}$ is the set of points of $Q^+(7,q)$ and $\mathcal{L}$ is the set of lines of $Q^+(7,q)$. An element $g_1 \in \mathcal{G}_1$ is incident with an element $g_2 \in \mathcal{G}_2$ if and only if $g_1 \cap g_2$ is a plane. Incidence between other elements is symmetrized containment. A *triality* of the geometry Ω is a map

$$\tau : \mathcal{L} \to \mathcal{L}, \mathcal{P} \to \mathcal{G}_1, \mathcal{G}_1 \to \mathcal{G}_2, \mathcal{G}_2 \to \mathcal{P}$$

preserving the incidence in Ω and such that τ^3 is the identity. Trialities of Ω exist [45].

The dualities and the triality described here, are used frequently to construct substructures of a polar space from different ones, as we will see in the next sections.

3. Ovoids, Spreads, m-systems and m-ovoids

"Ovoids" of polar spaces are inspired by ovoids of the projective space $PG(3,q)$ (see e.g. [28]), and are defined for the first time in [78]. Also "spreads" occurred first in projective spaces, and are transferred to polar spaces.

Let $\mathcal{P}$ be a finite classical polar space of rank $r \geq 2$. An *ovoid* is a set $\mathcal{O}$ of points of $\mathcal{P}$, which has exactly one point in common with each generator of $\mathcal{P}$. A *spread* is a set $\mathcal{S}$ of generators of $\mathcal{P}$ which constitute a partition of the point set of $\mathcal{P}$.

Theorem 3.1. *An ovoid in $Q^-(2n-1,q)$, $Q(2n,q)$, $Q^+(2n+1,q)$ or $W(2n-1,q)$ has q^n+1 points. An ovoid of $H(2n,q^2)$ or $H(2n+1,q^2)$ has $q^{2n+1}+1$ points.*

A spread of $Q^-(2n-1,q)$, $Q(2n,q)$, $Q^+(2n+1,q)$ or $W(2n-1,q)$ contains q^n+1 generators. A spread of $H(2n,q^2)$ or $H(2n+1,q^2)$ contains $q^{2n+1}+1$ generators.

Proof. We demonstrate the proof for $\mathcal{P} = Q^+(2n+1,q)$ as an example, the proof is analogous for the other polar spaces.

By Theorem 1.2, $Q^+(2n+1,q)$ has $2(q+1)(q^2+1)\cdots(q^n+1)$ generators. By Result 1.1, the quotient space of a point is a $Q^+(2n-1,q)$, hence, every point lies in $2(q+1)(q^2+1)\cdots(q^{n-1}+1)$ generators. Thus an ovoid must have $[2(q+1)(q^2+1)\cdots(q^n+1)]/[2(q+1)(q^2+1)\cdots(q^{n-1}+1)] = q^n+1$ elements.

By Theorem 1.2, $Q^+(2n+1,q)$ has $(q^n+1)\theta_n(q)$ points. Each generator is a projective space of dimension n, that contains $\theta_n(q)$ points. Thus a spread must contain q^n+1 elements. □

So in a polar space $\mathcal{P}$, the size of an ovoid equals the size of a spread, this number is denoted by $\mu_{\mathcal{P}}$.

3.1. Ovoids

Ovoids of finite classical polar spaces are rare, they seem to exist only in low rank, and for many polar spaces of high rank, a non-existence proof for ovoids is known. One important observation to show the non-existence of ovoids is the following lemma.

Lemma 3.2. *If $\mathcal{O}$ is an ovoid of a finite classical polar space $\mathcal{P}$ of rank $r \geq 3$, then $\mathcal{O}$ induces an ovoid of a finite classical polar space of the same type of rank $r-1$*

Proof. Consider any point $P \notin \mathcal{O}$ of the polar space $\mathcal{P}$. The quotient space on P is a polar space $\mathcal{P}'$ of rank $r-1$ of the same type. But each generator of $\mathcal{P}$ on P contains exactly one point of $\mathcal{O}$, so $\mathcal{O}$ induces an ovoid of $\mathcal{P}'$. □

Hence, if the non-existence of ovoids is proved for a polar space of a certain type in some rank, the contraposition of Lemma 3.2 shows the non-existence in higher rank. In the very rare cases where an ovoid of a polar space of rank r is induced by an ovoid of a polar space of rank $r+1$, applying Lemma 3.2 is called "slicing". We now prove the non-existence of ovoids of $W(3,q)$, q odd.

Lemma 3.3. *The polar space $W(3,q)$ has ovoids if and only if q is even.*

Proof. If q is even, then $W(3,q)$ is isomorphic to $Q(4,q)$, and an embedded quadric $Q^-(3,q)$ in $Q(4,q)$ yields an ovoid of $W(3,q)$. Conversely, suppose that $\mathcal{O}$ is an ovoid of $W(3,q)$. Consider a line l of the ambient projective space $PG(3,q)$ spanned by two points of $\mathcal{O}$. Since a generator of $W(3,q)$ contains exactly one point of $\mathcal{O}$, the line l is not a generator of $W(3,q)$. So $|l \cap \mathcal{O}| = c \geq 2$. We count the pairs $\{(P,Q)|P \in l, Q \in \mathcal{O} \setminus l\}$. For any point $P \in l \setminus \mathcal{O}$, the $q+1$ generators of $W(3,q)$ on P each meet $\mathcal{O}$ in exactly one point, while on each point of $\mathcal{O} \setminus l$, there is exactly one generator of $W(3,q)$ meeting l in a point not in $\mathcal{O}$. It follows that $(q+1-c)(q+1)+c = q^2+1$. This is a contradiction unless $c=2$. But then, for any point $P \in W(3,q) \setminus \mathcal{O}$, in the plane $P^\perp$ we see $q+1$ points of $\mathcal{O}$, which, together with P, constitute a set $\mathcal{H}$ of $q+2$ points such that each line of $P^\perp$ meets $\mathcal{H}$ in 0 or 2 points. So $\mathcal{H}$ is a hyperoval of $PG(2,q)$, and q must be even (see e.g. [28, §2.1]). □

The non-existence of ovoids of $Q^-(5,q)$, $W(5,q)$ and $H(4,q^2)$, all for general q, can be proved using the same technique.

Corollary 3.4. *The polar spaces $Q(2n,q)$, $n \geq 3$, q even, and $Q^-(2n+1,q)$, $H(2n,q^2)$, $W(2n+1,q)$, $n \geq 2$, for general q, have no ovoid.*

Proof. Use Lemma 3.2 and the result for $Q^-(5,q)$, $W(5,q)$ and $H(4,q^2)$, and use the isomorphism between $W(2n-1,q)$, q even and $Q(2n,q)$, q even. □

Table 2. Existence and non-existence results on ovoids

polar space	existence and/or known examples	references
$Q^-(2n+1,q)$	$n>1$: no	[81]
$Q(4,q)$	classical; q odd prime: every ovoid is classical	[4]
	$q=3^h$: $Q(6,q)$ slices; 1 other example ($q=3^5$)	[46, 86, 87]; [69]
	$q=3^h$, $h>2$: Thas-Payne ovoids	[86]
	q odd non prime: Kantor ovoids	[46] ([86])
$W(3,q)$	q even: classical; Tits ovoid for $q=2^{2h+1}$	[71]+ [78], [89]
$Q(6,q)$	q even; $q>3$, q prime: no in both cases	[81]; [67]+ [4]
	$q=3^h$: two infinite families known	[10, 87]; [80]
$Q(2n,q)$	$n\geq 4$: no (different proofs for q odd and even)	[38]; [81]
$Q^+(3,q)$	several examples	[44]
$Q^+(5,q)$	yes, equivalent with spreads of $PG(3,q)$	[44]
$Q^+(7,q)$	$q=3^h$: known examples: from $Q(6,q)$	
	$q=2^h$: 1 infinite family; 1 other example ($q=8$)	[46]; [30]
	$q=p^h$, $p\equiv 2 \mod 3$, p prime, h odd: yes	[46]
	$q\geq 5$ prime: yes	[12, 64]
$Q^+(2n+1,q)$	$q=p^h$, p prime, $p^n>\binom{2n+p}{2n+1}-\binom{2n+p-2}{2n+1}$: no	[31], [7]
$W(2n+1,q)$	q odd $n=1$: no; all q, $n>1$: no	[81]
$H(2n,q^2)$	$n\geq 2$: no	[81]
$H(3,q^2)$	classical and many others, see spreads of $Q^-(5,q)$	[68]
$H(5,4)$	no	[20]
$H(2n+1,q^2)$	$q=p^h$, p prime, $p^{2n+1}>\binom{2n+p}{2n+1}^2-\binom{2n+p-1}{2n+1}^2$: no	[65]

The non-existence of ovoids of $\mathcal{P} = Q(8,q)$, q odd, is proved in [38] by associating a two-graph Γ to a hypothetical ovoid of $\mathcal{P}$. It is shown that Γ is regular, and using known relations between eigenvalues of the adjacency matrix of Γ, a contradiction follows rapidly. Lemma 3.2 closes the case $Q(2n,q)$, q odd, $n\geq 4$.

Conditions for the non-existence of ovoids of $Q^+(2n+1,q)$, $H(2n+1,q^2)$ respectively, are shown in [7], [65] respectively, by computing the p-rank of the incidence matrix of the points of $Q^+(2n+1,q)$, $H(2n+1,q^2)$ respectively, and the tangent hyperplanes to $Q^+(2n+1,q)$, $H(2n+1,q^2)$ respectively. The submatrix corresponding with the points of a hypothetical ovoid is necessarily the identity matrix, so comparing the size of an ovoid with the computed p-rank yields immediately a condition for non-existence. These conditions, shown in Table 2, leave open an infinite number of cases. We mention that Dye [31] gave an upper bound on the size of partial ovoids of the polar spaces $Q(2n,2)$, $Q^+(2n+1,2)$ and $Q^-(2n+1,2)$, which implies the non-existence of ovoids in some cases, in particular for $Q^+(2n+1,2)$ for $n\geq 4$.

In [47], it is shown that the polar space $H(2n+1,q^2)$ has no ovoids if $n>q^3$, and, similarly, in [18], that $Q^+(2n+1,q)$ has no ovoids if $n>q^2$. This is weaker than the earlier known conditions, but the proofs only use geometrical and combinatorial arguments. Pushing a little bit further these arguments, it is shown in [20] that $H(5,4)$ has no ovoid.

In [67], it is shown that $Q(6,q)$, $q > 3$, has no ovoids if all ovoids of $Q(4,q)$ are elliptic quadrics. It is shown in [4] that this condition is satisfied for q odd prime. This leaves open the existence or non-existence of ovoids of $Q(6,q)$ when $q = p^h$, p an odd prime, $h > 1$, except for $p = 3$, where ovoids are known to exist, see below.

Ovoids of $Q(4,q)$ and $H(3,q^2)$ can be constructed easily. The intersection with a hyperplane of the ambient projective space containing no generator, yields an ovoid. We call such ovoids *classical*. For $Q(4,q)$, $H(3,q^2)$ respectively, this is an elliptic quadric $Q^-(3,q)$, a hermitian curve $H(2,q^2)$ respectively. However, in $Q(4,q)$, q non-prime, and in $H(3,q^2)$, also non-classical ovoids exist. It is shown in [71, 78] that ovoids of $W(3,q)$, q even, are equivalent to ovoids of $PG(3,q)$. So the Tits ovoid in $PG(3,q)$, q even ([89], see also e.g. [28, §3.1.3]) yields an ovoid of $W(3,q)$, q even, and hence yields an ovoid of $Q(4,q)$, q even, which is non-classical, [68]. For q odd non prime, infinite families of non-classical ovoids of $Q(4,q)$ are known. Ovoids of $H(3,q^2)$ are equivalent to spreads of $Q^-(5,q)$, of which many non-classical examples are known, see Section 3.2..

The *Klein correspondence* is a bijective map from the line set of $PG(3,q)$ to the point set of the polar space $Q^+(5,q)$. Two lines of $PG(3,q)$ have a point in common if and only if they are mapped to two points of $Q^+(5,q)$ being contained in a common generator. Hence a spread of $PG(3,q)$ is mapped to a set of q^2+1 points of $Q^+(5,q)$ two by two not contained in a common generator, so constituting necessarily an ovoid. Since many different families of spreads of $PG(3,q)$ are known (see e.g. [44]), there are many different examples of ovoids of $Q^+(5,q)$. We mention that a regular spread of $PG(3,q)$ corresponds to an elliptic quadric $Q^-(3,q) \subset Q^+(5,q)$.

Only two infinite families of ovoids of $Q(6,q)$ are known, for $q = 3^h$, $h \geq 1$. Embedding $Q(6,q)$ as a hyperplane section in $Q^+(7,q)$, it is easily observed that an ovoid of $Q(6,q)$ induces an ovoid of $Q^+(7,q)$, and all known ovoids of $Q^+(7,q)$, $q = 3^h$, arise from ovoids of $Q(6,q)$. But several (infinite families of) ovoids of $Q^+(7,q)$, $q \neq 3^h$, are known, and all of them are not contained in a hyperplane section.

We now refer to Table 2 for an overview, including references.

3.2. Spreads

From the definition, it follows that ovoids of a polar space of rank 2 are spreads of the dual of $\mathcal{P}$. This immediately yields some examples of spreads in the rank two case. But we start with a construction result in the symplectic polar space $W(2n+1,q)$.

Consider the projective space $PG(d,q)$. When $(t+1) \mid (d+1)$, the multiplicative group of $\mathbb{F}_{q^{d+1}}$ can be partitioned by cosets of the multiplicative group of $\mathbb{F}_{q^{t+1}}$. Each such coset is an $\mathbb{F}_q$ vector space, so we find a partition of $PG(d,q)$ by t-dimensional projective spaces. For $d = 2n+1$ and $t = n$, we find a spread of $PG(2n+1,q)$ consisting of n-dimensional subspaces. It is shown in [30] that there exists always a symplectic polarity ϕ of $PG(2n+1,q)$ such that all n-dimensional subspaces of this spread are totally isotropic with relation to ϕ. This yields a spread of the polar space $W(2n+1,q)$, $n \geq 1$, and, when q is even, a spread of the polar space $Q(2n+2,q)$, $n \geq 1$. The same result is also shown in [79] for $n = 2$, with a proof that is extendable to general n.

The polar space $Q^+(4n+1,q)$, $n \geq 1$, has no spread, since by Result 1.3, at most two generators can be skew. Consider now $Q(4n+2,q)$, $n \geq 1$, as a hyperplane intersection

Table 3. Existence and non-existence results on spreads

polar space	existence and/or known examples	references
$W(2n+1,q), n \geq 1$	yes; $n=1$: also see ovoids of $Q(4,q)$	[30], [79]; [68]
$Q(2n,q), n \geq 2$	q even: yes; $n=2$: also see ovoids of $Q(4,q)$	Section 2.; [68]
$Q(6,q)$	all known examples: see spreads of $Q^+(7,q)$	Result 3.5
$Q^-(2n+1,q), n \geq 2$	q even: yes	Result 3.5
$Q^-(5,q)$	yes, e.g. from spreads of $PG(3,q)$	[68], [82]
$Q^+(4n+3,q), n \geq 1$	q even: yes: see spreads of $Q(4n+2,q)$ and Theorem 1.3	
$Q^+(4n+1,q)$	no	[44]
$Q^+(3,q)$	yes	[44]
$Q^+(7,q)$	all known examples: see ovoids of $Q^+(7,q)$	[88]
$Q(4n,q)$	q odd: no	[78, 84]
$H(2n+1,q^2)$	no	[81, 84]
$H(4,4)$	no, unpublished computer result of A.E. Brouwer	

of $Q^+(4n+3,q)$. Suppose that $Q(4n+2,q)$ has a spread $\mathcal{S}$. Then each element $\pi \in \mathcal{S}$ is contained in two generators of $Q^+(4n+3,q)$, one of each class, meeting in π. By Result 1.3, the set $\mathcal{S}'$ of all generators of one class, meeting $Q(4n+2,q)$ in an element of $\mathcal{S}$, is a spread of $Q^+(4n+3,q)$. Also, using hyperplane sections, the following proposition is easy to see.

Result 3.5 ([44]). *If the polar space* $Q^+(2n+1,q), n \geq 2; Q(2n,q), n \geq 3; H(2n+1,q^2), n \geq 2$, *respectively, has a spread, then the polar space* $Q(2n,q), n \geq 2; Q^-(2n-1,q), n \geq 3; H(2n,q^2), n \geq 2$, *respectively, has a spread.*

It is shown in [68] that any spread of $PG(3,q)$ gives rise to a spread of $Q^-(5,q)$. Many spreads of $PG(3,q)$ are known, so this gives rise to many spreads of $Q^-(5,q)$, and, dually to ovoids of $H(3,q^2)$. Finally, using the existence of a triality of $Q^+(7,q)$, one observes easily that an ovoid of $Q^+(7,q)$ is equivalent to a spread of $Q^+(7,q)$. This has also consequences for $Q(6,q)$, since a spread of $Q^+(7,q)$ induces, using a hyperplane section, a spread of $Q(6,q)$. The non-existence of spreads of the polar spaces $Q(4n,q)$, q odd and $n > 1$, and $H(2n+1,q^2)$, $n > 1$ is proved for the first time in [84]. The proofs are purely geometric.

We refer now to Table 3 for an overview, including references.

3.3. m-Systems

Let $\mathcal{P}$ be a finite classical polar space of rank $r \geq 2$. A *partial m-system* of $\mathcal{P}$ is a set $\mathcal{M} = \{\pi_1, \ldots, \pi_k\}$ of m-dimensional subspaces of $\mathcal{P}$, such that no generator of $\mathcal{P}$ containing π_i has any point in common with an element of $\mathcal{M} \setminus \{\pi_i\}$, for all elements $\pi_i \in \mathcal{M}$. If $|\mathcal{M}| = \mu_P$, then the partial m-system is called an *m-system*. Remark that for $m=0$, an m-system is an ovoid of $\mathcal{P}$, while for $m = r-1$, an m-system is a spread of $\mathcal{P}$.

This definition is given by Shult and Thas in [73]. Within the scope of this article, it is not possible to survey all existence and non-existence results of m-systems in a detailed way. Therefore, we will give information on particular facts and refer to existing surveys.

Field reduction is an appropriate way to construct m'-systems from m-systems. Consider the hermitian variety $\mathrm{H}(3,q^{2e})$, e odd, with associated hermitian form κ. With T the trace map from $\mathbb{F}_{q^{2e}}$ into $\mathbb{F}_{q^2}$, it is easy to check that the map $T \circ \kappa$ induces a hermitian form on $V(3e,q^2)$, so there is a map from $\mathrm{H}(3,q^{2e})$ to $\mathrm{H}(3e-1,q^2)$, mapping points of $\mathrm{H}(3,q^{2e})$ to $(e-1)$-dimensional subspaces of $\mathrm{H}(3e-1,q^2)$. We have seen that $\mathrm{H}(3,q^{2e})$ has plenty of ovoids, and hence, $\mathrm{H}(3e-1,q^2)$ has plenty of $(e-1)$-systems.

The first examples of m-systems, those described in [73], are actually obtained by field reduction. Most known cases now are still found there, two cases are described in [74], and two cases are described in [40], and we refer to [85] for a survey. Mappings between finite classical polar spaces based on field reduction are studied comprehensively in [34].

We discuss three sources of non-existence results on m-systems. The oldest results are due to Shult and Thas, who obtain non-existence results on m-systems comparable with the non-existence results on ovoids of Blokhuis and Moorehouse in [7, 65]. The following results are shown in [75], and are essentially based on the computation of the p-rank of an incidence matrix in two ways.

Result 3.6 (see [75]). *If the finite classical polar space $\mathcal{P}$ admits an m-system, then*
(i) *for* $\mathcal{P} = \mathrm{Q}^+(2n+1,2^h)$, $2^n \leq \binom{2n+2}{m+1}$
(ii) *for* $\mathcal{P} = \mathrm{Q}(2n,q)$, $q=p^h$, *p an odd prime*, $p^n \leq \binom{\binom{2n+1}{m+1}+p-2}{p-1} - \binom{\binom{2n+1}{m+1}+p-4}{p-3}$
(iii) *for* $\mathcal{P} = \mathrm{Q}^+(2n+1,q)$, $q=p^h$, *p an odd prime*, $p^n \leq \binom{\binom{2n+2}{m+1}+p-2}{p-1} - \binom{\binom{2n+2}{m+1}+p-4}{p-3}$
(iv) *for* $\mathcal{P} = \mathrm{H}(2n+1,q^2)$, $q=p^h$, *p a prime*, $p^{2n+1} \leq \binom{\binom{2n+2}{m+1}+p-2}{p-1}^2 - \binom{\binom{2n+2}{m+1}+p-4}{p-3}^2$

Recently, Sin showed in [76] an upper bound on the number of elements of a partial m-system, using the p-rank approach of an incidence matrix in a more elaborate way. Let $N(n+1,r,p-1)$ be the number of monomials in $n+1$ variables of total degree r and with (partial) degree at most $p-1$ in each variable. This number is equal to the coefficient of x^r in $(1+x+\cdots+x^{p-1})^{n+1}$.

Result 3.7 (see [76]). *Let $\mathcal{M}$ be a partial m-system of a finite classical polar space $\mathcal{P}$ with ambient projective space* $\mathrm{PG}(n,q)$, $q=p^h$, *p prime. Then* $|\mathcal{M}| \leq 1+N(n+1,(m+1)(p-1),p-1)^h$.

If the right hand side is smaller than $\mu_{\mathcal{P}}$, then this implies the non-existence of m-systems in $\mathcal{P}$. It is hard to compare both bounds in general. Both bounds imply non-existence of m-systems for polar spaces of "high" rank, but for given m and q, Result 3.7 implies non-existence often for lower rank than Result 3.6. A careful analysis is done in [76].

To describe the third non-existence result, we first have to go back to [73]. Suppose that $\mathcal{P} \in \{\mathrm{W}(2n+1,q), \mathrm{Q}^-(2n+1,q), \mathrm{H}(2n,q^2)\}$ and that $\mathcal{M}$ is an m-system of $\mathcal{P}$. The point set $\widetilde{\mathcal{M}}$ is the union of the elements of $\mathcal{M}$ as point sets. In [73], it is shown that $\widetilde{\mathcal{M}}$ is a two intersection set with respect to the hyperplanes of the ambient projective space. This implies that a strongly regular graph can be associated to $\widetilde{\mathcal{M}}$. Hamilton and Mathon study this graph in [39] and compute its eigenvalues. This yields the following result.

Result 3.8. *m-Systems of* $\mathrm{W}(2n+1,q), \mathrm{Q}^-(2n+1,q), \mathrm{H}(2n,q^2)$ *do not exist for* $n > 2m+1$.

Hamilton and Mathon analyze their result and give examples for $W(2n+1,q)$, q even and n odd, and $Q^-(2n+1,q)$, n odd, showing that their bound is sharp in these cases. They also give an example that shows that their bound is better in some cases than the bound of [75]. Finally, the paper contains classification results for m-systems of $W(2n+1,2), Q^-(2n+1,2)$, and $Q^+(2n+1,2)$ for $m = 1,2,3$, and 4, and applications.

A recent paper providing a general classification result is [6]. Bamberg and Penttila give a complete classification of m-systems admitting an insoluble transitive collineation group. There is no restriction on m, so their classification also holds for ovoids and spreads satisfying the condition. This paper also contains a detailed overview of some construction methods mentioned here, and a long list of references.

3.4. m-Ovoids

Let $\mathcal{P}$ be a finite classical polar space of rank $r \geq 2$. An *m-ovoid* is a set $\mathcal{O}$ of points of $\mathcal{P}$, which has exactly m points in common with each generator of $\mathcal{P}$. Thas defined m-ovoids of generalized quadrangles in [83]. Before this introduction, Segre studied already m-ovoids of $Q^-(5,q)$, but in the dual setting, i.e. as sets of lines of $H(3,q^2)$ covering each point m times. Segre proved that $m = \frac{q+1}{2}$, when q is odd, [72]. A line sets of $H(3,q^2)$ covering each point exactly $\frac{q+1}{2}$ times is also called a *hemisystem* of $H(3,q^2)$. Segre also gives an example of a hemisystem for $q = 3$, and it is only in [14] that hemisystems of $H(3,q^2)$ are constructed for all odd q. In [13], m-ovoids of $W(3,q)$ are constructed, for q odd and $m = \frac{q+1}{2}$ and for q even and $m \in \{2,\dots q-1\}$.

Up to our knowledge, the first systematic treatise of m-ovoids of polar spaces is [5]. In this paper, m-ovoids are treated in a more general framework, related to i-tight sets and intriguing sets of polar spaces. It is shown that m-ovoids of a polar space $\mathcal{P}$, with $\mathcal{P} \in \{H(2n,q^2), Q^-(2n+1,q), W(2n+1,q)\}$ have two intersection numbers with relation to hyperplanes of the ambient projective space. This gives rise to a strongly regular graph. Expressing that one of the parameters must be larger than 0, yields the lower bound on m. The following result is obtained in this way.

Result 3.9. *Let $\mathcal{P}$ be $H(2r,q^2), Q^-(2r+1,q), W(2r-1,q)$ respectively. If an m-ovoid of $\mathcal{P}$ exists, then $m \geq b$, with $b = \frac{(-3+\sqrt{9+4q^{2r+1}})}{2q^2-2}, \frac{(-3+\sqrt{9+4q^{r+1}})}{2q-2}, \frac{(-3+\sqrt{9+4q^{r}})}{2q-2}$ respectively.*

The above bounds are larger than 1 for $H(2r,q^2)$ and $Q^-(2r+1,q)$ for $r \geq 2$ and for $W(2r-1,q)$ for $r > 2$, for all q. Using a slicing argument that is in fact comparable with Lemma 3.2, the authors obtain the following result.

Result 3.10. *The following polar spaces do not admit a 2-ovoid: $W(2r-1,q)$, q odd and $r > 2$; $Q^-(2r+1,q)$, $r > 2$; $H(2r,q^2)$, $r > 2$; and $Q(2r,q)$, $r > 4$.*

Proof. Suppose that $\mathcal{O}$ is a 2-ovoid of the polar space $\mathcal{P}$ of rank r which is one of the mentioned examples. Consider any point $P \in \mathcal{O}$. Then the quotient space on P is a polar space of rank $r-1$ of the same type. Since all generators of $\mathcal{P}$ on P meet $\mathcal{O} \setminus \{P\}$ in exactly one point, $\mathcal{O}$ induces an ovoid in this quotient space. The result now follows from the non-existence of ovoids in the polar spaces mentioned. □

Table 4. Lower bounds on the size of maximal partial ovoids

polar space	lower bound	references
$W(2n+1,q)$	$q+1$ (sharp)	[11], [17]
$Q(4,q)$, q odd $Q(6,q)$, $q \in \{3,5,7\}$; $q \geq 9$ odd $Q(2n,q)$, $n \geq 4$, q odd; $Q(8,3)$	$1.419q$ $2q$; $2q-1$ $2q+1$; $2q$	[17, Theorem 2.2 (b)]
$Q^-(5,q)$, $q=2$; 3; $q \geq 4$ $Q^-(2n+1,q)$	6; 16; $2q+2$ $2q+1$	[17, Theorem 2.2 (c)]
$Q^+(2n+1,q)$, $n=2$; $n \geq 3$	$2q$; $2q+1$	[17, Theorem 2.2 (a)]
$H(3,q^2)$, q odd; even $H(2n+1,q^2)$, $n \geq 2$	$q^2+1+\frac{4}{9}q$; q^2+1 (sharp) q^2+q+1	[61]; [2] [16, Theorem 2.3]
$H(2n,q^2)$, $n=2$; $n \geq 3$	q^2+q+1	[63]; [16, Theorem 2.2]

4. Partial Ovoids and Partial Spreads

Let $\mathcal{P}$ be a finite classical polar space. A *partial ovoid* of $\mathcal{P}$ is a set $\mathcal{O}$ of points of $\mathcal{P}$ with the property that every generator of $\mathcal{P}$ contains at most one point of $\mathcal{O}$. A partial ovoid is called *proper* if it is not an ovoid. A (proper) partial ovoid is called *maximal* if it is not contained in a partial ovoid of larger size. Clearly, a maximal proper partial ovoid is not an ovoid.

A *partial spread* of $\mathcal{P}$ is a set $\mathcal{S}$ of pairwise disjoint generators. A partial spread is called *proper* if it is not a spread. A (proper) partial spread is called *maximal* if it is not contained in a partial spread of larger size. Clearly, a maximal proper partial spread is not a spread.

Obviously, in the rank 2 case, (maximal) (proper) partial ovoids become (maximal) (proper) partial spreads in the dual space. After non-existence proofs for ovoids, spreads respectively, partial ovoids, partial spreads respectively, arise naturally, and then we are interested in an upper bound on their size. Secondly, we wish to derive a lower bound on the size in case of maximality. Finally, when ovoids, spreads respectively, exist, extendability of proper partial ovoids, proper partial spreads respectively, is studied.

4.1. Partial Ovoids

The first series of results we mention are based on the use of a combinatorial approach also found in [35], where Glynn derives a lower bound on the size of maximal partial spreads of $PG(3,q)$. Under the Klein correspondence, this is equivalent with a lower bound on the size of maximal partial ovoids of $Q^+(5,q)$. But not only the result translates, also the proof, and this proof can also be applied for partial ovoids of other polar spaces. This yields lower bounds on the size of maximal partial ovoids of $Q^+(2n+1,q)$, $n \geq 2$, $Q^-(2n+1,q)$, $n \geq 2$ and $Q(2n,q)$, $n \geq 3$ and q odd. A proof can be found in e.g. [17]. Lower bounds for other polar spaces obtained using a combinatorial approach, are also known. We refer to Table 4 for an overview.

Recall from Lemma 3.3 that the existence of ovoids of $W(3,q)$ is equivalent with the existence of $(q+2)$-arcs (hyperovals) of $PG(2,q)$. As we will see in the following lemma, a partial ovoid of $W(3,q)$ gives rise to k-arcs of $PG(2,q)$

Lemma 4.1. *Let $\mathcal{O}$ be a partial ovoid of* $W(3,q)$, *with* $|\mathcal{O}| > q^2 - q + 1$. *For any point* $P \notin \mathcal{O}$, *the set* $\mathcal{K} := \{P\} \cup (P^\perp \cap \mathcal{O})$ *is an arc in the plane* $P^\perp$.

Proof. Suppose that l is a line of the ambient projective space $PG(3,q)$ meeting $\mathcal{O}$ in $c \geq 2$ points. Necessarily, l is not a generator of $W(3,q)$. Counting pairs $\{(P,Q) | P \in l, Q \in \mathcal{O} \setminus l\}$ yields the inequality $(q+1-c)(q+1)+c \geq |\mathcal{O}|$. If there exists a line l meeting $\mathcal{O}$ in at least three points, it follows that $|\mathcal{O}| \leq q^2 - q + 1$. So we may conclude that every line of $PG(3,q)$ meets $\mathcal{O}$ in at most 2 points. Hence, if $P \notin \mathcal{O}$, then the set $\{P\} \cup (P^\perp \cap \mathcal{O})$ is an arc in the plane $P^\perp$. □

An upper bound on the size of a partial ovoid of $W(3,q)$, q odd, is obtained now, using that the size of an arc of $PG(2,q)$, q odd, is at most $q+1$ (see e.g. [28, §2.1]).

Lemma 4.2. *Let $\mathcal{O}$ be a partial ovoid of* $W(3,q)$, *q odd. Then* $|\mathcal{O}| \leq q^2 - q + 1$.

Proof. Suppose that $|\mathcal{O}| > q^2 - q + 1$. Consider a point $P \notin \mathcal{O}$. By Lemma 4.1, $\mathcal{K} = \{P\} \cup (P^\perp \cap \mathcal{O})$ is an arc of the projective plane $P^\perp$. Since we assumed that q is odd, necessarily $|\mathcal{K}| \leq q+1$, hence, $|P^\perp \cap \mathcal{O}| \leq q$. Consider now a generator g of $W(3,q)$ that meets $\mathcal{O}$ in the unique point S. Clearly $|S^\perp \cap \mathcal{O}| = 1$, and any plane $\pi \neq S^\perp$ on g meets $\mathcal{O}$ in at most $q-1$ points of $\mathcal{O}$ different from S. Hence $|\mathcal{O}| \leq 1 + q(q-1) = q^2 - q + 1$, a contradiction. □

To show an upper bound on the size of maximal proper partial ovoids of $W(3,q)$, q even, *extendability* of arcs of $PG(2,q)$ can be used.

Lemma 4.3. *Let $\mathcal{O}$ be a proper partial ovoid of* $W(3,q)$, *q even, of size $q^2 + 1 - \delta$. If $\delta < q$, then $\mathcal{O}$ can be extended.*

Proof. Assume that $|\mathcal{O}| = q^2 + 1 - \delta$, $0 < \delta < q$. Since $\mathcal{O}$ is a proper partial ovoid, there exists a generator g of $W(3,q)$ not meeting $\mathcal{O}$. Hence all planes through g meet $\mathcal{O}$ in at most q points. If all planes through g meet $\mathcal{O}$ in at most $q-2$ points, then $|\mathcal{O}| \leq (q-2)(q+1) = q^2 - q - 2$. So by the assumption on the size of $\mathcal{O}$, there exists a plane π on g containing $q-1$ or q points of $\mathcal{O}$. Define $P := \pi^\perp$, then $P \in g$, and $\{P\} \cup (P^\perp \cap \mathcal{O})$ is an k-arc $\mathcal{K}$ in the plane π, with $k = q$ or $k = q+1$. In any case, $\mathcal{K}$ can be extended to a hyperoval $\overline{\mathcal{K}}$, and $\overline{\mathcal{K}}$ contains a point $T \in g \setminus \{P\}$.

Consider now a generator $l \neq g$ of $W(3,q)$ on T. Suppose that l meets $\mathcal{O}$ in a point Q. Then the plane $Q^\perp$, which does not contain g, and which intersects π in a line through T, contains a point R of $\overline{\mathcal{K}} \setminus g$. Necessarily $R \notin \mathcal{O}$. Hence, if $|\pi \cap \mathcal{O}| = q$ then no generator of $W(3,q)$ on T can meet $\mathcal{O}$, but then $\mathcal{O}$ can be extended with the point T; if $|\pi \cap \mathcal{O}| = q-1$ then at most one generator of $W(3,q)$ on T can meet $\mathcal{O}$. In this case, count the number of points of $\mathcal{O}$ in the $q+1$ planes π_i of $PG(3,q)$ on g to find

$$\sum_{\pi_i} |\pi_i \cap \mathcal{O}| = q - 1 + 1 + \sum_{\pi_i \notin \{T^\perp, \pi\}} |\pi_i \cap \mathcal{O}| = q^2 + 1 + \delta > q^2 - q + 1,$$

since $\delta < q$. This yields a contradiction if $|\pi_i \cap \mathcal{O}| \leq q - 1$ for all $\pi_i \notin \{T^\perp, \pi\}$. So at least one of the planes π_i, $\pi_i \notin \{T^\perp, \pi\}$ contains exactly q points of $\mathcal{O}$. But then the above argument shows the existence of a point $T' \in g$ by which $\mathcal{O}$ can be extended. □

Table 5. Upper bounds on the size of maximal proper partial ovoids in low rank polar spaces

polar space	upper bound	references
$\mathrm{W}(3,q)$	q^2-q+1 (sharp for q even)	[9]
$\mathrm{W}(5,q)$	$1+\frac{q}{2}(\sqrt{5q^4+6q^3+7q^2+6q+1}-q^2-q-1)$	[17]
$\mathrm{Q}(4,q)$, q odd	q^2 (see description above)	
$\mathrm{Q}(6,q)$, $q>13$, q prime	q^3-2q+1	[17]
$\mathrm{Q}(8,q)$, q odd, q not a prime	$q^4-q\sqrt{q}$	[17]
$\mathrm{Q}^-(5,q)$	$\frac{1}{2}(q^3+q+2)$ (sharp for $q=2,3$)	[16] ([31,32])
$\mathrm{H}(3,q^2)$	q^3-q+1 (sharp)	[48]
$\mathrm{H}(5,q^2)$	$q^5+1-(q^2+\frac{1}{4}q-1)/\sqrt{2}$	[16]
$\mathrm{H}(4,q^2)$	$q^5-q^4+q^3+1$	[16]

Both cases can be formulated in one statement as follows.

Corollary 4.4. *Let $\mathcal{O}$ be a maximal proper partial ovoid of* $\mathrm{W}(3,q)$*. Then* $|\mathcal{O}|\leq q^2-q+1$.

The proofs of Lemma's 4.1, 4.2 and 4.3 are based on results from [9], where actually the dual problem is discussed, i.e. the extendability of partial spreads of $\mathrm{Q}(4,q)$. Also in [9], examples of maximal partial spreads of $\mathrm{Q}(4,q)$, q even, of size q^2-q+1 are given. So the obtained upper bound is sharp for q even. Remark that the result of Lemma 4.2 was first shown in [77], by exploiting the fact that a maximal partial spread of $\mathrm{Q}(4,q)$ is mapped to a blocking set with respect to the planes of $\mathrm{PG}(3,q)$ under the Klein correspondence.

In [48], results on blocking sets of $\mathrm{PG}(4,q)$ contained in cones over a quadric $\mathrm{Q}^-(3,q)$ are obtained (see Lemma 5.1). These results are obtained for general q, and can be applied to study extendability of partial spreads of $\mathrm{Q}(4,q)$ and partial spreads of $\mathrm{Q}^-(5,q)$. This approach yields an alternative proof for the dual of Corollary 4.4, see also Theorem 5.2 (a), and it yields an upper bound on the size of maximal proper partial spreads of $\mathrm{Q}^-(5,q)$, (see Theorem 5.2 (b)), or, dually, an upper bound on the size of maximal proper partial ovoids of $\mathrm{H}(3,q^2)$. The following example, also found in [48], shows that this bound is sharp.

Example 4.5. *Consider a hermitian spread $\mathcal{S}$ of* $\mathrm{Q}^-(5,q)$*, that is a spread translating to a classical ovoid of* $\mathrm{H}(3,q^2)$ *under the duality between* $\mathrm{Q}^-(5,q)$ *and* $\mathrm{H}(3,q^2)$*. Using this duality, it is easy to see that such a spread is the union of q^2 reguli R_i through a common line l. Consider two reguli R_1 and R_2 containing the line l. Let R_i^{opp} be the regulus opposite to R_i, $i=1,2$. Replacing the $2q+1$ lines of $\mathcal{S}$ in $R_1\cup R_2$ by $q+1$ lines in $R_1^{opp}\cup R_2^{opp}$ such that every point of l is covered exactly once, yields a partial spread $\mathcal{S}'$, with $|\mathcal{S}'|=q^3-q+1$. If at least one line from R_1^{opp} and one line from R_2^{opp} is chosen, then $\mathcal{S}'$ is maximal.*

An upper bound on the size of maximal proper partial ovoids of $\mathrm{H}(5,q^2)$ is obtained in [16], where also an upper bound on the size of partial ovoids of $\mathrm{H}(4,q^2)$ is obtained, which improves an earlier result of [37].

Table 6. Inductive bounds on the size of partial ovoids

polar space	recursion	references
$W(2n+1,q)$	$x_{n,q} \leq 2+(q-1)x_{n-1,q}$	[17]
$Q^-(2n+1,q)$	$x_{n,q} \leq 2+\frac{q^n+1}{q^{n-1}+1}(x_{n-1,q}-2)$	[47]
$Q(2n,q)$	$x_{n,q} \leq 1+q(x_{n-1,q}-1)$	[17]
$Q^+(2n+1,q)$	$x_{n,q} \leq 2+\frac{q^n-1}{q^{n-1}-1}(x_{n-1,q}-2)$	[17]
$H(2n,q^2)$	$x_{n,q^2} \leq q^2x_{n-1,q^2}-q^2+1$	[16]
$H(2n+1,q^2)$	$x_{n,q^2} \leq q^2x_{n-1,q^2}-q^2+1$	[16]

In [84], an upper bound on the size of partial ovoids in $W(2n+1,q)$, $n \geq 2$, is obtained. In [17], this bound is improved for $n=2$, and using an inductive argument, this yields an upper bound for general n that is better than the one in [84]. The inductive argument is valid in all polar spaces, so we continue with the low rank cases, and then give an overview of the inductive bounds.

The case $Q(4,q)$, q odd, seems to be hard. Currently, it is only known that partial ovoids of size q^2 always extend to ovoids and that maximal proper partial ovoids of $Q(4,q)$ of size q^2-1, $q=p^h$, p odd, do not exist for $h>1$, and that examples are known for $q \in \{3,5,7,11\}$. The non-existence result is shown in [15], and the proof is also presented in [3, Corollary 6.9]. Furthermore, in [37], it is shown that if a maximal proper partial ovoid of $Q(4,q)$, q odd, of size $q^2+1-\delta$ exists, $\delta < \sqrt{q}$, then δ is even. Projection arguments and the results known on proper partial ovoids of $Q(4,q)$ for different values of q, yield an upper bound on the size of maximal proper partial ovoids of $Q(6,q)$ in [17].

A recent treatment of the case $Q^-(5,q)$ can be found in [16], where in fact the dual, i.e. partial spreads of $H(3,q^2)$, are considered, and which is described below (Theorem 4.8).

Upper bounds on the size of maximal proper partial ovoids of $Q^+(5,q)$ are under the Klein correspondence equivalent with upper bounds on the size of maximal proper partial spreads of $PG(3,q)$. For q not prime and not a square, the best upper bound is found in [53]. A comprehensive survey, also including results for q square and for q prime, can be found in [62]. Improvements on parts of [62] can be found in [33]. Constructions of maximal partial spreads of $PG(3,q)$ can e.g. be found in the series of papers [41–43].

Suppose that $\mathcal{P}_r$ is a polar space of a given type of rank r. If it has no ovoid, and an upper bound on the size of a partial ovoid is known, then the argument used in Lemma 3.2 makes it possible to deduce an upper bound for a partial ovoid of a polar space $\mathcal{P}_{r+1}$. Inductive bounds described in [16] and [17] are presented in Table 6, where $x_{n,q}$ denotes the upper bound on the size of a partial ovoid in the corresponding classical finite polar space with ambient projective space $PG(2n,q)$ or $PG(2n+1,q)$.

4.2. Partial Spreads

Partial spreads require a different treatment than partial ovoids. On the one hand, counting techniques like the one of Glynn mentioned above for maximal partial ovoids, applied in rank 2 to obtain lower bounds, yield, dualizing, lower bounds on the size of maximal partial spreads. On the other hand, inductive bounds are not possible for spreads, so arguments must be found for general rank.

We first mention results on lower bounds on the size of maximal partial spreads of polar spaces. It is shown in [16] that any maximal partial spread of a polar space $\mathcal{P}$ has at least $t+1$ elements, where $t+1$ is the number of lines through a point in the polar space $\mathcal{P}'$ of rank 2 of the same type as $\mathcal{P}$. For hyperbolic quadrics, this theorem yields a lower bound of 2, which is improved in [16] for $\mathrm{Q}^+(4n+3,q)$ to $q+1$. Better lower bounds for polar spaces of rank 2 can, if applicable, be found in Table 4, by applying duality. For $\mathrm{H}(4,q^2)$, the following result is known.

Result 4.6 (see [63, Theorem 2.2]). *A maximal partial spread of* $\mathrm{H}(4,q^2)$ *contains at least* $\lceil q^3+q\sqrt{q}-\frac{q}{2}-\frac{3}{8}\sqrt{q}+\frac{7}{8}\rceil$ *elements.*

As indicated, we start our overview of upper bounds with the case $\mathrm{H}(3,q^2)$. The proof relies on a geometric property of hermitian varieties that is useful in several cases.

Result 4.7 (see [84]). *Let* π_1, π_2 *and* π *be mutually skew generators of* $\mathrm{H}(2n+1,q^2)$. *Then the points of* π *that lie on a line of* $\mathrm{H}(2n+1,q^2)$, *meeting* π_1 *and* π_2, *form a hermitian variety* $\mathrm{H}(n,q^2)$ *in* π.

Theorem 4.8 (see [16]). *A partial spread of* $\mathrm{H}(3,q^2)$ *has at most* $\frac{1}{2}(q^3+q+2)$ *elements.*

Proof. Suppose that $\mathcal{S}$ is a partial spread of $\mathrm{H}(3,q^2)$ and that $|\mathcal{S}|=q^3+1-\delta$. Then the number of points of $\mathrm{H}(3,q^2)$ not covered by lines of $\mathcal{S}$ is $h=\delta(q^2+1)$. We call these points *holes*.

Consider triples (l_1,l_2,P), where l_1 and l_2 are different elements of $\mathcal{S}$ and where P is a hole. We will estimate how many of these triples have the property that the unique line of $\mathrm{PG}(3,q^2)$ on P that meets l_1 and l_2 is a line of $\mathrm{H}(3,q^2)$.

To do so, we consider a hole P. Then P lies on $q+1$ lines of $\mathrm{H}(3,q^2)$. If x_i, $i=1,\dots,q+1$, is the number of points on the i-th line on P covered by an element of $\mathcal{S}$, then we have $\sum x_i=|\mathcal{S}|$ and hence

$$\sum x_i(x_i-1)\geq (q+1)\frac{|\mathcal{S}|}{q+1}\left(\frac{|\mathcal{S}|}{q+1}-1\right).$$

So we find a lower bound on the number of triples, using that the number of holes equals $\delta(q^2+1)$.

Now choose a pair (l_1,l_2) of distinct spread elements. There are q^2+1 lines of $\mathrm{H}(3,q^2)$ that meet l_1 and l_2. These lines cover $(q^2+1)(q^2-1)$ points of $\mathrm{H}(3,q^2)$ not on l_1 and l_2. By Result 4.7, every line of $\mathcal{S}\setminus\{l_1,l_2\}$ contains $q+1$ of these points. Thus there are $(q^2+1)(q^2-1)-(|\mathcal{S}|-2)(q+1)$ holes. Together with the lower bound, this gives

$$|\mathcal{S}|(|\mathcal{S}|-1)\left[(q^4-1)-(|\mathcal{S}|-2)(q+1)\right]\geq (q^3+1-|\mathcal{S}|)(q^2+1)|\mathcal{S}|\left(\frac{|\mathcal{S}|}{q+1}-1\right).$$

After simplification, we obtain $|\mathcal{S}|\leq\frac{1}{2}(q^3+q+2)$. □

Remarkably, this bound is sharp for $q = 2$ and $q = 3$, [30,32]. But for $q = 4,5$, exhaustive computer searches have shown that this bound is not sharp.

In [16], the proof of Theorem 4.8 is presented for partial spreads of $\mathrm{H}(4n+3,q^2)$ and also yields for $n \geq 1$ an upper bound.

Result 4.7 has an analogon for hyperbolic quadrics and symplectic polar spaces.

Result 4.9 (see [49]). (i) *Let g_1, g_2 and g_3 be three mutually skew generators of $\mathrm{Q}^+(4n+3,q)$. Then the lines of g_1 that lie in a totally isotropic 3-space intersecting g_2, g_3 in a line, form a symplectic space $\mathrm{W}(2n+1,q)$ in g_1.*

(ii) *Let g_1, g_2 and g_3 be three pairwise skew generators of $\mathrm{W}(2n+1,q)$, $n \geq 2$. Let $\mathcal{P}$ be the set of points P in g_1 such that there exists a line in $\mathrm{W}(2n+1,q)$ through P intersecting g_2 and g_3.*
For q even and n even, $\mathcal{P}$ forms a pseudo-polarity of g_1.
For q even and n odd, $\mathcal{P}$ is either a pseudo-polarity or a symplectic polarity (depending on the relative position of g_1, g_2 and g_3).
For q odd and n even, $\mathcal{P}$ is a parabolic quadric in g_1.
For q odd and n odd, $\mathcal{P}$ is either an elliptic or hyperbolic quadric (depending on the relative position of g_1, g_2 and g_3).

In [49], these results are used to derive lower bounds on the size of maximal partial spreads in these polar spaces.

Vanhove [90] obtained an upper bound on the size of partial spreads of $\mathrm{H}(4n+1,q^2)$. The proof relies on a remarkable link to *association schemes*, combinatorial structures consisting of a set Ω and a set of symmetric relations partitioning $\Omega \times \Omega$, with high regularity. In our case, if Ω is the set of generators of a polar space of rank d, and two generators g_1 and g_2 are i-related if the codimension of $g_1 \cap g_2$ in g_1 is i, then $(\Omega,(R_0,\dots,R_d))$ is an association scheme. A partial spread of the polar space is a clique of the relation R_d of this association scheme.

The real vector space $\mathbb{R}\Omega$, for an association scheme $(\Omega,(R_0,\dots,R_d))$ in general, can be decomposed orthogonally into $d+1$ subspaces V_i, such that each non-zero vector of V_i is an eigenvector of the relation R_j with eigenvalue P_{ij}. Define the matrix $P = (P_{ij})$. Then the *dual matrix of eigenvalues* is defined as $Q = |\Omega|P^{-1}$. Define the inner distribution vector of any non-empty subset X of Ω as $\mathbf{a} = (a_i)$ with $a_i = \frac{|(X\times X)\cap R_i|}{|X|}$. Then it is shown in e.g. [29] that every entry of $\mathbf{a}Q$ is non-negative. For X a clique of a non-trivial relation R_j, it is shown in [36, Lemma 2.4.1] or [66, Lemma 3.2.2] that $1-\frac{k}{\lambda}$ is an upper bound on the size of X, with k the valency of the relation R_j and λ the smallest eigenvalue of the relation R_j.

Applied to the specific case of $\mathrm{H}(4n+1,q^2)$, $q^{2n+1}+1$ is found as upper bound on the size of a partial spread. For other polar spaces, this method does not give non-trivial results.

Vanhove gives in [91] an alternative proof for this result, which is now purely geometric and based on a clever generalization of steps taken in [23]. As in [23], this method gives also some insight in case of equality.

Note that the upper bound $q^{2n+1}+1$ on the size of a partial spread in $\mathrm{H}(4n+1,q^2)$ is sharp. One sees easily that a spread of the symplectic polar space $\mathrm{W}(2n+1,q)$ embedded in $\mathrm{H}(2n+1,q^2)$ extends to a partial spread of $\mathrm{H}(2n+1,q^2)$. Maximality (proved earlier for $n = 2$ in [1], and for n even, $n \geq 2$ in [52]) now follows from the upper bound.

Table 7. Upper bounds on the size of partial spreads

polar space	upper bound	references
$Q(4n,q)$, q odd	$q^n+1-\delta$, $\delta \geq \varepsilon$, with $q+1+\varepsilon$ the size of the smallest non-trivial blocking set of $PG(2,q)$	[37]
$Q^+(4n+1,q)$	2	
$H(3,q^2)$	$\frac{1}{2}(q^3+q+2)$ (sharp for $q=2,3$)	[16] ([31,32])
$H(4n+1,q^2)$	$q^{2n+1}+1$ (sharp)	[23], [90], [91]
$H(4n+3,q^2)$	$q^{4n+3}-q^{3n+3}(\sqrt{q}-1)$	[16]

Only for $Q(4n,q)$, q odd, and $Q^+(4n+1,q)$, it is proved, without further assumptions on q, that spreads do not exist. This is clear for $Q^+(4n+1,q)$ by Result 1.3. An upper bound on the size of partial spreads of $Q(4n,q)$, q odd, is proved in [37]. The upper bound is related to the size blocking sets of $PG(2,q)$ (see e.g. [8]), and is obtained by analyzing the set of points of $Q(4n,q)$ not covered by any element of the partial spread, and describing this set using characterization results on multiple weighted blocking sets (minihypers) of the ambient projective space. Recent results on the latter objects can be found in [50].

Table 7 contains an overview of the cases where the non-existence of a spread is proved.

The existence of spreads of the polar space $Q(6,q)$ and $Q^+(7,q)$ is not known for all q. In this situation the difficulty is to find an upper bound on the size of a maximal partial spread, without any assumption on the existence of spreads. Using the results on (maximal) partial ovoids of $Q^+(7,q)$ and the triality map of $Q^+(7,q)$, the following result is derived in [17].

Result 4.10. *The polar space $Q^+(7,q)$ has no maximal proper partial spread of size $q^3+1-\delta$ with $0<\delta<q+1$.*

Embedding $Q(6,q)$ in $Q^+(7,q)$ as a hyperplane section, we find in [17] exactly the same result for $Q(6,q)$.

Recall that upper bounds on the size of maximal proper partial spreads of $Q(4,q)$ and $Q^-(5,q)$ respectively, are found in Corollary 4.4 and Theorem 5.2 (a), and Theorem 5.2 (b) respectively.

5. Covers and Blocking Sets

Let $\mathcal{P}$ be a classical finite polar space. A *cover* is a set $\mathcal{C}$ of generators such that every point of P lie in at least one generator of $\mathcal{C}$. A cover is *minimal* if it does not contain a smaller cover. A *blocking set* is a set $\mathcal{B}$ of points with the property hat every generator contains at least one point of $\mathcal{B}$. A blocking set is *minimal* if it does not contain a smaller blocking set.

If $\mathcal{P}$ has rank 2, then clearly a blocking set of $\mathcal{P}$ is mapped by a duality on a cover of the dual space of $\mathcal{P}$. So as in the ovoid-spread case, dualities, and other isomorphisms, can play a role in the construction of these objects from each other.

The study of blocking sets and covers is motivated in the same way as the study of partial ovoids and partial spreads. Non-existence of ovoids motivates the study of the sets of points blocking all generators. Existence of ovoids poses the question how large a blocking set must be if it does not contain an ovoid. The motivation for the study of covers is of course the same.

5.1. Covers

The study of covers is similar to the study of maximal partial spreads, but there are additional difficulties. We explain this with the following example.

Consider a minimal cover $\mathcal{C}$ (or maximal partial spread) of $Q(4,q)$ (or $Q^-(5,q)$) with $q^2+1\pm\delta$ (or $q^3+1\pm\delta$) lines. Let $w:\mathcal{P}\to\mathbb{N}$ be the function that assigns to every point of $Q(4,q)$ (or $Q^-(5,q)$) the number $w(P)$ of lines of $\mathcal{C}$ through P. Let $w'(P)=w(P)-1$ (or $w'(P)=1-w(P)$ if we start from a partial spread).

From now on, we work with the weight function w' and the arguments are the same for covers and spreads. The only difference is that in the case of partial spreads we know that w' has range $\{0,1\}$. Let π be a hyperplane, every line of $\mathcal{C}$ meets π either in 1 or $q+1$ points, so

$$\sum_{P\in\pi} w'(P)\equiv\delta \mod q\ .$$

This shows that for $1\le\delta<q$, the weight function w' defines a blocking set of the ambient projective space, completely contained in $\mathcal{P}$. For such blocking sets we have the following result.

Lemma 5.1 (see [48], Lemma 2.1). *Consider in* $PG(4,q)$ *a quadric that is a cone with vertex a point P over a non-degenerate elliptic quadric* $Q^-(3,q)$. *Suppose that B is a set of at most 2q points contained in the quadric. If every solid of* $PG(4,q)$ *meets B, then one of the following occurs:*

(a) Some line of the quadric is contained in B.

(b) $|B|>\frac{9}{5}q+1$, $P\in B$, *and there exists a unique line l of the quadric that meets B in at least* $2+\frac{1}{2}(3q-|B|)$ *points. This line has at most* $|B|-1-q$ *points in B.*

Applying this lemma to the weight function w' shows immediately that for $\delta\le\frac{4}{5}q$ the corresponding blocking set contains a line. In the case of partial spreads this result is exactly what we want, and with some extra work one can use the information that w' is at most 1 to extend the result to all $\delta\le q$. Thus we get the following theorem.

Theorem 5.2 (see [48]). *(a) Every partial spread of* $Q(4,q)$ *of size* $q^2+1-\delta$, $\delta<q$, *extends to a spread.*

(b) Every partial spread of $Q^-(5,q)$ *of size* $q^3+1-\delta$, $\delta<q$, *extends to a spread.*

(c) Let $\mathcal{C}$ *be a cover of* $Q(4,q)$ *of size* $q^2+1+\delta$, $\delta\le\frac{4}{5}q$. *For every point P let* $w'(P)+1$ *the number of lines of C through P. Then there exists lines* $l_1,\dots,l_\delta$ *of* $Q(4,q)$ *such that* $w'(P)$ *is equal to the number of lines* l_i *through P.*

(d) Let $\mathcal{C}$ *be a cover of* $Q^-(5,q)$ *of size* $q^3+1+\delta$, $\delta\le\frac{4}{5}q$. *For every point P let* $w'(P)+1$ *the number of lines of* $\mathcal{C}$ *through P. Then there exists lines* $l_1,\dots,l_\delta$ *of* $Q^-(5,q)$ *such that* $w'(P)$ *is equal to the number of lines* l_i *through P.*

In the case of covers, it is unclear if the lines $l_1, \ldots, l_\delta$ belong to the cover, so it is not shown that the cover can be reduced to a smaller cover. For $Q(4,q)$, q odd, this was done for small δ using a long and complicated algebraic argument.

Result 5.3 (see [48, Theorem 1.3]). *Let q be odd. Then a cover of $Q(4,q)$ contains at least $q^2 - q - \frac{3}{2} + \frac{\sqrt{8q^2+20q+25}}{2} \approx q^2 + 0.414q$ lines.*

For $Q^-(5,q)$ this is however not possible as $Q^-(5,q)$ has small minimal covers, constructed in the following example. The construction uses, as in Example 4.5, hermitian spreads of $Q^-(5,q)$.

Example 5.4. *Consider a hermitian spread $\mathcal{S}$ of $Q^-(5,q)$. Recall that such a spread is the union of q^2 reguli R_i through a common line. Let R_i^{opp} be the regulus opposite to R_i. Define $\mathcal{S}' := (\mathcal{S} \cup R_1^{opp}) \setminus R_1$. Then $\mathcal{S}'$ is again a spread. But this procedure can be repeated, and now $\mathcal{S}'' := (\mathcal{S}' \cup R_2^{opp}) \setminus R_2$ will be a minimal cover of size q^3+2. Clearly, one can construct minimal covers of any size in the range $q^3+2, \ldots, q^3+q^2$ using this method.*

This is quite typical for covers and blocking sets of finite polar spaces. Using arguments from the partial spread and partial ovoid case yield results similar to Theorem 5.2. Deciding if the extra lines (or points) are already inside the cover (or blocking set) is the hard part.

5.2. Blocking Sets

Suppose that $\mathcal{P}_r$ is a polar space of rank r of a given type. In most cases where the non-existence of ovoids of $\mathcal{P}_{r-1}$ is proved, the smallest minimal blocking sets of $\mathcal{P}_r$ are known. To describe the examples, we introduce a *truncated cone*. Suppose that π is any subspace in $PG(n,q)$, and $\mathcal{O}$ any point set contained in π', a subspace skew to π. The truncated cone $\pi^*\mathcal{O}$, is the set of all points on all lines connecting a point of π and a point of $\mathcal{O}$, minus the points of π. Usually, for polar spaces, $\pi \subseteq \mathcal{P}_r$, $\pi' \subseteq \pi^\perp$ and $\mathcal{O} \subseteq \mathcal{P}_r \cap \pi'$.

Table 8 lists the smallest minimal blocking sets of polar spaces of which the non-existence of ovoids is proved.

The result on blocking sets of the polar spaces $W(2n+1,q)$, $n \geq 2$ is found by Metsch [59]. It classifies the smallest minimal blocking sets when q is even, and shows a lower bound on the size when q is odd. Apart from this lower bound, nothing is known. Independently, De Beule and Storme treated the case $n = 2$ and q even in [25].

Another interesting open case is to determine the smallest minimal blocking sets of $Q(2n,q)$, q odd, q not prime and $q \neq 3$. It is conjectured (see e.g. [67]) that $Q(2n,q)$ has ovoids if and only if $q = 3^h$, and it is expected that the smallest minimal blocking sets always are truncated cones $\pi_{n-3}^*\mathcal{O}$, $\mathcal{O}$ an ovoid of $Q(4,q)$, when $q \neq 3^h$.

In the spaces $Q^+(2n+1,q)$, $q \in \{2,3\}$, $n \geq 4$, not only the smallest minimal blocking sets are known. In [27] and [24], the geometrical arguments used to study blocking sets enable to classify the two smallest minimal blocking sets of $Q^+(2n+1,2)$, $n \geq 3$, and the three smallest minimal blocking sets of $Q^+(2n+1,3)$, $n \geq 3$.

For $Q(4,q)$, the smallest blocking sets are ovoids. Clearly, a truncated cone $\pi_0^*\mathcal{C}$, $\mathcal{C}$ a conic, is a minimal blocking set of $Q(4,q)$ different from an ovoid. But up to now, for q even, minimal blocking sets different from an ovoid of size s, $s < q^2+1+\frac{q+4}{6}$, are excluded [70]. For q odd, q prime, only minimal blocking sets of size q^2+2 are excluded [21]. The

Table 8. Smallest minimal blocking sets

polar space	example	references
$W(2n+1,q)$, q even, $n>2$	$\pi^*_{n-2}O$, O an ovoid of $W(3,q)$	[59], [25]
$Q(2n,p)$, $p>3$ odd prime, $n>2$	$\pi^*_{n-3}Q^-(3,q)$	[19]
$Q^-(2n+1,q)$, $n\geq 2$	$\pi^*_{n-2}Q^-(3,q)$	[54]
$H(2n,q^2)$, $n\geq 2$	$\pi^*_{n-2}H(2,q^2)$	[22]
$Q(2n,3)$, $n\geq 4$	$\pi^*_{n-4}O$, O an ovoid of $Q(6,3)$	[26]
$Q^+(2n+1,q)$, $q\in\{2,3\}$, $n\geq 4$	$\pi^*_{n-4}O$, O an ovoid of $Q^+(7,q)$	[27], [24]

smallest minimal blocking sets of $Q(6,3)$ different from an ovoid are truncated cones $\pi^*_0 O$, O an ovoid of $Q(4,3)$ [26]. Blocking sets of $W(3,q)$, q odd, are dually the same as covers of $Q(4,q)$, q odd, so we refer to Result 5.3. Finally, minimal blocking sets of $H(3,q^2)$, dually minimal covers of $Q^-(5,q)$, are constructed with size in the range $q^3+2,\ldots,q^3+q^2$ in Example 5.4.

Let now $\mathcal{P}$ be a finite classical polar space of rank r. A *blocking set with respect to the s-dimensional spaces of $\mathcal{P}$* is a set of points of $\mathcal{P}$ blocking all s-dimensional spaces, $s\leq r-1$, contained in $\mathcal{P}$. When $s=r-1$, we are considering blocking sets. We have seen that in some cases the smallest blocking sets are truncated cones with base an ovoid of a polar space of low rank, so the existence or non-existence of ovoids, which is not completely known for all polar spaces, complicates the work. However, more can be done for blocking sets with respect to s-spaces for $1\leq s<r-1$. The basic observation is that s-dimensional spaces of $\mathcal{P}$ also are s-dimensional subspaces of the ambient projective space, and these are all blocked by a subspace of the ambient projective space of codimension s. In many cases, it can be shown that a blocking set with respect to s-dimensional subspaces of $\mathcal{P}$ can be constructed from an intersection of $\mathcal{P}$ with a subspace of the ambient projective space of codimension s . All results we describe here, are based on results found in the series of papers [51, 54–58, 60]. The following general result for quadrics is proved in [58].

Result 5.5. *Let $\mathcal{Q}$ be a non-degenerate quadric and d the dimension of its generators. Assume that $s<d$ when $\mathcal{Q}$ is not elliptic, and assume $s\leq d$ otherwise. Then the smallest (minimal) blocking sets with respect to the s-dimensional spaces of $\mathcal{Q}$ have the form $(T\setminus T^\perp)\cap\mathcal{Q}$ for a suitable subspace T of the ambient projective space of codimension s.*

The *suitability* of the subspace T refers to its intersection type with $\mathcal{Q}$, which we will describe in detail below. The size of the constructed blocking set is dependent on the intersection type, hence it is not surprising that in some cases minimal blocking sets are obtained that are not the smallest. Result 5.5 leads to the classification of blocking sets with respect to s-dimensional spaces below a given size. Table 9 surveys known results for quadrics. Each line must be interpreted as: *a blocking set B of the space $\mathcal{P}$ with respect to its s-dimensional spaces, with size smaller than the given size, contains (one of) the given example(s).* Only for $Q^-(2n+1,q)$ the shown result includes the result for the smallest minimal blocking sets with respect to its generators.

The following result and corollary for $H(2n+1,q^2)$ are proved in [60].

Table 9. Blocking sets with respect to s-spaces

polar space	dimension s	given size	given example(s)
$Q^+(2n+1,q)$	$2 \leq s \leq n-1$	$(q^n+q^{s-2}+1)\theta_{n-s}(q)$	$\pi^*_{s-3}Q^-(2(n-s)+3,q)$
$Q^+(2n+1,q)$	1	$(q^{n-1}+1)\theta_{n-1}(q)$	$Q(2n,q)$ or $P^*Q^+(2n-1,q)$
$Q(2n,q)$	$1 \leq s \leq n-2$	$(q^n+q^{s-1}+1)\theta_{n-s-1}(q)$	$\pi^*_{s-2}Q^-(2(n-s)+1,q)$
$Q^-(2n+1,q)$	$2 \leq s \leq n-2$	$(q^{n+1}+q^s+1)\theta_{n-s-1}(q)$	$\pi^*_{s-1}Q^-(2(n-s)+1,q)$

Result 5.6. *Consider* $H(2n+1,q^2)$ *and an integer s,* $1 \leq s < n$. *Concerning the cardinalities of the minimal blocking sets of* $H(2n+1,q^2)$ *with respect to s-spaces, the sets* $(T \setminus T^\perp) \cap H(2n+1,q^2)$, *T a subspace of the ambient projective space* $PG(2n+1,q^2)$ *of codimension s, provide the two smallest cardinalities when* $s \in \{1,2\}$ *and the* $s-2$ *smallest cardinalities when* $s \geq 3$.

Corollary 5.7. *The smallest blocking sets of* $H(2n+1,q^2)$ *with respect to s-spaces,* $1 \leq s < n$, *are truncated cones* $\pi^*_{s-2}H(2n+2-2s,q^2)$.

Consider the embedding of $H(2n,q^2)$ in $H(2n+1,q^2)$ as a hyperplane section. It is clear that a point set $B \subset H(2n,q^2)$ is a blocking set of $H(2n,q^2)$ with respect to s-spaces, $1 \leq s \leq n-1$, if and only if B is a blocking set of $H(2n+1,q^2)$ with respect to $(s+1)$-spaces. So by Corollary 5.7 we know the smallest blocking sets of $H(2n,q^2)$ with relation to s-spaces, $1 \leq s < n-1$. Recall that the case $s = n-1$ for $H(2n,q^2)$ is described in the fourth line of Table 8. Finally, the case $W(2n+1,q)$, q odd, is completely open. Even the smallest blocking sets with respect to lines of $W(2n+1,q)$, q odd, are not known.

Acknowledg ments

The authors would like to thank Frédéric Vanhove for his help with summarizing [90].

References

[1] A. Aguglia, A. Cossidente, and G. L. Ebert, *Complete spans on Hermitian varieties*, Des. Codes Cryptogr., 29 (2003), pp. 7–15.

[2] A. Aguglia, G. L. Ebert, and D. Luyckx, *On partial ovoids of Hermitian surfaces*, Bull. Belg. Math. Soc. Simon Stevin, 12 (2005), pp. 641–650.

[3] S. Ball, *The polynomial method in Galois geometries*, in Current research topics in Galois geometry, Nova Sci. Publ., New York, 2011, ch. 5, pp. 103–128.

[4] S. Ball, P. Govaerts, and L. Storme, *On ovoids of parabolic quadrics*, Des. Codes Cryptogr., 38 (2006), pp. 131–145.

[5] J. Bamberg, S. Kelly, M. Law, and T. Penttila, *Tight sets and m-ovoids of finite polar spaces*, J. Combin. Theory Ser. A, 114 (2007), pp. 1293–1314.

[6] J. Bamberg and T. Penttila, *A classification of transitive ovoids, spreads, and m-systems of polar spaces*, Forum Math., 21 (2009), pp. 181–216.

[7] A. Blokhuis and G. E. Moorhouse, *Some p-ranks related to orthogonal spaces*, J. Algebraic Combin., 4 (1995), pp. 295–316.

[8] A. Blokhuis, P. Sziklai, and T. Szőnyi, *Blocking sets in projective spaces*, in Current research topics in Galois geometry, Nova Sci. Publ., New York, 2011, ch. 3, pp.61–84.

[9] M. R. Brown, J. De Beule, and L. Storme, *Maximal partial spreads of $T_2(O)$ and $T_3(O)$*, European J. Combin., 24 (2003), pp. 73–84.

[10] P. J. Cameron, J. A. Thas, and S. E. Payne, *Polarities of generalized hexagons and perfect codes*, Geometriae Dedicata, 5 (1976), pp. 525–528.

[11] M. Cimráková, S. De Winter, V. Fack, and L. Storme, *On the smallest maximal partial ovoids and spreads of the generalized quadrangles $W(q)$ and $Q(4,q)$*, European J. Combin., 28 (2007), pp. 1934–1942.

[12] J. H. Conway, P. B. Kleidman, and R. A. Wilson, *New families of ovoids in O_8^+*, Geom. Dedicata, 26 (1988), pp. 157–170.

[13] A. Cossidente, C. Culbert, G. L. Ebert, and G. Marino, *On m-ovoids of* $\mathrm{W}_3(q)$, Finite Fields Appl., 14 (2008), pp. 76–84.

[14] A. Cossidente and T. Penttila, *Hemisystems on the Hermitian surface*, J. London Math. Soc. (2), 72 (2005), pp. 731–741.

[15] J. De Beule and A. Gács, *Complete arcs on the parabolic quadric* $\mathrm{Q}(4,q)$, Finite Fields Appl., 14 (2008), pp. 14–21.

[16] J. De Beule, A. Klein, K. Metsch, and L. Storme, *Partial ovoids and partial spreads in Hermitian polar spaces*, Des. Codes Cryptogr., 47 (2008), pp. 21–34.

[17] ——, *Partial ovoids and partial spreads in symplectic and orthogonal polar spaces*, European J. Combin., 29 (2008), pp. 1280–1297.

[18] ——, *Partial ovoids and partial spreads of classical finite polar spaces*, Serdica Math. J., 34 (2008), pp. 689–714.

[19] J. De Beule and K. Metsch, *Small point sets that meet all generators of $Q(2n,p)$, $p > 3$ prime*, J. Combin. Theory Ser. A, 106 (2004), pp. 327–333.

[20] ——, *The Hermitian variety $H(5,4)$ has no ovoid*, Bull. Belg. Math. Soc. Simon Stevin, 12 (2005), pp. 727–733.

[21] ——, *Minimal blocking sets of size q^2+2 of $Q(4,q)$, q an odd prime, do not exist*, Finite Fields Appl., 11 (2005), pp. 305–315.

[22] ——, *The smallest point sets that meet all generators of $H(2n,q^2)$*, Discrete Math., 294 (2005), pp. 75–81.

[23] ——, *The maximum size of a partial spread in* $H(5,q^2)$ *is* q^3+1, J. Combin. Theory Ser. A, 114 (2007), pp. 761–768.

[24] J. De Beule, K. Metsch, and L. Storme, *Characterization results on small blocking sets of the polar spaces* $Q^+(2n+1,2)$ *and* $Q^+(2n+1,3)$, Des. Codes Cryptogr., 44 (2007), pp. 197–207.

[25] J. De Beule and L. Storme, *The smallest minimal blocking sets of* $Q(6,q)$, *q even*, J. Combin. Des., 11 (2003), pp. 290–303.

[26] ——, *On the smallest minimal blocking sets of* $Q(2n,q)$, *for q an odd prime*, Discrete Math., 294 (2005), pp. 83–107.

[27] ——, *Blocking all generators of* $Q^+(2n+1,3)$, $n \geq 4$, Des. Codes Cryptogr., 39 (2006), pp. 323–333.

[28] F. De Clerck and N. Durante, *Constructions and characterizations of classical sets in* $\mathrm{PG}(n,q)$, in Current research topics in Galois geometry, Nova Sci. Publ., New York, 2011, ch. 1, pp. 1–32.

[29] P. Delsarte, *An algebraic approach to the association schemes of coding theory*, Philips Res. Rep. Suppl., (1973), pp. vi+97.

[30] R. H. Dye, *Partitions and their stabilizers for line complexes and quadrics*, Ann. Mat. Pura Appl. (4), 114 (1977), pp. 173–194.

[31] ——, *Maximal sets of nonpolar points of quadrics and symplectic polarities over* $\mathrm{GF}(2)$, Geom. Dedicata, 44 (1992), pp. 281–293.

[32] G. L. Ebert and J. W. P. Hirschfeld, *Complete systems of lines on a Hermitian surface over a finite field*, Des. Codes Cryptogr., 17 (1999), pp. 253–268.

[33] S. Ferret and L. Storme, *Results on maximal partial spreads in* $\mathrm{PG}(3,p^3)$ *and on related minihypers*, Des. Codes Cryptogr., 29 (2003), pp. 105–122.

[34] N. Gill, *Polar spaces and embeddings of classical groups*, New Zealand J. Math., 36 (2007), pp. 175–184.

[35] D. G. Glynn, *A lower bound for maximal partial spreads in* $\mathrm{PG}(3,q)$, Ars Combin., 13 (1982), pp. 39–40.

[36] C. Godsil, *Association schemes.*, tech. rep., University of Waterloo, 2001.

[37] P. Govaerts, L. Storme, and H. Van Maldeghem, *On a particular class of minihypers and its applications. III. Applications*, European J. Combin., 23 (2002), pp. 659–672.

[38] A. Gunawardena and G. E. Moorhouse, *The non-existence of ovoids in* $O_9(q)$, European J. Combin., 18 (1997), pp. 171–173.

[39] N. Hamilton and R. Mathon, *Existence and non-existence of m-systems of polar spaces*, European J. Combin., 22 (2001), pp. 51–61.

[40] N. Hamilton and C. T. Quinn, *m-systems of polar spaces and maximal arcs in projective planes*, Bull. Belg. Math. Soc. Simon Stevin, 7 (2000), pp. 237–248.

[41] O. Heden, *Maximal partial spreads and the modular n-queen problem*, Discrete Math., 120 (1993), pp. 75–91.

[42] ——, *Maximal partial spreads and the modular n-queen problem. II*, Discrete Math., 142 (1995), pp. 97–106.

[43] ——, *Maximal partial spreads and the modular n-queen problem. III*, Discrete Math., 243 (2002), pp. 135–150.

[44] J. W. P. Hirschfeld, *Finite projective spaces of three dimensions*, Oxford Mathematical Monographs, The Clarendon Press Oxford University Press, New York, 1985. Oxford Science Publications.

[45] J. W. P. Hirschfeld and J. A. Thas, *General Galois geometries*, Oxford Mathematical Monographs, The Clarendon Press Oxford University Press, New York, 1991. Oxford Science Publications.

[46] W. M. Kantor, *Ovoids and translation planes*, Canad. J. Math., 34 (1982), pp. 1195–1207.

[47] A. Klein, *Partial ovoids in classical finite polar spaces*, Des. Codes Cryptogr., 31 (2004), pp. 221–226.

[48] A. Klein and K. Metsch, *New results on covers and partial spreads of polar spaces*, Innov. Incidence Geom., 1 (2005), pp. 19–34. See also *Corrections to "New results on covers and partial spreads of polar spaces"*, Innov. Incidence Geom., to appear.

[49] A. Klein, K. Metsch, and L. Storme, *Small maximal partial spreads in classical finite polar spaces*, Adv. Geom., 10 (2010), pp. 379–402.

[50] I. Landjev and L. Storme, *Galois geometries and coding theory*, in Current research topics in Galois geometry, Nova Sci. Publ., New York, 2011, ch. 8, pp. 185–212.

[51] D. Luyckx, *Blocking $(n-2)$-dimensional subspaces on $Q(2n,q)$*, European J. Combin., 22 (2001), pp. 521–528.

[52] ——, *On maximal partial spreads of $H(2n+1,q^2)$*, Discrete Math., 308 (2008), pp. 375–379.

[53] K. Metsch, *Improvement of Bruck's completion theorem*, Des. Codes Cryptogr., 1 (1991), pp. 99–116.

[54] ——, *The sets closest to ovoids in $Q^-(2n+1,q)$*, Bull. Belg. Math. Soc. Simon Stevin, 5 (1998), pp. 389–392. Finite geometry and combinatorics (Deinze, 1997).

[55] ——, *A Bose-Burton theorem for elliptic polar spaces*, Des. Codes Cryptogr., 17 (1999), pp. 219–224.

[56] ——, *Bose-Burton type theorems for finite projective, affine and polar spaces*, in Surveys in combinatorics, 1999 (Canterbury), vol. 267 of London Math. Soc. Lecture Note Ser., Cambridge Univ. Press, Cambridge, 1999, pp. 137–166.

[57] ——, *On blocking sets of quadrics*, J. Geom., 67 (2000), pp. 188–207. Second Pythagorean Conference (Pythagoreion, 1999).

[58] ——, *A Bose-Burton type theorem for quadrics*, J. Combin. Des., 11 (2003), pp. 317–338.

[59] ——, *Small point sets that meet all generators of* $W(2n+1,q)$, Des. Codes Cryptogr., 31 (2004), pp. 283–288.

[60] ——, *Blocking structures of Hermitian varieties*, Des. Codes Cryptogr., 34 (2005), pp. 339–360.

[61] ——, *Small maximal partial ovoids of* $H(3,q^2)$, Innov. Incidence Geom., 3 (2006), pp. 1–12.

[62] K. Metsch and L. Storme, *Partial t-spreads in* $\mathrm{PG}(2t+1,q)$, Des. Codes Cryptogr., 18 (1999), pp. 199–216. Designs and codes—a memorial tribute to Ed Assmus.

[63] ——, *Maximal partial ovoids and maximal partial spreads in Hermitian generalized quadrangles*, J. Combin. Des., 16 (2008), pp. 101–116.

[64] G. E. Moorhouse, *Ovoids from the* E_8 *root lattice*, Geom. Dedicata, 46 (1993), pp. 287–297.

[65] ——, *Some p-ranks related to Hermitian varieties*, J. Statist. Plann. Inference, 56 (1996), pp. 229–241. Special issue on orthogonal arrays and affine designs, Part II.

[66] M. Newman, *Independent Sets and Eigenspaces*, PhD thesis, University of Waterloo, 2004.

[67] C. M. O'Keefe and J. A. Thas, *Ovoids of the quadric* $Q(2n,q)$, European J. Combin., 16 (1995), pp. 87–92.

[68] S. E. Payne and J. A. Thas, *Finite generalized quadrangles*, EMS Series of Lectures in Mathematics, European Mathematical Society (EMS), Zürich, second ed., 2009.

[69] T. Penttila and B. Williams, *Ovoids of parabolic spaces*, Geom. Dedicata, 82 (2000), pp. 1–19.

[70] V. Pepe, L. Storme, and G. Van de Voorde, *On codewords in the dual code of classical generalised quadrangles and classical polar spaces*, Discrete Math., 310 (2010), pp. 3132–3148.

[71] B. Segre, *On complete caps and ovaloids in three-dimensional Galois spaces of characteristic two*, Acta Arith., 5 (1959), pp. 315–332 (1959).

[72] ——, *Forme e geometrie hermitiane, con particolare riguardo al caso finito*, Ann. Mat. Pura Appl. (4), 70 (1965), pp. 1–201.

[73] E. E. Shult and J. A. Thas, *m-Systems of polar spaces*, J. Combin. Theory Ser. A, 68 (1994), pp. 184–204.

[74] ——, *Constructions of polygons from buildings*, Proc. London Math. Soc. (3), 71 (1995), pp. 397–440.

[75] ——, *m-Systems and partial m-systems of polar spaces*, Des. Codes Cryptogr., 8 (1996), pp. 229–238. Special issue dedicated to Hanfried Lenz.

[76] P. Sin, *The p-rank of the incidence matrix of intersecting linear subspaces*, Des. Codes Cryptogr., 31 (2004), pp. 213–220.

[77] G. Tallini, *Blocking sets with respect to planes in* $\mathrm{PG}(3,q)$ *and maximal spreads of a nonsingular quadric in* $\mathrm{PG}(4,q)$, in Proceedings of the First International Conference on Blocking Sets (Giessen, 1989), no. 201, 1991, pp. 141–147.

[78] J. A. Thas, *Ovoidal translation planes*, Arch. Math. (Basel), 23 (1972), pp. 110–112.

[79] ——, *Two infinite classes of perfect codes in metrically regular graphs*, J. Combinatorial Theory Ser. B, 23 (1977), pp. 236–238.

[80] ——, *Polar spaces, generalized hexagons and perfect codes*, J. Combin. Theory Ser. A, 29 (1980), pp. 87–93.

[81] ——, *Ovoids and spreads of finite classical polar spaces*, Geom. Dedicata, 10 (1981), pp. 135–143.

[82] ——, *Semipartial geometries and spreads of classical polar spaces*, J. Combin. Theory Ser. A, 35 (1983), pp. 58–66.

[83] ——, *Interesting pointsets in generalized quadrangles and partial geometries*, Linear Algebra Appl., 114/115 (1989), pp. 103–131.

[84] ——, *Old and new results on spreads and ovoids of finite classical polar spaces*, in Combinatorics '90 (Gaeta, 1990), vol. 52 of Ann. Discrete Math., Amsterdam, 1992, North-Holland, pp. 529–544.

[85] ——, *Ovoids, spreads and m-systems of finite classical polar spaces*, in Surveys in combinatorics, 2001 (Sussex), vol. 288 of London Math. Soc. Lecture Note Ser., Cambridge Univ. Press, Cambridge, 2001, pp. 241–267.

[86] J. A. Thas and S. E. Payne, *Spreads and ovoids in finite generalized quadrangles*, Geom. Dedicata, 52 (1994), pp. 227–253.

[87] J. Tits, *Les groupes simples de Suzuki et de Ree.* Sem. Bourbaki 13 (1960/61), No. 210, 18 p. (1961).

[88] ——, *Sur la trialité et certains groupes qui s'en déduisent*, Inst. Hautes Études Sci. Publ. Math., (1959), pp. 13–60.

[89] ——, *Ovoïdes et groupes de Suzuki*, Arch. Math., 13 (1962), pp. 187–198.

[90] F. Vanhove, *The maximum size of a partial spread in* $\mathrm{H}(4n+1,q^2)$ *is* $q^{2n+1}+1$, Electron. J. Combin., 16 (2009). Note 13.

[91] ——, *A geometric proof of the upper bound on the size of partial spreads in* $\mathrm{H}(4n+1,q^2)$, Adv. Math. Comm., (to appear).

In: Current Research Topics in Galois Geometry
Editors: J. De Beule and L. Storme
ISBN: 978-1-61209-523-3

Chapter 3

BLOCKING SETS IN PROJECTIVE SPACES

***Aart Blokhuis*[1,*], *Péter Sziklai*[2,†] and *Tamás Szőnyi*[2,3,‡]**
[1] Eindhoven University of Technology,
Department of Mathematics and Computer Science,
P.O. Box 513, 5600 MB Eindhoven, the Netherlands
[2] Eötvös Loránd University, Institute of Mathematics,
Pázmány Péter sétány 1/C, H–1117 Budapest, Hungary
[3] Computer and Automation Research Institute,
Hungarian Academy of Sciences, Lágymányosi u. 11,
H–1111 Budapest, Hungary

Abstract

In this paper we collect results on the possible sizes of k-blocking sets. Since previous surveys focused mainly on blocking sets in the plane, we concentrate our attention on blocking sets in higher dimensions. Lower bounds on the size of the smallest non-trivial k-blocking set are surveyed in detail. The linearity conjecture and known results supporting the conjecture (e.g. proofs in particular cases) are collected. The known constructions are also presented. In case of planar minimal blocking sets we only discuss the constructions briefly. In case of higher dimensions the situation is not satisfactory, there are more open questions than known constructions.

Key Words: blocking set, (semi-)ovoid, Rédei type, minihyper.

AMS Subject Classification: 51E21

1. Introduction and Definitions

In a projective or affine space, a *blocking set* with respect to k-dimensional subspaces is a point set meeting every k-dimensional subspace. As a blocking set plus a point is still a blocking set, we are interested in minimal ones (with respect to set-theoretical inclusion) mainly. In the literature this is sometimes called a k-blocking set, we prefer the dual notion:

*E-mail address: aartb@win.tue.nl
†E-mail address: sziklai@cs.elte.hu
‡E-mail address: szonyi@cs.elte.hu

Definition 1.1. *A blocking set with respect to $(n-k)$-dimensional (so k-codimensional) subspaces of an n-dimensional projective or affine space is called a* k-blocking set. *A 1-blocking set is also called just blocking set.*

A k-dimensional subspace of $\mathrm{PG}(n,q)$ is a k-blocking set, a blocking set containing one is called *trivial*. We will see that a k-dimensional subspace is the smallest possible k-blocking set.

A point P of a k-blocking set B is *essential* if $B \setminus \{P\}$ is no longer a k-blocking set, i.e. there is an $(n-k)$-dimensional ("tangent") subspace meeting B in P only. Hence a k-blocking set is *minimal* if and only if all of its points are essential.

We say that a blocking set B is *d-dimensional* if the subspace generated by B has dimension d; in particular, for $d=2$, we say that B is *planar*.

Definition 1.2. *A blocking set is called* d-proper *if it is d-dimensional and does not contain a $(d-1)$-dimensional blocking set. The minimum size of a d-proper blocking set is denoted by $f_d(q)$. A blocking set of* $\mathrm{PG}(m,q)$ *is called* proper, *if it is m-proper.*

So a 2-proper blocking set is a blocking set in the plane, not containing a line. We also use the notation $f_d(q) = q+1+r_d(q)$. The dimension n of the space containing the blocking set does not play a role (see Proposition 2.3), so we may think of a d-proper blocking set as a set in $\mathrm{PG}(d,q)$. See also the remark after Theorem 2.6.

Another place where blocking sets occur are 'good' 2-colourings of the points of a projective space. Here good means that there is no monochromatic $(n-k)$-subspace. Each colour class is a blocking set w.r.t. $(n-k)$-subspaces, not containing an $(n-k)$-subspace. This motivates the following more general definition of Huber [39, 40].

Definition 1.3. *A* (k,s)-blocking set *of* $\mathrm{PG}(n,q)$, *$(n>k)$ is a k-blocking set that does not contain an s-subspace.*

In many cases the smallest blocking set is a cone: a *cone with vertex V and base B* is the union of (V and) the subspaces $\langle V, X\rangle$, with $X \in B$; here B is a set of points and V is a subspace such that $V \cap \langle B \rangle = \emptyset$. The cone is denoted by VB. Note that the definition allows B to be empty in which case the cone is just V.

Blocking sets, or intersection sets, are a central concept in the study of hypergraphs [29]. A hypergraph $\mathcal{H} = (H,E)$ consists of a set H of points and a collection E of subsets of H called edges. A blocking set is a subset of H intersecting every edge. A fractional blocking set is a map $b: H \to \mathbf{R}_{\geq 0}$, with $\sum_{v \in E} b(v) \geq 1$ for every edge. It's size is $\sum_{v \in H} b(v)$. If all edges have the same cardinality $\mathcal{H}$ is called *uniform*, if every point is in the same number of edges it is called *regular*. The fractional blocking number (which is a lower bound for the ordinary blocking number) of an e-uniform, regular hypergraph equals $|H|/e$. In all our problems we are dealing with uniform, regular hypergraphs. A general upper bound $(1+\log(r))|H|/e$ for the blocking number follows from a theorem by Lovász [50] (here r is the degree).

In the affine space $\mathrm{AG}(n,q)$ it is typically not possible to block k-subspaces with few points, here few means close to the fractional blocking number. For instance if $k=n-1$ then one may take n concurrent lines spanning $\mathrm{AG}(n,q)$, this set of $n(q-1)+1$ points will block all the hyperplanes, and we will see that you can't do better. Here the fractional blocking number is only q.

The blocking problem can be generalized to the following: *What is the minimum cardinality of a set T of t-subspaces such that every s-subspace is incident with at least one element of T and what is the structure of the corresponding sets T?* This is the main problem considered in Metsch [56], where one can also find several results on blocking sets.

We use the notation

$$\theta_k = \begin{bmatrix} k+1 \\ 1 \end{bmatrix}_q = \frac{q^{k+1}-1}{q-1} = q^k + q^{k-1} + \cdots + q + 1.$$

for the number of points in an k-dimensional (sub)space $\mathrm{PG}(k,q)$ (or the number of one-spaces in an $k+1$-dimensional vector space).

2. History and Basic Bounds

In this section we collect results about (non-trivial) k-blocking sets having the least number of points.

By combinatorial and counting arguments one gets the following result.

Theorem 2.1 (Bose-Burton [21]). *In $\mathrm{PG}(n,q)$ a k-blocking set has at least θ_k points. In case of equality the blocking set is the point set of a k-dimensional subspace.*

There is a nice and non-trivial generalization of this result by Klaus Metsch:

Theorem 2.2 (Metsch [57]). *For a set of $\theta_k - e \geq 0$ points in $\mathrm{PG}(n,q)$ there are at least $eq^{(k-1)(n-k+1)}$ disjoint $(n-k)$-spaces.*

To further motivate the terminology "k-blocking set" we mention the following fact.

Proposition 2.3. *In a projective space, a k-blocking set of a subspace is also a k-blocking set of the whole space.*

The converse is of course not always true, but the intersection of a k-blocking set with a hyperplane is a $(k-1)$-blocking set (of that hyperplane). For non-trivial planar blocking sets Bruen [23, 24] proved the following bound.

Theorem 2.4 (Bruen [23,24]). *$f_2(q) \geq q + \sqrt{q} + 1$. In case of equality the blocking set is a Baer subplane (i.e. a subplane of order $\sqrt{q}$).*

A slightly weaker bound was found by Pelikán [60]. We remark that Bruen's and Pelikán's proofs are combinatorial and valid for all (so also non-Desarguesian) planes. The bound on $f_2(q)$ (in the Desarguesian case) was substantially improved by Blokhuis [10, 11], for non-square q. We recall here the result for q prime or the cube of a prime.

Theorem 2.5 (Blokhuis [10]). *$f_2(q) \geq 3(q+1)/2$ if q is a prime. For $q = p^3$ we have $f_2(q) \geq p^3 + p^2 + 1$.*

This result is sharp (see the projective triangle later in Definition 5.4), so for $r_2(q)$ the previous results mean $\sqrt{q} \leq r_2(q) \leq (q+1)/2$. For more details on $f_2(q)$, see [11] or [76] and also Section 4.. The above results can be extended to higher dimensional spaces.

Theorem 2.6. *(1)* (Beutelspacher [9]). *A non-trivial blocking set of* PG(n,q) *has at least* $q+\sqrt{q}+1$ *points. In case of equality the blocking set is planar.*

(2) (Heim [34]). *A non-trivial blocking set of* PG(n,q) *has at least* $f_2(q)$ *points. In case of equality the blocking set is planar.*

Actually, Heim [34] also proved that $f_{e+1}(q) > f_e(q)$ if $f_e(q) \leq 2q$, and $f_3(q) \geq f_2(q)+2$ (if $q > 3$). The following table lists the values of $f_3(q)$.

q	2	3	4	5	7	8	9	11
$f_3(q)$	5	7	9	11	≥ 14	15	≥ 15	≥ 20

Before discussing the general case of k-blocking sets, let us see what to expect for 2-blocking sets in PG$(3,q)$. Take a planar blocking set B^* and a point P not in $\langle B^*\rangle$ and form the cone $B = PB^*$. Then B blocks the lines and has size $q|B^*|+1$. Taking a Baer subplane as B^* we get $|B| = q^2+q\sqrt{q}+q+1$, which should be compared with the size of the trivial blocking set which is q^2+q+1. The same phenomenon occurs in the general case.

Theorem 2.7 (Heim [34]). *Let B be a non-trivial k-blocking set in* PG(n,q)*, $n > k$, $q > 2$. Then $|B| \geq \theta_k + q^{k-1}r_2(q)$. In case of equality B is a cone VB^*, where V is a subspace of dimension $(k-2)$ and B^* is a planar blocking set of size $q+1+r_2(q)$.*

The special case when q is a square was proved earlier by Beutelspacher [9]. In this case the blocking set is a Baer cone. In general, a *Baer cone of type* (d,e) is a cone VB with vertex a d-subspace V and with base B which is a Baer subspace of an e-subspace. Baer cones were characterized by Huber in [39, 40].

The case $q = 2$ was treated by Govaerts and Storme:

Theorem 2.8 (Govaerts and Storme [30]). (1) *In* PG$(n,2), n \geq 3$*, the smallest non-trivial blocking sets are elliptic quadrics in solids in* PG$(n,2)$*: five points in a 3-space, no four in a plane. Up to isomorphism, this is the only non-trivial minimal blocking set in* PG$(3,2)$.

(2) *Up to isomorphism, there is only one non-trivial minimal 2-blocking set in* PG$(3,2)$*. It consists of ten points not on an elliptic quadric, i.e. the complement of the set in (1).*

(3) *In* PG$(n,2), n \geq 3$*, the smallest non-trivial k-blocking sets, $2 \leq k \leq n-1$, have size $2^{k+1}+2^{k-1}+2^{k-2}-1$ and are cones with vertex a $(k-3)$- space π_{k-3} and base the set of ten points not on an elliptic quadric in a solid skew to π_{k-3}.*

The case $q = 2$ is special: the complement of a (minimal) $(n-1)$-blocking set in PG$(n,2)$ is a (complete) cap. Caps in binary spaces were studied by several authors, see [27, 45] and the references in them. For instance, the structure result by Davydov and Tombak gives a description of large caps, i.e. ones of size $\geq 2^{n-1}+1$.

Let us return to the characterization of Baer cones.

Theorem 2.9 (Huber [40]). *Let B be a (k,s)-blocking set of* PG(n,q) *, $q \geq 5$, $k \geq s-1$. Then $|B| \geq \theta_k + \sqrt{q}(q^{k-1}+\cdots+q^{s-1})$, with equality if and only if B is a Baer cone of type $(s-2, 2(k-s+1))$.*

Again, this is a generalization of earlier results proved for the case $s = 1$ by Beutelspacher [9].

The fact that not only Baer subplanes but also Baer subspaces can be used to construct k-blocking sets follows from the next result.

Proposition 2.10. *Let* $\mathrm{PG}(t, q_1)$ *be a subgeometry of* $\mathrm{PG}(n, q)$, $q = q_1^e$, $t \leq n$. *A k-blocking set of* $\mathrm{PG}(t, q_1)$ *is a also a* $\lfloor k/e \rfloor$*-blocking set of the whole space.*

In particular, $\mathrm{PG}(n, q_1)$ is a $\lfloor n/e \rfloor$-blocking set of $\mathrm{PG}(n, q = q_1^e)$, of course if e does not divide n then the size is not close to the minimum value.

There is an even more general form of the above theorem of Heim, namely when $r_d(q) \leq (q+1)/2$, then for a d-proper k-blocking set B one has $|B| \geq \theta_k + q^{k-1} r_d(q)$ and also in this case the structure of B can be described. For more details, see 3.1.2 Hauptsatz in [34].

Let us continue with an easy example for q square. In $\mathrm{PG}(4, q)$ a Baer subspace $\mathrm{PG}(4, \sqrt{q})$ is a 2-blocking set of size $q^2 + q\sqrt{q} + q + \sqrt{q} + 1$. It does not contain lines, so it is a $(2, 1)$-blocking set. Another example is a cone PB, where P is a point, B is a Baer subplane. This is a $(2, 2)$-blocking set of size $q^2 + q\sqrt{q} + q + 1$. If the base B is replaced by another planar blocking set then by Theorem 2.6, or rather by its improvements mentioned after the theorem, the resulting cone will have much larger size. This shows that there is also hope for characterizing not only the smallest (non-trivial) k-blocking sets but also minimal k-blocking sets whose size is close to the minimum. Such results were obtained by Bokler [17–19] and also follow from the 1 modulo p results to be discussed later in Section 4.. Before going into detail about Bokler's result, let us briefly recall the result of Heim [34], Satz 4.1.4. about proper 2-blocking sets in $\mathrm{PG}(4, q)$: such a 2-blocking set has at least $q^2 + qr_2(q) + q + r_2(q) + 1$ points and equality only occurs for a Baer subspace.

Bokler's results are about the first few numbers of the set

$$S_{n,k,q} = \{|B| : B \text{ is a minimal } k\text{-blocking set of } \mathrm{PG}(n, q)\}$$

and the structure of the corresponding blocking sets.

Theorem 2.11 (Bokler [17]). *The smallest minimal k-blocking sets in* $\mathrm{PG}(n, q)$, *q square,* $q \neq 4$ *are cones with a vertex of dimension* $k - 1 - s$ *over a Baer subspace* $\mathrm{PG}(2s, \sqrt{q})$, *where* $s = \min\{k, n - k\}$.

An earlier proof of this result, using also that $q \neq 9$, appeared in [18]. The special cases $k = 2$, $q \neq 9$ and $k = 3$ are due to Metsch and Storme [58] and Bokler and Metsch [20], respectively.

Bokler [19] has also investigated minimal k-blocking sets in spaces of non-square order. He proved the following.

Theorem 2.12 (Bokler [19]). *Let B be a minimal k-blocking set of* $\mathrm{PG}(n, q)$, $q > 2$. *Suppose that for an integer* n' *with* $k \leq n' \leq 2k$ *we have* $|B| \leq \theta_k + \theta_{n'-k-1} r_2(q) q^{2k-n'}$. *Then the dimension of B is at most* n'. *If it is exactly* n' *then we have equality in the bound for* $|B|$ *(and also the structure of B can be described).*

As an illustration, for $n' = k$, the set B is a k-dimensional subspace. When $n' = k+1$, then B is a cone with a k-dimensional vertex and with base a non-trivial planar blocking set of size $f_2(q)$.

Regarding the $n+1-k$ smallest numbers $c_0 < c_1 < \ldots < c_{n-k}$ in $S_{n,k,q}$, Metsch [56] proves that for a minimal k-blocking set B either $|B| \geq \theta_k + q^k$ or $|B| \geq c_{\dim\langle B\rangle - k}$.

Note that some of the above results can be somewhat improved by using 1 modulo p results from Section 4. and the results of Bokler for q square were improved substantially by Weiner [80].

There are also upper bounds for the size of a minimal blocking set. By a result of Bruen–Thas, minimal blocking sets can have at most $q\sqrt{q}+1$ points. This was recently improved for non-square q's, see [75].

Theorem 2.13. *Suppose B is a minimal blocking set in* PG$(2,q)$, $q \neq 5$, *and denote by s the fractional part of $\sqrt{q}$. Then $|B| \leq q\sqrt{q} + 1 - \frac{1}{4}s(1-s)q$.*

The 'vertexless' triangle in PG$(2,5)$ shows that $q = 5$ really is an exception. The results are not only valid for blocking sets but also for sets having a tangent line at each point. This can be generalized to higher dimensions; a *semi-ovoid* is a set of points that has a tangent hyperplane at each of its points. In 3 dimensions ovoids have this property, and they have the maximum number of points. Recently, it was shown by Metsch and Storme [59] that there are no minimal blocking sets of size q^2 and q^2-1 in PG$(3,q)$. Similar results for the planar case were obtained by Blokhuis and Metsch, see e.g. in [76]. In general, we have the following upper bounds for the size of a semi-ovoid.

Theorem 2.14. 1. (Bruen-Thas [25]) *If S is a semi-ovoid in* PG(n,q) *then* $|S| \leq q^{(n+1)/2}+1$.
2. (Blokhuis-Moorhouse [15]) *If S is a semi-ovoid in* PG(n,q), $q = p^e$, *then*

$$|S| \leq \binom{p+n-1}{n}^e + 1\,.$$

Here we refer to the Techniques 54 of [47] by Landjev and Storme.

3. Natural Constructions

3.1. Subgeometry

We want to construct a k-blocking set in PG(n,q). Let $q = q_1^e$, so GF(q_1) is a subfield of GF(q), assume that $ke \leq n$. Take a subgeometry $B = $ PG(ke, q_1). The vector space dimension of the whole space is $n+1$ over GF(q), so $(n+1)e$ over GF(q_1). An $(n-k)$-subspace has dimension $(n-k+1)e$ while B is $ke+1$-dimensional as vector spaces over GF(q_1), hence they intersect.

Note that $|B| = q^k + q^k/q_1 + q^k/q_1^2 + \cdots$. One also sees that choosing $q_1 = q$ results in a subspace of dimension k.

3.2. Cone and Projection

Let $B^* \subset$ PG$(n^*,q) = \Pi$ be a k^*-blocking set. In PG$(n,q) \supset \Pi$, we can build a cone with vertex $V = $ PG(m,q) and base B^* (where $V \cap \Pi = \emptyset$), which will be a (k^*+m+1)-blocking set.

One can proceed in the opposite way in a sense, using projection. Let $B \subset \mathrm{PG}(n,q)$ be a k-blocking set, and $C = \mathrm{PG}(n-r-1,q)$, $(C \cap B = \emptyset)$ the center of the projection. Projecting B from C onto $\Pi = \mathrm{PG}(r,q)$, $(\mathrm{PG}(r,q) \subset \mathrm{PG}(n,q)$, $C \cap \mathrm{PG}(r,q) = \emptyset)$ we get a k-blocking set of Π. Note that the projection is not necessarily one-to-one, and the way C is chosen determines the structure of the projected image. This results in a large variety of *linear blocking sets*, defined as follows:

Definition 3.1. *A* linear point set *in* $\mathrm{PG}(r,q)$ *is the projection of a subgeometry* $\mathrm{PG}(n^*,q_1)$ *of* $\mathrm{PG}(n,q)$ *onto* $\mathrm{PG}(r,q)$ *for suitable* n, n^* *and* q_1 *(*q *a power of* q_1*).*

3.3. Directions and the Generalized Rédei Construction

Definition 3.2. We say that a set of points $U \subset \mathrm{AG}(n,q)$ *determines the direction* $d \in H_\infty$ (where H_∞ is the hyperplane at infinity), if there is an affine line with direction d meeting U in at least two points. Let D denote the set of determined directions.

We will always suppose that $|U| = q^k$, the number of $(n-k)$-spaces in a parallel class. Now we show the connection between directions and blocking sets.

Proposition 3.3. *If* $U \subseteq \mathrm{AG}(n,q)$, $|U| = q^k$, *then* U *together with the infinite points corresponding to directions in* D *forms a* k*-blocking set in* $\mathrm{PG}(n,q)$. *If the set* D *does not form a* k*-blocking set in* H_∞ *then all the points of* U *are essential.*

Proof. Any $(n-k)$-subspace at infinity $H_{n-k} \subset H_\infty$ is blocked by D: there are q^{k-1} (disjoint) affine $(n-k+1)$-spaces through H_{n-k}, so at least one of them has at least two points in U.

Consider next an $(n-k-1)$-space $H_{n-k-1} \subset H_\infty$. The $(n-k)$-subspaces through it determine an affine parallel class. If $D \cap H_{n-k-1} \neq \emptyset$ then they are all blocked by an infinite point. If H_{n-k-1} does not contain any point of D, then every affine $(n-k)$-subspace through it must contain exactly one point of U (as if one contained at least two then the direction determined by them would fall into $D \cap H_{n-k-1}$), so again they are blocked. Hence $U \cup D$ blocks all affine $(n-k)$-subspaces and all the points of U are essential unless maybe D is a k-blocking set in H_∞. □

It may happen of course that some points of D are non-essential, but it follows from the above that if D is not too big (i.e. $|D| < \theta_k$, roughly q^k), then this is not the case.

Proposition 3.4. *If* $|D| < \theta_k$, *then all points of* D *are essential.*

Proof. The smallest k-blocking set is the trivial one of size θ_k by the theorem of Bose-Burton (Theorem 2.1). □

A k-blocking set B arising in this way has the property that it meets a particular hyperplane in $|B| - q^k$ points. On the other hand, if a minimal k-blocking set B (of size $< q^k + \theta_k$) meets a hyperplane in $|B| - q^k$ points then, after deleting this hyperplane, we find a set of points in the affine space determining these $|B| - q^k$ directions, so the following notion is more or less equivalent to a point set plus its directions: a k-blocking set B is of *Rédei type* if there is a hyperplane meeting it in $|B| - q^k$ points. We remark that the theory developed by Rédei in his book [64] (or [65]) is highly related to these blocking sets. Minimal

k-blocking sets of Rédei type are in a sense extremal examples, as for any (non-trivial) minimal k-blocking set B and hyperplane H, where H intersects B in a set $H \cap B$ which is not a k-blocking set in H, $|B \setminus H| \geq q^k$ holds.

Since the k-blocking set that results has size $q^k + |D|$, in order to find small k-blocking sets, we will have to look for sets determining a small number of directions. Hence classifying sets determining few directions is an important problem, since it is equivalent with classifying small k-blocking sets of Rédei type.

4. Linear Blocking Sets

Until now we used the adjective "small" in a general sense. Here we specify what a small blocking set means.

Definition 4.1. *A k-blocking set in* $\mathrm{PG}(n,q)$ *is* small *when its size is smaller than* $\frac{3}{2}(q^k+1)$.

This size is easier to work with than the slightly bigger 'natural bound':

$$\theta_k + (1/2)(q+1)q^{k-1},$$

the size of the relevant cone over the blocking set of Rédei type called projective triangle, to be defined in Definition 5.4 of size $|B| = \frac{3}{2}(q+1)$ in $\mathrm{PG}(2,q)$ for each q odd. None of these blocking sets is a linear point set in the sense of Definition 3.1.

The *exponent* of a k-blocking set B is the maximal e such that each $(n-k)$-dimensional subspace intersects B in 1 mod p^e points. For a linear blocking set it is easy to see that $e \geq 1$. Szőnyi, Szőnyi-Weiner and Sziklai proved strong conditions in general on the exponents of *small* k-blocking sets, see below.

Conjecture 4.2 ([71]). **The Linearity Conjecture.** *In* $\mathrm{PG}(n,q)$ *every small k-blocking set is a linear point set.*

There are some cases of the Conjecture that are proved already.

Theorem 4.3. *For* $q = p^h$, *every small minimal non-trivial k-blocking set is linear, if*

(a) $n = 2$, $k = 1$ *(so we are in the plane) and*

- *(i)* (Blokhuis [10]) $h = 1$ *(i.e. there is no small non-trivial blocking set at all);*
- *(ii)* (Szőnyi [74]) $h = 2$ *(the only non-trivial example is a Baer subplane with* $p^2 + p + 1$ *points);*
- *(iii)* (Polverino [62]; Polverino, Storme [63]) $h = 3e$, *where* $e \geq 1$ *is the* exponent *of the blocking set (there are two or three examples (of Rédei type), one with* $q + p^{2e} + 1$ *and another with* $q + p^{2e} + p^e + 1$ *points and possibly the Baer subplane of size* $q + \sqrt{q} + 1$*);*
- *(iv)* (Blokhuis, Ball, Brouwer, Storme, Szőnyi [12], Ball [3]) *if* $p > 2$ *and there exists a line* ℓ *intersecting* B *in* $|B \cap \ell| = |B| - q$ *points (so a blocking set of Rédei type);*

(b) for general k:

(i) (Szőnyi-Weiner [77] $h = 1$ *(i.e. there is no small non-trivial blocking set at all);*

(ii) (Szőnyi and Weiner [77]) *if* $hk \leq n$, $p > 2$ *and* B *is not contained in an* $(hk-1)$-*dimensional subspace;*

(iii) (Storme-Weiner [70] (*for* $k = 1$), Bokler [18] and Weiner [80]) $h = 2$, $q \geq 16$*;*

(iv) (Storme-Sziklai [69]) *if* $p > 2$ *and there exists a hyperplane* H *intersecting* B *in* $|B \cap H| = |B| - q^k$ *points (so a blocking set of Rédei type);*

(v) (Lavrauw, Storme, Van de Voorde [48], [49], Harrach, Metsch, Szőnyi, Weiner [33], [32]) $h = 3e$, $p \geq 7$, *and every* $(n-k)$-*subspace intersects* B *in* 1 *modulo* p^e *points.*

The following result serves as a main tool in the proof of many particular cases of the Conjecture.

Theorem 4.4. *(i)* (Szőnyi [74]) *In* PG$(2,q)$, $q = p^h$, *if* B *is a minimal blocking set of size less than* $3(q+1)/2$, *then each line intersects it in 1 modulo* p^e *points for some* $e \geq 1$*;*

(ii) (Sziklai) *if* e *is maximal then here* $e|h$, *so* GF(p^e) *is a subfield of* GF(q). *Moreover, most of the secant lines intersect* B *in a point set isomorphic to* PG$(1,p^e)$, *i.e. in a linear point set.*

(iii) *If in (i)* e *was chosen to be maximal then, with* $E := p^e + 1$,

$$q+1+p^e\lceil\frac{q/p^e+1}{E}\rceil \leq |B| \leq \frac{1}{2}(1+E(q+1)-\sqrt{(1+E(q+1))^2-4E(q^2+q+1)}).$$

The bounds are due to Blokhuis, Polverino and Szőnyi, see [62,74], asymptotically they give

$$q+\frac{q}{p^e}-\frac{q}{p^{2e}}+\frac{q}{p^{3e}}-\cdots \leq |B| \leq q+a_0\frac{q}{p^e}+a_1\frac{q}{p^{2e}}+a_2\frac{q}{p^{3e}}+\cdots+a_{h/e-2}p^e,$$

where $a_0, a_1, \ldots$ are the Motzkin numbers 1,1,2,4,9,21,... . For this upper bound see [28]. Note that for $q = p^{2s}$ and $q = p^{3s}$, where s is a prime, the lower bound is sharp: $|B| \geq q + q/p^s + 1$.

To formulate the results in higher dimensions, let $S(q)$ denote the set of possible sizes of small minimal blocking sets in PG$(2,q)$. So $S(q) = S_{2,1,q}$ with the notation of Section 2.

Corollary 4.5. *Let* B *be a minimal* k-*blocking set of* PG(n,q), $q = p^h$, *of size* $|B| < \frac{3}{2}(q^k+1)$, *and of size* $|B| < \sqrt{2}q^k$ *if* $p = 2$. *Then*

- $|B| \in S(q^k)$*;*
- *if* $p > 2$ *then* $(|B|-1)q^{k(n-2)}+1 \in S(q^{k(n-1)})$.

If $p > 2$ *then there exists an integer* e, *called the* exponent *of* B, *such that* $1 \leq e|h$, *and every subspace that intersects* B, *intersects it in 1 modulo* p^e *points. Also* $|B|$ *lies in an interval belonging to some* $e' \leq e$, $e'|h$. *Most of the* $n-k$-*dimensional subspaces intersecting* B *in more than one point, intersect it in precisely* (p^e+1) *points, and each of these* (p^e+1)-*sets is a collinear point set isomorphic to* PG$(1,p^e)$.

Most of this was proved by Szőnyi and Weiner in [77] and Sziklai [71]. Consider the line determined by any two points in a (p^e+1)-secant $(n-k)$-subspace, this line should contain p^e+1 points. Then the technique of [77] can be used to derive a planar minimal blocking set (in a plane of order q^k) with the same exponent e: first embed $\mathrm{PG}(n,q)$ into $\mathrm{PG}(n,q^k)$ where the original blocking set B becomes a blocking set w.r.t. hyperplanes, then choose an $(n-3)$-dimensional subspace $\Pi \subset \mathrm{PG}(n,q^k)$ not meeting any of the secant lines of B and project B from Π onto a plane $\mathrm{PG}(2,q^k)$ to obtain a planar minimal blocking set, for which the planar results can be applied, implying $e|hk$.

Now in $\mathrm{PG}(n+1,q) \supseteq \mathrm{PG}(n,q)$ build a cone B^* with base B and vertex $V \in \mathrm{PG}(n+1,q) \setminus \mathrm{PG}(n,q)$; then B^* will be a (small, minimal) $k+1$-blocking set in $\mathrm{PG}(n+1,q)$. The argument above gives $e|h(k+1)$, so $e \mid \gcd(hk, h(k+1)) = h$.

There is an even more general version of the Linearity Conjecture. A t-fold k-blocking set is a point set which intersects each $(n-k)$-subspace in at least t points. Multiple points may be allowed as well.

Conjecture 4.6 ([71]). The Linearity Conjecture for multiple blocking sets. *In* $\mathrm{PG}(n,q)$ *a t-fold k-blocking set B is the union of some (not necessarily disjoint) linear point sets $B_1,\dots,B_s$, where B_i is a t_i-fold k-blocking set and $t_1+\dots+t_s=t$, provided that t and $|B|$ are small enough ($t \le T(n,q,k)$ and $|B| \le S(n,q,k)$ for suitable functions T and S).*

This conjecture is supported by the following theorem from [16] describing roughly what we know in this case.

Theorem 4.7. *Let B be a t-fold blocking set in* $\mathrm{PG}(2,q)$, *$q=p^h$, p prime, of size $t(q+1)+c$. Let $c_2=c_3=2^{-1/3}$ and $c_p=1$ for $p>3$.*

(1) (Ball [1,2,5]) *When $q=p>3$ is a prime and $t<p/2$, then $c \ge \frac{1}{2}(p+1)$.*

(2) *If h is odd and $t<q/2-c_pq^{2/3}/2$, then $c \ge c_pq^{2/3}$, unless $t=1$ in which case B, with $|B|<q+1+c_pq^{2/3}$, contains a line.*

(3) *If q is a square, $t<q^{1/4}/2$ and $c<c_pq^{2/3}$, then $c \ge t\sqrt{q}$ and B contains the union of t pairwise disjoint Baer subplanes, except for $t=1$ in which case B contains a line or a Baer subplane.*

(4) *If $q=p^2$, p prime, and $t<q^{1/4}/2$ and $c<p\lceil\frac{1}{4}+\sqrt{\frac{p+1}{2}}\rceil$, then $c \ge t\sqrt{q}$ and B contains the union of t pairwise disjoint Baer subplanes, except for $t=1$ in which case B contains a line or a Baer subplane.*

In [14] a (much) more detailed description of the special cases that q is a square, a cube, a fourth power or a sixth power is given. This paper also gives a t mod p result which we mention here.

Theorem 4.8. *Let B be a minimal t-fold blocking set in* $\mathrm{PG}(2,q)$, *$q=p^h$, p prime, $h \ge 1$, $|B|=tq+t+c$ and $2c+3t+t^2<q+5$. Then every line intersects B in $t \pmod{p}$ points.*

There exists a $(\sqrt[4]{q}+1)$-fold blocking set in $\mathrm{PG}(2,q)$, constructed by Ball, Blokhuis and Lavrauw [6], which is *not* the union of smaller blocking sets. (This multiple blocking set is a linear point set.) This shows that in general one cannot hope that a multiple blocking set can always be dismantled into 1-fold blocking sets.

5. More Constructions

Since blocking sets were first studied in projective planes, most of the known constructions are planar, now we present some of them. Later we describe some sporadic examples in higher dimensions. Then recursive constructions of Heim [35] and Mazzocca, Polverino, Storme [53] are discussed. Of course, starting from planar or sporadic examples one can use the recursive constructions combined with the general constructions (building cones and projecting onto a subspace) to obtain various k-blocking sets in spaces. As there are several survey papers about the spectrum problem for minimal planar blocking sets, the planar constructions will only be discussed briefly.

5.1. Planar Constructions

Let us recall the planar version of Rédei's construction. Take a subset U of size q in $\mathrm{AG}(2,q)$. An infinite point, or direction (d) is *determined by* U if there are two points $P_1, P_2 \in U$, such that $P_1, P_2, (d)$ are collinear. Let D denote the set of directions determined by U. The following is a special case of Proposition 3.3:

Proposition 5.1. *The set $B = U \cup D$ is a minimal blocking set, if D is not the entire line at infinity. Conversely, if B is a minimal blocking set of size $q+m$, $m \leq q$, and there is a line ℓ so that $|B \setminus \ell| = q$, then the blocking set can be obtained by Rédei's construction.*

Since $|U| = q$, $|D| \leq q$, we obtain blocking sets of size at most $2q$ this way. Originally, the set U is the graph of a function f from $\mathrm{GF}(q)$ to $\mathrm{GF}(q)$, that is, $U = \{(x, f(x)) : x \in \mathrm{GF}(q)\}$. In this case (d) is determined when there are x, u such that $(f(x) - f(u))/(x-u) = d$.

Since Rédei's construction can be found in several survey papers on blocking sets, we just give some examples: from $f(x) = x^{(q+1)/2}$ we get a blocking set of size $3(q+1)/2$. It is conjecture that for $q \neq 2$ prime, $q \neq 7, 13$, this is the unique example of this size. This has been proved for $q < 41$ [13]. From the trace function from $\mathrm{GF}(q)$ to a subfield $\mathrm{GF}(q_1)$, that is from $f(x) = x + x^{q_1} + \ldots + x^{q/q_1}$, we get a blocking set of size $q + 1 + q/q_1$. This is minimal if q is a square or if $q = p^3$ for a prime p. Finally, if $q = q_1^e$ then $f(x) = x^{q_1}$ gives a blocking set of size $q + (q-1)/(q_1 - 1)$. For more details, we refer to Blokhuis [11], [76], and Sections 13.1 and 13.4 of the second edition of Hirschfeld's book [36].

The following examples are also connected to Rédei's construction. The *index* of a blocking set is the minimum number of lines that cover the blocking set. It is straightforward that the index of a non-trivial blocking set is at least three. If it is three then the lines covering the set can either be concurrent or form a triangle. In the former case the common point of the three lines must belong to the blocking set, while in the latter case if the three intersection points are not in the blocking sets then the blocking set either contains a line or has $3(q-1)$ points and consists of the points on 3 non-concurrent lines with the three intersection points removed. This is called the *vertexless triangle*.

Megyesi's construction. We shall consider minimal blocking sets and assume that one of the three lines is the line at infinity, and that the affine part of the blocking set is

$$\{(0, -a) : a \in A\} \cup \{(1, b) : b \in B\}$$

in case of concurrent and

$$\{0,a) : a \in A\} \cup \{(1/b,0) : b \in B\} \cup \{(0,0)\}$$

in case of non-concurrent lines. Here A and B are nonempty subsets of the additive or multiplicative group of $\mathrm{GF}(q)$, respectively.

If the blocking set comes from Rédei's construction, the infinite part D consists of the set of determined directions. Besides the point (∞) (or (∞) and (0)), $(d) \in D$ iff $d = a + b$ (or $d = ab$). If we take a subgroup H of G and $A = G \setminus H (= -A)$, $B = H$ in the description above, then $D = H$, hence the blocking set consists of $2q + 1 - |H|$ points.

Theorem 5.2 ([72], [26]). *A non-trivial minimal blocking set of index three, whose size is at most $2q$, has size $2q+1-d$ for some divisor d of q or $q-1$. Conversely, for any divisor d of q or $q-1$ there is a blocking set of index three whose size is $2q+1-d$.*

The second part comes from Megyesi's construction. The proof of the first part is a straightforward application of Kneser's theorem ([51]).

Theorem 5.3. *Let $(G, \oplus)$ be an abelian group, $\emptyset \neq A, B$ be finite subsets of G. Then there is a subgroup H of G such that $A \oplus B = A \oplus B \oplus H$ and $|A \oplus B| \geq |A+H| + |B+H| - |H|$.*

Definition 5.4. *If $d = (q-1)/2$, then the Rédei type blocking set arising from Megyesi's construction is called the* projective triangle, *while for $d = q/2$, it is called the* projective triad.

Actually, after a suitable change of coordinates they come from $f(x) = x^{(q+1)/2}$ (q odd) and $f(x) = x + x^2 + \ldots + x^{q/2}$ (q even), respectively. This description was presented earlier.

The case when the index of the blocking set is 4 is less elaborated, but there are some recent results by Harrach and Mengyán [31]. They place cosets on the lines $x = 0$, $y = 0$, and $y = x$, and then use Rédei's construction.

Proposition 5.5. *Let $3 \leq s | q-1$ and consider the multiplicative subgroup G of $\mathrm{GF}(q)^*$ of index s. Let α be a generator of $\mathrm{GF}(q)^*$, so G, αG, $\alpha^2 G$, $\ldots$, $\alpha^{s-1} G$ are the cosets of G. Form three non-empty subsets $I, J, K \subset \mathbb{Z}_s$ such that $|I| + |J| + |K| = s$. Let*

$$U = \{(0,x) : x \in \alpha^i G, i \in I\} \cup \{(x,0) : x \in \alpha^j G, j \in J\} \cup$$
$$\cup \{(x,x) : x \in \alpha^k G, k \in K\} \cup \{(0,0)\}$$

Using this construction one can obtain minimal blocking sets of sizes $(2 - \frac{t}{s^2})q + C\sqrt{q}$, where $t \in \{1, 2, k, kl\}$ and $k|s, l|s$ such that $kl < s$, and $|C| \leq 2t$.

The examples for the given t's are the following:

$t = 1$: $I = \{0,2\}$, $J = \{1\}$, $K = \{3, 4, \ldots, s-1\}$.

$t = 2$: $I = \mathbb{Z}_s \setminus \{u, v\}$, $J = \{u\}$, $K = \{v\}$.

$t = k$: Let H be a proper subgroup of $\mathbb{Z}_s$, $|H| = k$ (note that $1 \notin H$) and let $I = \{1\}$, $J = H$, $K = \mathbb{Z}_s \setminus (I \cup J)$.

$t = kl$: Let H_1 and H_2 be proper subgroups of $\mathbb{Z}_s$, $H_1 \neq H_2$, such that there is an element $h \in \mathbb{Z}_s$ such that $H_1 \cap H_2 + h = \emptyset$ and let $I = \mathbb{Z}_s \setminus (I \cup J)$, $J = H_1$, $K = H_2$.

Here we recall very briefly some constructions giving typically larger minimal blocking sets than those obtained by Rédei's construction. Note that there are blocking sets of index 3 that are larger than $2q$ but it is difficult to control their size if they are minimal. In particular, they cannot have size close to $3q-3$.

The following construction gives minimal blocking sets of index four. (In case of $k = 3q-3$ it gives the vertexless triangle.)

IMI construction: In $\mathrm{PG}(2,q)$, $q \geq 4$, there exists a minimal blocking set of index four having k points, for every k with $2q-1 \leq k \leq 3q-3$. For more details, see Innamorati and Maturo [42] and Illés, Szőnyi, Wettl [41].

Even larger minimal blocking set can be obtained by the next construction.

Parabola construction ([73]): This construction best works for $q \equiv 1 \pmod 4$. Let C be a maximal independent set in the Paley-graph and let $\mathcal{P}_c$ be the parabola with equation $Y = X^2 + c$. Then $\cup_{c \in C} \mathcal{P}_c$ is a minimal blocking set of size $|C|q+1$ in $\mathrm{PG}(2,q)$.

Using this construction one can get minimal blocking sets of size $cq\log q$. A slight modification gives similar results for $q \equiv 3 \pmod 4$.

The largest examples are unitals for q square, see Theorem 2.14.

Whenever we have a blocking set, we may try to modify it locally. If B is a minimal blocking set in $\mathrm{PG}(2,q)$ and $P \notin B$ lies on at least one tangent of B, then we may add the point P and delete some points of B lying on the tangents through P. We need accurate information on the structure of the blocking set to control the size and minimality of the resulting blocking set. For example, if one starts from a classical unital, then the tangents through a point outside intersect the unital in collinear points, so we can delete all of them but one. Note also that in case of classical unitals the construction can be repeated several times and this gives a lot of non-isomorphic minimal blocking sets. Together with a careful analysis of the above constructions, Mengyán [54] proved that in many intervals there are exponentially many pairwise non-isomorphic blocking sets (in some cases even having the same number of points). Combining the constructions discussed in the present paper with the above mentioned local modifications, one can more or less determine the spectrum of minimal blocking sets for planes of small order. For the details, see [76], Chapter 13 of [36], [37,38].

5.2. Sporadic Constructions in Higher Dimensions

Probably the most important construction giving large minimal blocking set is related to ovoids. The result for quadrics is due to Ball [4], and Ball, Govaerts, Storme [7].

Theorem 5.6. *Ovoids of* $\mathrm{PG}(3,q)$ *are blocking sets. Every ovoid of a non-singular parabolic quadric* $Q(2n,q)$, $n = 2,3$ *is a minimal blocking set in* $\mathrm{PG}(2n,q)$.

The known examples of ovoids in $\mathrm{PG}(3,q)$ are elliptic quadrics, which exist for any q, and Suzuki–Tits ovoids, which exist for $q = 2^{2h+1}$. They are also ovoids of $Q(4,q)$. Besides them there are examples due to Kantor [44] (for $q = 3^{2h+1}$, Thas–Payne [79] (for $q = 3^e \geq 27$, and Penttila–Williams [61] (for $q = 3^5$). The known ovoids of $Q(6,q)$ are the Thas-Kantor ovoids, with $q = 3^e$ and $e \geq 1$, and the Ree-Tits ovoids, with $q = 3^{2h+1}$, $h > 0$.

The next sporadic examples in $\mathrm{PG}(3,q)$ were given by Tallini [78].

Proposition 5.7 (Tallini). *The following examples are minimal blocking sets of* $\mathrm{PG}(3,q)$.

1. *Let* $q>2$. $B_1=(r\setminus\{N_1,N_2\})\cup(K_1\cup K_2)$, *where* r *and* r' *are skew lines* N_1,N_2 *are distinct points on* r *and* K_i, $i=1,2$, *is a* $(q+1)$*-set in the plane* $\pi_i=\langle N_i,r'\rangle$, *having* N_i *as a nucleus and every point of* r' *is on at least one line of* π_i *different from* r' *and disjoint from* K_i. *We have* $|B_1|=3q+1$.

2. *Let* $q>2$ *be even.* $B_2=(r\setminus\{N_1,N_2,N_3\})\cup(K_1\cup K_2\cup K_3)$, *where* r *and* r' *are skew lines* N_1,N_2,N_3 *are distinct points on* r, K_1 *is a* $(q+1)$*-set in the plane* $\pi_1=\langle N_1,r'\rangle$ *having* N_1 *as a nucleus, and* K_i $(i=2,3)$ *is the projection of* K_1 *on the plane* $\pi_i=\langle N_i,r'\rangle$ *from the point* N_j, *with* $\{i,j\}=\{2,3\}$. *We have* $|B_2|=4q+1$.

3. *Let* $q>2$. $B_3=(\ell_1\cup\ell_2\cup\ell_3)\setminus(r_1\cup r_2)\cup\{P_1,P_2\}$, *where* ℓ_1,ℓ_2,ℓ_3 *are distinct lines in a regulus,* r_1,r_2 *are distinct lines of the opposite regulus,* $P_i\in r_i$, $P_i\notin\ell_1\cup\ell_2\cup\ell_3$, $(i=1,2)$. *We have* $|B_3|=3q-1$.

4. *Let* $q>2$ *be even.* $B_4=(\mathcal{O}\setminus(\cup_{i=1}^{h}C_i))\cup\{N_1,\dots,N_h\}$, $1\le h\le q-2$, *where* $\mathcal{O}$ *is an ovoid of* $\mathrm{PG}(3,q)$, $C_i\subset\mathcal{O}$ *are disjoint plane sections of* $\mathcal{O}$ *with nuclei* N_i *(i.e.* $N_i=\pi_i^{\perp}$, *where* π_i *is the plane of* C_i*),* $1\le i\le h$. *We have* $|B_4|=q(q-h)+1$.

Rößing and Storme [66, 67] proved that the spectrum of minimal blocking sets in $\mathrm{PG}(3,q)$ contains a long interval. For q odd, it is roughly $[q^2/4,3q^2/4]$ while for q even it is roughly $[q^2/10,9q^2/10]$.

5.3. More Constructions in Higher Dimensions

Let us begin with some trivial ways of putting together lower dimensional blocking sets to obtain higher dimensional ones.

Proposition 5.8. (1) *Let* Π_1 *and* Π_2 *be disjoint subspaces of* $\Pi=\mathrm{PG}(n,q)$, *with* $\langle\Pi_1,\Pi_2\rangle=\Pi$. *Let* B_i *be a blocking set of* Π_i, *and* $P_i\in B_i$, $i=1,2$. *Finally, let* P *be a point on the line* $\langle P_1,P_2\rangle$, $P\neq P_i$, $i=1,2$. *Then* $B=B_1\cup B_2\cup\{P\}\setminus\{P_1,P_2\}$ *is a blocking set of* Π.

(2) *Let* Π_1 *and* Π_2 *be subspaces of* $\Pi=\mathrm{PG}(n,q)$ *meeting in a point, that is* $\Pi_1\cap\Pi_2=\{P\}$. *Let* B_1 *be a blocking set in* Π_1, $P\in B_1$. *Let* Π_2' *be a hyperplane of* Π_2 *not through* P, *and* B_2' *be a blocking set in* Π_2'. *Denote by* K_2' *the cone with vertex* P *and base* B_2', *and let* $P\notin K_2$ *be any subset of* K_2' *which is projected bijectively onto* B_2' *from* P. *Then* $B=B_1\setminus\{P\}\cup K_2$ *is a blocking set of* Π.

Of course, these trivial constructions can easily be generalized to subspaces intersecting in more than a point but it is more complicated to analyse properties of the resulting blocking sets. The special case when the two subspaces are a hyperplane and a plane was studied in detail by Heim [35]. Here we just recall some of his results. Recall that a blocking set is called *proper* if no hyperplane intersects it in a blocking set.

Theorem 5.9 (Heim). *Let H_1 be a hyperplane, E_2 be a plane, not in H_1, of* $\mathrm{PG}(n,q)$. *Put $g = H_1 \cap E_2$. Let B_1 be a proper minimal blocking set in H_1 and let B_2 be a minimal blocking set of Rédei type in E_2 such that g is a Rédei line of B_2. Suppose that $S_g = B_1 \cap B_2 \cap q \neq \emptyset$ and $B_i \cap g \neq S_g$. Then $B = (B_1 \cup B_2 \setminus g) \cup S_g$ is a proper blocking set.*

The minimality of the resulting blocking set is not straightforward. If one starts in three dimensions from Rédei type blocking sets B_1, B_2, and B_2 has a second Rédei line through a point of S_g, and then repeats the same construction (keeping the same extra property of B_2), then the resulting blocking sets are minimal. Note that the blocking sets having two Rédei lines are characterized, see Korchmáros, Mazzocca [46] and Sherman [68].

Theorem 5.10 (Heim [35]). *Let $q = p^h$, p prime. One can obtain proper minimal blocking sets in* $\mathrm{PG}(n,q)$, *of size $(n-1)q-(n-3)q/p+1$, if $h>2$ and of size $(n-1)q-(n-3)(q+1)/2+n-1$, if $h=1$ and p is odd.*

Using a similar recursive construction starting from subgeometries $\mathrm{PG}(n,q)$ in $\mathrm{PG}(n,q^n)$, Heim [35] obtained further families of proper minimal blocking sets.

Theorem 5.11. *In* $\mathrm{PG}(ce-(c-1)-i,q)$ *with $q=q_1^e$, c a positive integer and $i=1,\ldots,e-2$ there exists a proper minimal blocking set of size*

$$c(q+q/q_1+\cdots+q_1^2)+(c-1)-(q_1^i+\cdots+q_1).$$

5.4. The Mazzocca, Polverino, Storme Constructions

By the Mazzocca, Polverino, Storme (MPS) construction, starting from a blocking set in a projective space, one can construct blocking sets in spaces whose order is a power of the original one. The idea of the construction comes from [75] and it generalizes the planar version in Mazzocca, Polverino [52].

Let $\mathcal{S}$ be a Desarguesian $(e-1)$-spread of $\Sigma = \mathrm{PG}(ne-1,q_1)$. It defines a projective space $\mathrm{PG}(\mathcal{S})$ isomorphic to $\mathrm{PG}(n-1,q=q_1^e)$ in which the points are the elements of $\mathcal{S}$, and the subspaces of dimension $te-1$ ($2 \leq t \leq n$) whose points are partitioned by elements of $\mathcal{S}$. Such subspaces will be called $\mathcal{S}$*-subspaces*. Embed Σ as a hyperplane in $\Sigma' = \mathrm{PG}(ne,q_1)$ and define a point-line geometry $\Pi_n = \Pi_n(\Sigma',\Sigma,\mathcal{S})$ in the following way:

- the points of Π_n are the points of $\Sigma' \setminus \Sigma$ and the elements of $\mathcal{S}$;
- the lines of Π_n are the e-subspaces of Σ' intersecting Σ in an element of $\mathcal{S}$ and the lines of $\mathrm{PG}(\mathcal{S})$;
- the point-line incidences are inherited from Σ and Σ'.

The incidence structure Π_n is isomorphic to the projective space $\mathrm{PG}(n,q)$, where $q=q_1^e$ and we say that Π_n is *the Barlotti-Cofman representation* of $\mathrm{PG}(n,q)$ (see [8]). The points of Π_n in $\Sigma' \setminus \Sigma$ will be called *affine*.

Let Y be a fixed element of $\mathcal{S}$ and let $\Omega = \Omega_{e-2}$ be a hyperplane of Y. Let $\Gamma' = \Gamma'_{(n-1)e+1}$ be an $((n-1)e+1)$-subspace of Σ' disjoint from Ω. Also, denote by $\Gamma = \Gamma_{(n-1)r}$ the $(n-1)e$-subspace intersection of Γ' and Σ, and by T the intersection point of Γ and Y.

Let $\bar{B}$ be a blocking set of Γ' such that $\bar{B} \cap \Gamma = \{Q\}$, Q a point, with the following property:

(α) $\ell \setminus \{T\} \not\subset \bar{B}$, *for every line* ℓ *of* Γ' *through* T.

Denote by K the cone with vertex Ω and base $\bar{B}$. Note that, since $\Gamma' \cap \Omega = \emptyset$, $\langle \bar{P}, \Omega \rangle \cap \langle \bar{P'}, \Omega \rangle = \Omega$, for any distinct points $\bar{P}, \bar{P'} \in \bar{B}$. Let B be the subset of Π_n defined by

$$B = \big(K \setminus \Sigma\big) \cup \{X \in \mathcal{S} : X \cap K \neq \emptyset\},$$

and note that if $Q \in Y$ (i.e. $T = Q$), then $|B| = q_1^{e-1}(|\bar{B}|-1)+1$ and $B \cap \mathrm{PG}(\mathcal{S}) = \{Y\}$; while if $Q \notin Y$, then $|B| = q_1^{e-1}|\bar{B}|+1$ and $|B \cap \mathrm{PG}(\mathcal{S})| = q_1^{e-1}+1$.

Then B is a blocking set of the projective space Π_n.

MPS Construction A ([53])

Suppose that $\bar{B}$ is a minimal blocking set of Γ' such that $T = Q$, in other words, suppose that Γ is a tangent hyperplane of $\bar{B}$ at the point Q. In this case, $B = \big(K \setminus \Sigma\big) \cup \{Y\}$ and $|B| = |K \setminus \Sigma| + 1 = q_1^{e-1}(|\bar{B}|-1)+1$.

Then B is a minimal blocking set of Π_n if and only if $\bar{B}$ is a minimal blocking set of Γ'. If $\bar{B}$ is a d-dimensional blocking set of Γ', then $\dim\langle B \rangle \leq \min\{n,d\}$.

Theorem 5.12. *From a minimal d-dimensional blocking set* $\bar{B}$ *in a projective space of order* q_1, *it is possible to obtain via Construction A minimal n-dimensional blocking sets B in* $\Pi_n \cong \mathrm{PG}(n, q = q_1^e)$ *for any n such that* $\max\{2, \frac{d-1}{e}+1\} \leq n \leq d$.

MPS Construction B ([53])

Suppose that $Q \notin Y$, let Z be the unique element of $\mathcal{S}$ such that $Q \in Z$ and let $\Gamma \cap Y = \{T\}$. In this case, the size of B is given by

$$|B| = q_1^{e-1}(|\bar{B}|-1) + q_1^{e-1} + 1 = q_1^{e-1}|\bar{B}|+1;$$

also $B \cap \mathrm{PG}(\mathcal{S})$ is contained in the line $\langle Y, Z \rangle$ of $\mathrm{PG}(\mathcal{S})$ and $|B \cap \mathrm{PG}(\mathcal{S})| = q_1^{e-1}+1$. The intersection numbers of B with respect to the hyperplanes can be determined as in Construction A. However, in this construction the minimality of the blocking set is not automatic and needs extra care particularly in the case of infinite points (that is points of $\mathrm{PG}(\mathcal{S})$).

We say that $\bar{B}$ satisfies Condition $(*)$ with respect to the point T if:

(*) for each point $\bar{P} \in \bar{B} \setminus \{Q\}$, there exists a tangent hyperplane to $\bar{B}$ passing through $\bar{P}$, but not containing T.

The affine points of B are essential points if and only if $\bar{B}$ satisfies Condition $(*)$ with respect to the point T. If $\bar{B}$ satisfies Condition $(*)$ with respect to the point T, then the size of B' can be determined, and it is roughly $|B'| = q_1^{e-1}(|\bar{B}|-1)$, see Theorem 5.13.

Similarly to MPS Construction A, if $\bar{B}$ is a d-dimensional blocking set, then $\dim\langle B \rangle \leq \min\{n, d+1\}$.

Theorem 5.13. *From a minimal d-dimensional blocking set* $\bar{B}$ *in a projective space of order* q_1, *we can obtain via Construction B minimal n-dimensional blocking sets* B' *in* $\mathrm{PG}(n, q = q_1^e)$ *for any n such that* $\max\{2, \frac{d}{e}+1\} \leq n \leq d+1$. *Also,*

(a) $|B'| = q_1^{e-1}|\bar{B}| + 1$ (i.e. $B' = B$) if $d \leq (n-2)e+1$;
(b) $|B'| = q_1^{e-1}|\bar{B}| + \varepsilon$ ($\varepsilon = 0,1$) if $d = (n-2)e+2$ and there exist at least two tangent hyperplanes to $\bar{B}$ at the point Q;
(c) $|B'| \geq q_1^{e-1}(|\bar{B}|-1) + q_1^{(n-1)e-d} + \varepsilon$ ($\varepsilon = 0,1$) if $(n-2)e+1 < d \leq (n-1)e$;
(d) $|B'| \geq q^{e-1}(|\bar{B}|-1) + q^{(n-1)e-d+1} + \varepsilon$ ($\varepsilon = 0,1$) if $(n-2)e+2 < d \leq (n-1)e$ and there exist at least two tangent hyperplanes to $\bar{B}$ at the point Q.

5.5. Some Interesting Examples Obtained by the MPS Construction

The next two tables contain some applications of the MPS constructions. In the table n denotes the dimension of the space, the column "Starting BS" gives a reference to the construction given in this paper, while "Reference" refers to the original paper(s). The first table is about MPS construction A, $q = q_1^e$.

Table 1. Blocking sets obtained by MPSA construction

Size	Starting BS	q_1, e, n	Reference
$2q+1$	$P5.8(2)$	$n = 2,3$	[52,53]
$3q+1$	$P5.7(1)$	$n = 2,3, q_1 > 2$	[52,53]
$4q+1$	$P5.7(2)$	$n = 2,3, q_1 > 2$, q even	[52,53]
$3q - 2q/q_1 + 1$	$P5.7(3)$	$n = 2,3, q_1 > 2$,	[52,53]
$kq+1$	$P5.7(4)$	$n = 2,3, q_1 > 2$, q even, $2 \leq k \leq q_1 - 1$	[52,53]
$(d-1)q - (d-3)q/p + 1$	$T5.9(1)$	$q_1 = p^h$, $h \geq 2$, $\max\{2, \frac{d-1}{e}+1\} \leq n \leq d$	[52,53]
$\frac{(d+1)q - (d-1)q/q_1}{2} + 1$	$T5.9(2)$	$q_1 = p$, odd prime	[52,53]
$q \cdot q_1 + 1$	ovoid	$n = 2,3,4$, $q_1 = p^h, p$ prime, $h \geq 1$	[75], [52,53]
$q_1^{e+1} + q_1^{e-3} + 1$	ovoid	$n = 2, e \geq 3$	[75]
$q_1^{e+2} + 1$	ovoid in Q	$2 \leq n \leq 6, e \geq 3$ and q_1 a power of 3	[52,53]
$q_1^{1+e/2} + 1$	ovoid in Q	$2 \leq n \leq 6$, q_1 an even power of 3	[52,53]

Note that the blocking sets in the first seven rows are not isomorphic to blocking sets obtained by the original construction giving the starting examples, if applied for $PG(n,q)$. In all cases the blocking sets are n-dimensional. When the blocking set comes from an ovoid, also the intersection sizes with hyperplanes are determined.

When $n = 2$, q a power of 3 and $e = 5$, $n = 2$, we get blocking sets of size $q^{7/5}$ from ovoids of $Q(4,q_1)$. Here the exponent is close to $3/2$, which is the upper bound by Bruen, Thas [25], see Theorem 2.14 here.

The next table is about MPS construction B.

Table 2. Blocking sets obtained by MPSB construction

Size	Starting BS	q,e,n	Reference
$q\cdot q_1+q/q_1+1$	ovoid	$n=2$	[52]
$kq+q/q_1+1$	$P5.7(4)$	$n=2, q$ even, $2\le k\le q_1-1$	[52]
$q_1^{e+1}+q_1^{e-4}+$ $+l_Q(q_1^{e-3}-q_1^{e-4})+1$	ovoid	$e\ge 4, q_1=p^h, h\ge 1$, p an odd prime, or $q_1=2^{2h+1}, h\ge 1$	[52]
$q_1^{e+2}+q_1^{e-6}+$ $+l_Q(q_1^{e-5}-q_1^{e-6})+1$	ovoid	$e\ge 6, q_1=3^h, h\ge 1$	[52]
$q\cdot q_1+q/q_1+1$	ovoid	$n=3,4$	[53]
$q\cdot q_1+q/q_1+1$	ovoid	$3\le n\le 5, e\ge 3$, $q_1=p^h, p$ an odd prime , $h>1$	[53]
$q_1'^{e+1}+q_1'^{e-1}+1$	ovoid	$4\le n\le 5, q_1=q_1'^2$, p an odd prime , $h>1$	[53]
$q\cdot q_1+q/q_1+1$	ovoid	$4\le n\le 5, q_1=2^{2f+1}, f\ge 1$	[53]
$q^2+\sqrt{q}+1$	ovoid	$5\le n\le 7, q_1=9$	[53]
q^2+2	ovoid	$q_1=9$	[53]

Note that we used the notation $q=q_1^e$ also in this table. In the second row $0<l_Q\le q_1^2+q_1+1$, in the third $0<l_Q\le q_1^4+q_1^3+q_1^2+q_1+1$.

6. Affine Blocking Sets

So far we have been focusing on k-blocking sets for projective spaces. The answers usually depend heavily on the structure of the underlying field. For affine spaces the situation is different. First of all there is no trivial k-blocking set, quite the opposite, the only general sharp theorem is the Jamison/Brouwer-Schrijver result mentioned in the introduction. The result is sharp for all n, is independent of the structure of the field, and is much larger than the fractional blocking number. Note that in contrast to the projective case the bound depends also on the dimension n of the ambient space.

Theorem 6.1 ([22,43]). *The size of a (1-)blocking set in* $\mathrm{AG}(n,q)$ *is at least* $1+n(q-1)$.

A k-blocking set of $\mathrm{AG}(n,q)$ intersects every hyperplane in a $(k-1)$-blocking set and since $\mathrm{AG}(n,q)$ can be partitioned into hyperplanes we get by induction that the size $a_k(n)$ of a k-blocking set of $\mathrm{AG}(n,q)$ is at least $a_k(n)\ge qa_{k-1}(n-1)\ge q^{k-1}(1+(n-k+1)(q-1))$. We can also get a recursive upper bound of the form $a_k(n)\le a_k(n-1)+(q-1)a_{k-1}(n-1)$ by taking a $k-1$-blocking set in a hyperplane, together with a k-blocking set in every parallel one.

If instead of fixing k and taking n variable, we fix $n-k$, we get a problem that has been studied a bit more. If $n-k=1$ we want to block lines. For $q=2$ the problem becomes trivial, but for $q=3$ this problem shows up in different ways. Looking at the complement of the blocking set we get a cap in AG$(n,3)$. Tables for the size of the largest cap in (low-dimensional) affine and projective spaces can be found in [37, 38]. A cap in AG$(n,3)$ is also an example of a sum free set in $\mathbb{Z}_3^n$, since in this space three points are collinear precisely when they sum to zero. If c_n denotes the size of the largest cap in AG$(n,3)$ then $\lim_{n\to\infty}\sqrt[n]{c_n}=c$ exists, and it is a famous open problem whether $c<3$. We know that $c>2.14$ from the existence of the Hill Cap [37, 38] and that $c_n<2\cdot(3^n/n)$ [55]

Acknowledg ments

The third author thanks the Technical University of Eindhoven for the kind hospitality during his visit, where parts of the present paper were written. He gratefully acknowledges the financial support of NWO, including the support of the DIAMANT and Spinoza projects. The second and third authors were partly supported by OTKA Grants T 49662 and NK 67867. The third author was also supported partly by the Hungarian-Slovenian bilateral project.

References

[1] S. Ball, *Multiple blocking sets and arcs in finite planes*, J. London Math. Soc. (2), 54 (1996), pp. 581–593.

[2] ——, *On the size of a triple blocking set in* PG$(2,q)$, European J. Combin., 17 (1996), pp. 427–435.

[3] ——, *The number of directions determined by a function over a finite field*, J. Combin. Theory Ser. A, 104 (2003), pp. 341–350.

[4] ——, *On ovoids of* O$(5,q)$, Adv. Geom., 4 (2004), pp. 1–7.

[5] S. Ball and A. Blokhuis, *On the size of a double blocking set in* PG$(2,q)$, Finite Fields Appl., 2 (1996), pp. 125–137.

[6] S. Ball, A. Blokhuis, and M. Lavrauw, *Linear* $(q+1)$*-fold blocking sets in* PG$(2,q^4)$, Finite Fields Appl., 6 (2000), pp. 294–301.

[7] S. Ball, P. Govaerts, and L. Storme, *On ovoids of parabolic quadrics*, Des. Codes Cryptogr., 38 (2006), pp. 131–145.

[8] A. Barlotti and J. Cofman, *Finite Sperner spaces constructed from projective and affine spaces*, Abh. Math. Sem. Univ. Hamburg, 40 (1974), pp. 231–241.

[9] A. Beutelspacher, *Blocking sets and partial spreads in finite projective spaces*, Geom. Dedicata, 9 (1980), pp. 425–449.

[10] A. Blokhuis, *On the size of a blocking set in* PG$(2,p)$, Combinatorica, 14 (1994), pp. 111–114.

[11] ——, *Blocking sets in Desarguesian planes*, in Combinatorics, Paul Erdős is eighty, Vol. 2 (Keszthely, 1993), D. Miklós, V. Sós, and T. Szőnyi, eds., vol. 2 of Bolyai Soc. Math. Stud., János Bolyai Math. Soc., Budapest, 1996, pp. 133–155.

[12] A. Blokhuis, S. Ball, A. E. Brouwer, L. Storme, and T. Szőnyi, *On the number of slopes of the graph of a function defined on a finite field*, J. Combin. Theory Ser. A, 86 (1999), pp. 187–196.

[13] A. Blokhuis, A. E. Brouwer, and H. A. Wilbrink, *Blocking sets in* $\mathrm{PG}(2,p)$ *for small p, and partial spreads in* $\mathrm{PG}(3,7)$, Adv. Geom., (2003), pp. S245–S253. Special issue dedicated to Adriano Barlotti.

[14] A. Blokhuis, L. Lovász, L. Storme, and T. Szőnyi, *On multiple blocking sets in Galois planes*, Adv. Geom., 7 (2007), pp. 39–53.

[15] A. Blokhuis and G. E. Moorhouse, *Some p-ranks related to orthogonal spaces*, J. Algebraic Combin., 4 (1995), pp. 295–316.

[16] A. Blokhuis, L. Storme, and T. Szőnyi, *Lacunary polynomials, multiple blocking sets and Baer subplanes*, J. London Math. Soc. (2), 60 (1999), pp. 321–332.

[17] M. Bokler, *Blockierende Mengen in endlichen projektiven Räumen*, PhD thesis, University of Giessen, 2001.

[18] ——, *Minimal blocking sets in projective spaces of square order*, Des. Codes Cryptogr., 24 (2001), pp. 131–144.

[19] ——, *Lower bounds for the cardinality of minimal blocking sets in projective spaces*, Discrete Math., 270 (2003), pp. 13–31.

[20] M. Bokler and K. Metsch, *On the smallest minimal blocking sets in projective space generating the whole space*, Beiträge Algebra Geom., 43 (2002), pp. 43–53.

[21] R. C. Bose and R. C. Burton, *A characterization of flat spaces in a finite geometry and the uniqueness of the Hamming and the MacDonald codes*, J. Combinatorial Theory, 1 (1966), pp. 96–104.

[22] A. E. Brouwer and A. Schrijver, *The blocking number of an affine space*, J. Combinatorial Theory Ser. A, 24 (1978), pp. 251–253.

[23] A. A. Bruen, *Baer subplanes and blocking sets*, Bull. Amer. Math. Soc., 76 (1970), pp. 342–344.

[24] ——, *Blocking sets in finite projective planes*, SIAM J. Appl. Math., 21 (1971), pp. 380–392.

[25] A. A. Bruen and J. A. Thas, *Blocking sets*, Geometriae Dedicata, 6 (1977), pp. 193–203.

[26] P. J. Cameron, *Four lectures on projective geometry*, in Finite geometries (Winnipeg, Man., 1984), C. Baker and L. Batten, eds., vol. 103 of Lecture Notes in Pure and Appl. Math., Dekker, New York, 1985, pp. 27–63.

[27] A. A. Davydov, S. Marcugini, and F. Pambianco, *Minimal 1-saturating sets and complete caps in binary projective spaces*, J. Combin. Theory Ser. A, 113 (2006), pp. 647–663.

[28] V. Fack, Sz. L. Fancsali, L. Storme, G. Van de Voorde, and J. Winne, *Small weight codewords in the codes arising from Desarguesian projective planes*, Des. Codes Cryptogr., 46 (2008), pp. 25–43.

[29] Z. Füredi, *Matchings and covers in hypergraphs*, Graphs Combin., 4 (1988), pp. 115–206.

[30] P. Govaerts and L. Storme, *The classification of the smallest nontrivial blocking sets in* $\mathrm{PG}(n,2)$, J. Combin. Theory Ser. A, 113 (2006), pp. 1543–1548.

[31] N. V. Harrach and Cs. Mengyán, *Minimal blocking sets in* $\mathrm{PG}(2,q)$ *arising from a generalized construction of Megyesi*, Innov. Incidence Geom., 6/7 (2007/08), pp. 211–226.

[32] N. V. Harrach and K. Metsch, *Small point sets of* $\mathrm{PG}(n,q^3)$ *intersecting each k-subspace in 1 mod q points*, Des. Codes Cryptogr., 56 (2010), pp. 235–248.

[33] N. V. Harrach, K. Metsch, T. Szőnyi, and Zs. Weiner, *Small point sets of* $\mathrm{PG}(n,p^{3h})$ *intersecting each line in 1 mod* p^h *points*, J. Geom., 98 (2010), pp. 59–78.

[34] U. Heim, *Blockierende Mengen in endlichen projektiven Räumen*, Mitt. Math. Sem. Giessen, (1996), pp. 1–82. Dissertation, Justus-Liebig-Universität Giessen, Giessen, 1995.

[35] ——, *Proper blocking sets in projective spaces*, Discrete Math., 174 (1997), pp. 167–176. Combinatorics (Rome and Montesilvano, 1994).

[36] J. W. P. Hirschfeld, *Projective geometries over finite fields*, Oxford Mathematical Monographs, The Clarendon Press Oxford University Press, New York, second ed., 1998.

[37] J. W. P. Hirschfeld and L. Storme, *The packing problem in statistics, coding theory and finite projective spaces*, J. Statist. Plann. Inference, 72 (1998), pp. 355–380. R. C. Bose Memorial Conference (Fort Collins, CO, 1995).

[38] ——, *The packing problem in statistics, coding theory and finite projective spaces: update 2001*, in Finite geometries, A. Blokhuis, J. W. P. Hirschfeld, D. Jungnickel, and J. A. Thas, eds., vol. 3 of Dev. Math., Kluwer Acad. Publ., Dordrecht, 2001, pp. 201–246.

[39] M. Huber, *A characterization of Baer cones in finite projective spaces*, Geom. Dedicata, 18 (1985), pp. 197–211.

[40] ——, *Baer cones in finite projective spaces*, J. Geom., 28 (1987), pp. 128–144.

[41] T. Illés, T. Szőnyi, and F. Wettl, *Blocking sets and maximal strong representative systems in finite projective planes*, Mitt. Math. Sem. Giessen, (1991), pp. 97–107.

[42] S. Innamorati and A. Maturo, *On irreducible blocking sets in projective planes*, Ratio Math., 2 (1991), pp. 151–155.

[43] R. E. Jamison, *Covering finite fields with cosets of subspaces*, J. Combinatorial Theory Ser. A, 22 (1977), pp. 253–266.

[44] W. M. Kantor, *Ovoids and translation planes*, Canad. J. Math., 34 (1982), pp. 1195–1207.

[45] M. Khatirinejad and P. Lisoněk, *Classification and constructions of complete caps in binary spaces*, Des. Codes Cryptogr., 39 (2006), pp. 17–31.

[46] G. Korchmáros and F. Mazzocca, *Nuclei of point sets of size* $q+1$ *contained in the union of two lines in* $\mathrm{PG}(2,q)$, Combinatorica, 14 (1994), pp. 63–69.

[47] I. Landjev and L. Storme, *Galois geometries and coding theory*, in Current research topics in Galois geometry, Nova Sci. Publ., New York, 2011, ch. 8, pp. 185–212.

[48] M. Lavrauw, L. Storme, and G. Van de Voorde, *A proof of the linearity conjecture for k-blocking sets in* $\mathrm{PG}(n,p^3)$*, p prime*, J. Combin. Theory Ser. A, 118 (2011), pp. 808–818.

[49] M. Lavrauw and G. Van de Voorde, *On linear sets on a projective line*, Des. Codes Cryptogr., 56 (2010), pp. 89–104.

[50] L. Lovász, *On the ratio of optimal integral and fractional covers*, Discrete Math., 13 (1975), pp. 383–390.

[51] H. B. Mann, *Addition theorems: The addition theorems of group theory and number theory*, Interscience Publishers John Wiley & Sons New York-London-Sydney, 1965.

[52] F. Mazzocca and O. Polverino, *Blocking sets in* $\mathrm{PG}(2,q^n)$ *from cones of* $\mathrm{PG}(2n,q)$, J. Algebraic Combin., 24 (2006), pp. 61–81.

[53] F. Mazzocca, O. Polverino, and L. Storme, *Blocking sets in* $\mathrm{PG}(r,q^n)$, Des. Codes Cryptogr., 44 (2007), pp. 97–113.

[54] Cs. Mengyán, *On the number of pairwise non-isomorphic minimal blocking sets in* $\mathrm{PG}(2,q)$, Des. Codes Cryptogr., 45 (2007), pp. 259–267.

[55] R. Meshulam, *On subsets of finite abelian groups with no* 3*-term arithmetic progressions*, J. Combin. Theory Ser. A, 71 (1995), pp. 168–172.

[56] K. Metsch, *Blocking sets in projective spaces and polar spaces*, J. Geom., 76 (2003), pp. 216–232. Combinatorics, 2002 (Maratea).

[57] ——, *How many s-subspaces must miss a point set in* $\mathrm{PG}(d,q)$, J. Geom., 86 (2006), pp. 154–164 (2007).

[58] K. Metsch and L. Storme, *2-blocking sets in* $\mathrm{PG}(4,q)$*, q square*, Beiträge Algebra Geom., 41 (2000), pp. 247–255.

[59] ——, *Tangency sets in* $\mathrm{PG}(3,q)$, J. Combin. Des., 16 (2008), pp. 462–476.

[60] J. Pelikán, *Properties of balanced incomplete block designs*, in Combinatorial theory and its applications, III (Proc. Colloq., Balatonfüred, 1969), North-Holland, Amsterdam, 1970, pp. 869–889.

[61] T. Penttila and B. Williams, *Ovoids of parabolic spaces*, Geom. Dedicata, 82 (2000), pp. 1–19.

[62] O. Polverino, *Small minimal blocking sets and complete k-arcs in* $\mathrm{PG}(2,p^3)$, Discrete Math., 208/209 (1999), pp. 469–476. Combinatorics (Assisi, 1996).

[63] O. Polverino and L. Storme, *Small minimal blocking sets in* $\mathrm{PG}(2,q^3)$, European J. Combin., 23 (2002), pp. 83–92.

[64] L. Rédei, *Lückenhafte Polynome über endlichen Körpern*, Birkhäuser Verlag, Basel, 1970. Lehrbücher und Monographien aus dem Gebiete der exakten Wissenschaften, Mathematische Reihe, Band 42.

[65] ——, *Lacunary polynomials over finite fields*, North-Holland Publishing Co., Amsterdam, 1973. Translated from the German by I. Földes.

[66] C. Rößing and L. Storme, *A spectrum result on maximal partial ovoids of the generalized quadrangle* $Q(4,q)$*, q even*, European J. Combin., 31 (2010), pp. 349–361.

[67] ——, *A spectrum result on minimal blocking sets with respect to the planes of* $PG(3,q)$*, q odd*, Des. Codes Cryptogr., 55 (2010), pp. 107–119.

[68] B. F. Sherman, *Minimal blocking sets in finite planes*, J. Geom., 43 (1992), pp. 178–187.

[69] L. Storme and P. Sziklai, *Linear point sets and Rédei type k-blocking sets in* $\mathrm{PG}(n,q)$, J. Algebraic Combin., 14 (2001), pp. 221–228.

[70] L. Storme and Zs. Weiner, *On 1-blocking sets in* $\mathrm{PG}(n,q)$*,* $n \geq 3$, Des. Codes Cryptogr., 21 (2000), pp. 235–251. Special issue dedicated to Dr. Jaap Seidel on the occasion of his 80th birthday (Oisterwijk, 1999).

[71] P. Sziklai, *On small blocking sets and their linearity*, J. Combin. Theory Ser. A, 115 (2008), pp. 1167–1182.

[72] T. Szőnyi, *Combinatorial problems for abelian groups arising from geometry*, in Proceedings of the Second International Mathematical Miniconference, Part II (Budapest, 1988), Periodica Polytechnica (Transportation Engineering), vol. 19, 1991, pp. 91–100.

[73] ——, *Note on the existence of large minimal blocking sets in Galois planes*, Combinatorica, 12 (1992), pp. 227–235.

[74] ——, *Blocking sets in Desarguesian affine and projective planes*, Finite Fields Appl., 3 (1997), pp. 187–202.

[75] T. Szőnyi, A. Cossidente, A. Gács, Cs. Mengyán, A. Siciliano, and Zs. Weiner, *On large minimal blocking sets in* $\mathrm{PG}(2,q)$, J. Combin. Des., 13 (2005), pp. 25–41.

[76] T. Szőnyi, A. Gács, and Zs. Weiner, *On the spectrum of minimal blocking sets in* $\mathrm{PG}(2,q)$, J. Geom., 76 (2003), pp. 256–281. Combinatorics, 2002 (Maratea).

[77] T. Szőnyi and Zs. Weiner, *Small blocking sets in higher dimensions*, J. Combin. Theory Ser. A, 95 (2001), pp. 88–101.

[78] G. Tallini, *Blocking sets with respect to planes in* $\mathrm{PG}(3,q)$ *and maximal spreads of a nonsingular quadric in* $\mathrm{PG}(4,q)$, Mitt. Math. Sem. Giessen, (1991), pp. 141–147.

[79] J. A. Thas and S. E. Payne, *Spreads and ovoids in finite generalized quadrangles*, Geom. Dedicata, 52 (1994), pp. 227–253.

[80] Zs. Weiner, *Small point sets of* $\mathrm{PG}(n,q)$ *intersecting each k-space in 1 modulo* $\sqrt{q}$ *points*, Innov. Incidence Geom., 1 (2005), pp. 171–180.

In: Current Research Topics in Galois Geometry
Editors: J. De Beule and L. Storme
ISBN: 978-1-61209-523-3

Chapter 4

LARGE CAPS IN PROJECTIVE GALOIS SPACES

Jürgen Bierbrauer[1,*] ***and Yves Edel***[2,†]
[1] Department of Mathematical Sciences, Michigan Technological University,
Houghton, Michigan 49931, USA
[2] Ghent University, Department of Mathematics,
Krijgslaan 281 S22, B–9000 Gent, Belgium

Key Words: Caps, Galois geometries, codes.

AMS Subject Classification: 51E22, 94B05.

1. What Is a Cap?

A **cap** in a projective or affine geometry over a finite field is a set of points no three of which are collinear. The most natural question to ask is:

```
What is the maximum size of a cap in the given space?
```

This is also known as the **packing problem.** In this paper, $m_2(r,q)$ denotes the size of the largest caps in $\mathrm{PG}(r,q)$.

2. Classical Examples

If the underlying field is $\mathbb{F}_2$, the answer is easy: $\mathrm{AG}(n,2)$ is itself a cap of 2^n points and it forms up to projective equivalence the unique largest cap in $\mathrm{PG}(n,2)$.

Assume therefore we work in $\mathrm{PG}(n,q)$ or $\mathrm{AG}(n,q)$ for $q>2$. The canonical models for caps are quadrics of Witt index 1. They yield $(q+1)$-caps in $\mathrm{AG}(2,q)$ (and in $\mathrm{PG}(2,q)$) and obviously these **ovals** are maximal for odd q. In odd characteristic each oval in $\mathrm{PG}(2,q)$ is a conic section (Segre [49,50]). This is not true in characteristic 2, where moreover each oval $\mathcal{O}$ is embedded in a unique **hyperoval** $\mathcal{O}\cup\{N\}$. Here N is the **nucleus**, the intersection of all the tangents to $\mathcal{O}$. A hyperoval is a $(q+2)$-cap and this is maximal. The hyperovals are

*E-mail address: jbierbra@mtu.edu
†E-mail address: yedel@cage.ugent.be

described by a special kind of permutation polynomials. This is an active line of research, see the survey [38]. In PG$(3,q)$ an elliptic quadric is a (q^2+1)-cap. This is maximal for all $q>2$ (see Bose [13] and Qvist [47]). Its affine part is a q^2-cap in AG$(3,q)$ and this is maximal. In characteristic 2 the Tits ovoids form another family of (q^2+1)-caps in PG$(3,q)$, see [55]. They may be considered classical as they admit a family of classical groups, the Suzuki groups ${}^2B_2(q)$ for $q=2^{2m+1}$, as groups of automorphisms.

3. Exceptional Caps

For projective dimensions $d>3$ and fields $\mathbb{F}_q, q>2$, the basic question seems to be hard to answer. Only for some small dimensions and fields the answer is known. In those cases, the corresponding maximal caps tend to be exotic, in particular more or less uniquely determined and very symmetric.

The Ternary Case

In PG$(4,3)$ and AG$(4,3)$, the maximum is 20 (see Pellegrino [45]), with the 20-cap in AG$(4,3)$ (the **Pellegrino cap**) uniquely determined. In PG$(5,3)$ and AG$(5,3)$, the maximum is 56 and 45 respectively. In both cases, the caps are uniquely determined, the **Hill cap** in PG$(5,3)$ and the **affine Hill cap** (contained in the Hill cap) in AG$(5,3)$. The unicity was shown by Hill [36] in the projective, in [6, 25] in the affine case. The automorphism group of the Hill cap is an extension of the simple group PSL$(3,4)$ by a group of order 2. There are numerous links to other exceptional mathematical structures, see Hill [37]. The points of the elliptic quadric in PG$(5,3)$ can be chosen to be the one-dimensional subspaces of $\mathbb{F}_3^6$ generated by the vectors of weights 3 or 6. This indicates how those 112 points can be split into two halves each of which forms a cap (for details, see [6]). The automorphism group of the Hill cap is a rank 3 permutation group on the points of the Hill cap, the stabilizer of a point having orbits of lengths $1,10,45$. The points of the long orbit form a copy of the affine Hill cap whose automorphism group is the stabilizer PGL$(2,9)$. The remaining 11 points form a block of the uniquely determined $(56,11,2)$-symmetric design (a biplane).

There is a general doubling construction, see [42].

Theorem 1. *An n-cap in* PG(d,q) *allows the construction of a* $2n$*-cap in* AG$(d+1,q)$.

This also explains the Pellegrino cap in AG$(4,3)$. It follows from the doubling construction applied to the elliptic quadric in PG$(3,3)$. When applied to the Hill cap, doubling yields a 112-cap in AG$(6,3)$. Potechin [46] showed that this cap is maximal and uniquely determined. Starting from PG$(6,3)$ we are in uncharted territory. Most of the known constructions of large caps in higher dimensional spaces over $\mathbb{F}_3$ make use of the Hill cap. This starts with the Calderbank-Fishburn 236-cap [14] in AG$(7,3)$ and a 248-cap in PG$(7,3)$ (see [21]). Those caps have as automorphism groups semidirect products of E_{32} by S_5 and of E_{64} by S_5, respectively. Exceptions are a recently discovered 541-cap in PG$(8,3)$ and a 2744-cap in PG$(10,3)$ which resulted from a computer search. The game of SET can be used as a playful motivation to study caps in affine ternary spaces, see [7, Section 3.6] and [16]. The 81 cards of the game correspond to the points of AG$(4,3)$ and a cap is a point set not containing a SET. Thanks to Pellegrino, Hill, and Potechin, we now know what are the largest cardinalities of SET-free collections of cards in the d-dimensional generalizations of the game where $d \leq 6$.

When $q > 3$

The maximum sizes of caps in PG$(4,4)$ and AG$(4,4)$ are 41 and 40, respectively. The 40-cap in AG$(4,4)$ is uniquely determined [22]. It is complete in PG$(4,4)$. Its automorphism group is a semidirect product of E_{16} and A_5. It can be shown that the two 41-caps given in [19] are in fact the only 41-caps in PG$(4,4)$. There is a relation of duality between one of the two 41-caps in PG$(4,4)$ and the 40-cap K in AG$(4,4)$: embed AG$(4,4)$ in PG$(4,4)$. There are 40 hyperplanes of PG$(4,4)$ meeting K in 4 points. Those hyperplanes together with the empty hyperplane form the dual of a 41-cap. The other 41-cap in PG$(4,4)$ had in fact been found earlier, by Tallini [54]. Its automorphism group is solvable of order 240. Hill [36] observes: *For each of the known values of $m_2(r,q)$, there is a cap K in* PG(r,q) *of that size on which* Aut(K) *acts as a transitive permutation group.* Unfortunately, this is no longer true as none of the two 41-caps in PG$(4,4)$ admits a transitive automorphism group. Still the metarule that extremal objects tend to be very symmetric is verified also here: the more symmetric 41-cap has a large automorphism group which is transitive on all but one of its points.

Another exceptional object is the Glynn cap [33], a 126-cap in PG$(5,4)$. It contains a 120-cap in AG$(5,4)$ and admits PGL$(3,4)$ as an automorphism group. Observe that this is the second time we encounter the simple group PSL$(3,4)$. We saw it acting on the ternary Hill cap as well.

4. The Link to Linear Codes

Let K an n-cap in PG$(k-1,q)$ and G a $k \times n$ matrix whose columns are representatives of the points of K. Then G is a check-matrix of a $[n, n-k, 4]_q$-code $C^{\perp}$ and this is an equivalent description of the cap property. Its dual $C = C(K)$ may be called a cap-code. It is a projective $[n,k,d]_q$-code where d is the largest number such that outside every hyperplane H of PG$(k-1,q)$ there are at least d points of K. Good caps often yield good cap-codes as well. For example, Pellegrino's result implies directly that the code of the Hill cap is an $[56,6,36]_3$-code and this is a code meeting the Griesmer bound with equality (see [7, Theorem 5.7]).

5. General Bounds

The best known general upper bound on the size of a cap uses a version of the Fourier transform (see [10], Meshulam [41] and [7, Section 16.3]). Let $C_k(q)$ be the maximum size of a cap in AG(k,q) and $c_k(q) = C_k(q)/q^k$. Then

$$c_k(q) \leq (q^{-k} + c_{k-1}(q))/(1 + c_{k-1}(q)) \text{ for } q > 2, k \geq 3.$$

A weak form states

$$c_k(q) \leq (k+1)/k^2 \text{ for } q > 2, k \geq 3.$$

Together with the doubling construction (Theorem 1) this also yields bounds on caps in projective spaces. In fact, if there is an n-cap in $PG(k-1,q)$ then there is a $2n$-cap in $AG(k,q)$, hence $n \leq C_k(q)/2$. In low dimensions, the bounds of [53] are better.

6. Recursive Constructions

The archetype of all recursive cap constructions is Mukhopadhyay's product construction from [42]. Here is a generalization, [7, Theorem 16.62]:

Theorem 2. *If there is an n-cap $K_1 \subset AG(k,q)$ and an m-cap $K_2 \subset PG(l,q)$, then there is a cap (the product cap) of nm points in* $PG(k+l,q)$.

If A is avoided by $i \geq 1$ hyperplanes in general position and B by $j \geq 0$ hyperplanes in general position, then the product cap is avoided by $i+j-1$ hyperplanes in general position.

The doubling construction Theorem 1 is a special case of Theorem 2. Here is a generalization, [7, Theorem 16.63]:

Theorem 3. *Assume the following exist:*

- *An n-cap in* $AG(k,q)$ *which can be extended to an $(n+w)$-cap by some w points in the hyperplane at infinity, and*

- *An m-cap in* $PG(l,q)$.

Then $PG(k+l,q)$ *contains an $(nm+w)$-cap.*

An application to the elliptic quadric in $PG(3,q)$ yields a classical construction of B. Segre [51]: an m-cap in $PG(l,q)$ leads to an (q^2m+1)-cap in $PG(l+3,q)$. A **tangent hyperplane** to a given point set in a projective space is a hyperplane which meets the point set in precisely one point. The following is [20, Theorem 10].

Theorem 4. *Assume the following exist:*

- *An n-cap in* $PG(k,q)$ *possessing a tangent hyperplane, and*

- *An m-cap in* $PG(l,q)$ *possessing a tangent hyperplane.*

Then $PG(k+l,q)$ *contains an $(nm-1)$-cap.*

Application to the elliptic quadric in $PG(3,q)$ yields a (q^4+2q^2)-cap in $PG(6,q)$. For $q \geq 4$, this is the largest known cap in $PG(6,q)$. This leads to the natural question if larger caps can be constructed in $PG(6,q)$ for $q > 3$. Many of the best known caps, even in moderately small dimensions, have been constructed by applying some version of the product construction to caps from lower-dimensional spaces. Rather sophisticated product constructions are used in [17] to construct a 1216-cap in $PG(9,3)$ (whose automorphism group is an extension of a normal subgroup of order 2^8 by S_5) and a 6464-cap in $PG(11,3)$.

7. Families of Caps in Fixed Dimension

We are interested in families of caps in $\mathrm{PG}(d,q)$ for all q, or at least for an infinite family of fields $\mathbb{F}_q$, whose number of points is $cq^\alpha +$ lower terms. What is the largest exponent α and, for this α, what is the largest constant c? We then speak of a family of order cq^α. Clearly $(\alpha,c)=(2,1)$ for $d=3$.

The Case of Projective Dimension $d=4$

This is the smallest interesting dimension and it is difficult. It is not known if an exponent $\alpha>2$ can be reached. Choosing elliptic quadrics in two solids shows that order $2q^2$ can always be reached. A family of order $2.5q^2$ for arbitrary odd characteristic is constructed in [9]. In characteristic 2, only one family of order cq^2 for $c>2$ is known. This is a family of $(3q^2+4)$-caps $K_q\subset AG(4,q)$, $q=2^{odd}$ constructed in [23]. For $q=2^{even}$ the existence of a family of caps of order cq^2, for $c>2$, remains an open problem.

Definition 1. *Let $q=2^f$. For $0\neq v\in\mathbb{F}_q$, let p_v be the number of elements $0\neq x\in\mathbb{F}_q$ such that*

$$tr(x)=tr(v/x)=1$$

where $tr:\mathbb{F}_q\longrightarrow\mathbb{F}_2$ is the absolute trace.

The elliptic curve with affine equation $y^2+y=x+v/x$ has precisely $4p_v$ rational points. The weight distribution of the binary Kloosterman and Mélas codes are determined by the numbers p_v. Those numbers were determined by Schoof and van der Vlugt [48]. In [23], it is shown how the weight distribution of the cap-codes C_q corresponding to K_q is determined by the numbers p_v as well. In particular the minimum distance follows from the Hasse bound on the number of rational points of an elliptic curve. In the smallest case, C_8 is a $[196,5,164]_8$-code which can be extended to a $[200,5,168]_8$-code. This relation is one illustration of the use of algebraic geometry in coding theory. The most prominent such link is the construction of algebraic-geometric codes due to Goppa and Manin [34,40]. However there are many examples for the use of algebraic curves in determining the structure of classical codes as well.

The family K_q has more interesting structure. There is a special point P_0 such that $K_q\setminus\{P_0\}$ is a dual BCH-code, and those $3q+3$ points are distributed on three parabolic quadrics. This raises the general question to determine the cyclic codes of dual distance 4.

Projective Dimension $d\leq 5$ over $\mathbb{F}_5$

A 66-cap in $\mathrm{PG}(4,5)$ was found in [21] using a complicated recursive construction based on the ovoid in $\mathrm{PG}(3,5)$. This 66-cap is rather symmetric. Its automorphism group is a direct product of A_5 and the dihedral group D_8. This indicates a rich geometric structure. In fact, the 66-cap in $\mathrm{PG}(4,5)$ and its partner, a newly discovered 195-cap in $\mathrm{PG}(5,5)$, turn out to be closely related to the conic section in $\mathrm{PG}(2,5)$ and a classical geometric structure associated to it, the **Barlotti arcs** (see [2]). In the following we sketch the construction.

Start from the conic $C \subset PG(2,5)$ defined by the equation $Y^2 = XZ$. Its points are $P_\infty = (0:0:1)$ and $P_y = (1:y:y^2), y \in \mathbb{F}_5$. The tangents are

$$t_\infty = [1:0:0] \text{ and } t_y = [y^2 : -2y : 1].$$

These are the lines $[u:v:w]$, where $v^2 + uw = 0$. The interior points (those not on a tangent to C) are $(x:y:z)$, where $y^2 - xz$ is a non-square. These are the points $(1:y:z)$ where $y^2 - z = \pm 2$. The interior points are therefore $(1:y:y^2 \pm 2)$, where y is arbitrary. The secants are the lines $[u:v:w]$, where $v^2 + uw$ is a non-zero square and consequently the exterior lines are $[u:v:w]$, where $v^2 + uw$ is a non-square. The exterior line $[-v^2+2:v:1]$ contains the interior points

$$(1:2v-1:-v^2+v+3),\ (1:2v+1:-v^2-v+3) \text{ and } (1:2v:-v^2+3).$$

The exterior line $[-v^2-2:v:1]$ contains the interior points

$$(1:2v-2:-v^2+2v+2),\ (1:2v+2:-v^2-2v+2) \text{ and } (1:2v:-v^2+2).$$

Definition 2. *A* **half-point** *is a pair* $\pm v$ *of non-zero vectors. The* **parity** *of a non-zero element of* $\mathbb{F}_5$ *is its quadratic remainder symbol.*

Observe that each point of $PG(d,5)$ is the union of two half-points.

Definition 3. *Let* $G \subset GL(3,5)$ *the stabilizer of the set of vectors in* $\mathbb{F}_5^3$ *that represent the points of the conic* C. *Let*

$$K_\infty = (0,0,1) \text{ and } K_y = (1,y,y^2) \text{ for } y \in \mathbb{F}_5.$$

Define $K = \{\pm K_\tau \mid \tau \in PG(1,5)\}$, *a system of half-points representing the points of* C.

The group G acts on the two-element set $\{K, 2K\}$. The stabilizer of the system K of half-points is a subgroup $G_0 \subset G$ of index 2, where $G_0/\langle -1 \rangle \cong S_5$ and $G/\langle -1 \rangle \cong S_5 \times \mathbb{Z}_2$.

We turn to the action of G on vectors generating interior points.

Definition 4. *Let* $I(y,1) = (1,y,y^2+2)$ *and* $I(y,2) = 2(1,y,y^2-2)$ *for* $y \in \mathbb{F}_5$. *Then the union* I *of the* $\pm I(y,1)$ *and* $\pm I(y,2)$ *is a system of* 10 *half-points which generate the interior points of* C.

Here are those vectors:

$$(1,0,2),(1,1,3),(1,2,1),(1,3,1),(1,4,3)$$

$$(2,0,1),(2,1,4),(2,2,3),(2,3,3),(2,4,4).$$

Lemma 1. *The group* G *acts on the two-element set* $\{I, 2I\}$. *The stabilizer* G_1 *of the system* I *of half-points representing interior points satisfies* $G_1/\langle -1 \rangle \cong S_5$.

The points $(1:I)$ form a 20-cap in $AG(3,5)$. The stabilizer G_2 of K and I in G satisfies $G_2/\langle -1 \rangle \cong A_5$. Most important is the following lemma:

Lemma 2. *The half-points in $K \cup I$ have the following property: Let K_1, K_2, I_1, I_2 be non-zero vectors in $\mathbb{F}_5^3$ such that K_1, K_2 belong to different half-points from K and I_1, I_2 belong to different half-points from I.*

- *If $c_1K_1 + c_2K_2 + dI_1 = 0$, where $c_1, c_2, d \in \mathbb{F}_5$, not all $= 0$, then c_1, c_2 are non-zero of different parity.*
- *If $cK_1 + d_1I_1 + d_2I_2 = 0$, where $c, d_1, d_2 \in \mathbb{F}_5$, not all $= 0$, then d_1, d_2 are non-zero of different parity.*

This leads to a recursive construction procedure:

Theorem 5. *Let $l \geq 2$ and $A, B \subset \mathbb{F}_5^l$ such that the following are satisfied:*

1. *$0 \notin A = -A, 0 \notin B = -B$. In other words, A is the union of $|A|/2$ half-points, likewise for B. Denote by C_A, C_B the corresponding point sets in $\mathrm{PG}(l-1,5)$.*
2. *The set C_B is a $|B|/2$-cap in $\mathrm{PG}(l-1,5)$.*
3. *The points $(1 : a), a \in A$, form a cap in $\mathrm{AG}(l,5)$ (equivalently: $(A + A) \cap 2A = \emptyset$).*
4. *$C_A \cap C_B = \emptyset$.*
5. *The points represented by $A + 2A$ are disjoint from the points represented by B, and symmetrically with the roles of A, B exchanged.*

Then the points (P, a) and (Q, b) where $P \in K, Q \in I$ and $a \in A, b \in B$ represent a cap M of size $6|A| + 10|B|$ in $\mathrm{PG}(l+2,5)$.

In case $l = 2$, let

$$A = \pm\{(1,0),(1,2)\},\ B = \pm\{(0,1),(1,1)\}.$$

Then the conditions of Theorem 5 are satisfied. It follows that M is a 64-cap in $\mathrm{PG}(4,5)$. Points $(0:0:0:1:3)$ and $(0:0:0:1:4)$ are extension points. This yields a 66-cap. Each of the extension points is on an obvious tangent hyperplane. At this point, we have reconstructed the 66-cap in $\mathrm{PG}(4,5)$. A similar process works one dimension higher.

Let $l = 3$. Choose $A = K$ the union of the representatives of conic half-points. It is possible to find B with the same structure as K for a conic disjoint from C. One choice is the quadric $Q(X,Y,Z) = X^2 + Z^2 - 2(XY + XZ + YZ)$ and its half-points

$$B = \pm\{010, 101, 012, 210, 112, 211\}.$$

This yields a 192-cap in $\mathrm{PG}(5,5)$ by Theorem 5. Observe that B consists entirely of exterior points with respect to A. There are three extension points yielding a 195-cap in $\mathrm{PG}(5,5)$.

Higher Dimensions

In characteristic 2, the product construction applied to hyperovals and elliptic quadrics yields $(q+2)(q^2+1)$-caps in $PG(5,q)$. Recently, Kroll-Vincenti [39] constructed $((q+2)(q^2+2)-1)$-caps in $PG(5,q)$ for even $q \geq 8$. In odd characteristic a rather specialized version of the product construction (see [20]) applied to elliptic quadrics and conic sections yields $(q+1)(q^2+3)$-caps in $PG(5,q)$. The (q^4+2q^2)-caps in $PG(6,q)$ for arbitrary q have been mentioned before as an application of Theorem 4. An application of Theorem 3 to this cap produces $q^2(q^2+1)^2$-caps in $PG(9,q)$.

8. Concrete Bounds

Here is a list of the currently best known lower bounds on large caps in $PG(d,q)$, for $d \leq 11$ and $q \leq 9$. The superscript c indicates that the cap is known to be complete.

Table 1. Lower bounds

$d\backslash q$	3	4	5	7	8	9
2	4^c	6^c	6^c	8^c	10^c	10^c
3	10^c	17^c	26^c	50^c	65^c	82^c
4	20^c	41^c	66^c	132^c	208^c	212^c
5	56^c	126^c	195^c	434^c	695^c	840^c
6	112^c	288^c	675^c	2499^c	4224^c	6723^c
7	248^c	756^c	1715^c	6472^c	13520^c	17220^c
8	541^c	2110^c	5069^c	21555^c	45174	68070
9	1216^c	5040^c	17124^c	122500	270400	544644
10	2744^c	15423^c	43876	323318	878800	1411830
11	6464^c	34566	130951	1067080	2931457	5580100

The lower bounds are known to agree with the upper bound only when $d \leq 3$, for $d \leq 5$ in the ternary and for $d = 4$ in the quaternary case. The upper bound in $PG(6,3)$ currently is 136 [1]. In $PG(4,5)$ the upper bound is 88 [26].

9. The Atoms of Cap Theory

Most of the known large caps in larger dimensions result from applications of some recursive construction to exceptionally large caps in lower dimensions. This raises the question what the elementary building blocks are, the large caps which do not result from recursive constructions themselves and which are used as ingredients for the constructions in higher dimensions. We call them the atoms of cap theory. Clearly, the classical models have that status, see Section 2.. Also large caps possessing a large group of automorphisms will be considered to be atoms. This leads to the following list of atoms:

1. The ovals and hyperovals in $AG(2,q)$.

2. The elliptic quadrics in $PG(3,q)$.

3. The Tits ovoids in $PG(3,2^{2m+1}), m \geq 1$.

4. The Hill cap in $PG(5,3)$.

5. The highly symmetric 41-cap in $PG(4,4)$ and its dual partner, the 40-cap in $AG(4,4)$.

6. The Glynn cap in $PG(5,4)$.

Now we present some more examples of caps that have the potential to be regarded as atoms.

The Complete 14-cap in PG(3,4)

This object is uniquely determined. Its group of automorphisms is the semidirect product of an elementary abelian group of order 8 and $GL(3,2)$ (see [19]). Here is a construction using only hyperovals: there is a configuration in $PG(3,4)$ consisting of three collinear planes and a hyperoval in each plane, where the line of intersection is a secant for all three hyperovals. The union of those hyperovals is our 14-cap. We will encounter it in Section 11. as a quantum cap. It is also used in the construction of a quantum 38-cap in $PG(4,4)$. The complete 14-cap in $PG(3,4)$ is a special case of a result of Segre [52] who constructed complete $(3q+2)$-caps in $PG(3,q)$ for all even $q \geq 4$. The construction was further generalized by Pambianco-Storme [44].

A 66-cap in PG(4,5)

It has been mentioned in Section 7. that its group of automorphisms is $A_5 \times D_8$. It possesses a tangent hyperplane and therefore can be used in Theorem 4. This produces a 1715-cap in $PG(7,5)$.

A 132-cap in PG(4,7)

This cap resulted from a computer search. Its automorphism group has order 192.

A 208-cap in PG(4,8)

This is the largest cap known in $PG(4,8)$. It resulted from a computer construction based on a cyclic group of order $8^2-1=63$. We raise the problem if this construction can be generalized in the following way: a $(3q^2+2q)$-cap in $PG(4,q)$, $q=2^{2m+1}$, admitting the action of a certain cyclic group of order q^2-1, consisting of 3 regular orbits, one orbit of length $q+1$, one orbit of length $q-1$, and three fixed points. The conjecture is true for $q=8$ and for $q=32$.

A 195-cap in PG(5,5)

This cap was constructed as an application of Theorem 5. It possesses tangent hyperplanes and therefore can be used in Theorem 4. With the elliptic quadric in PG(3,5) as second ingredient, this yields a 5069-cap in PG(8,5). Application of Theorem 3 to the 195-cap in in PG(5,5) and the 675-cap in PG(6,5) yields a cap with $194 \times 675 + 1 = 130,951$ points in PG(11,5). We saw in an earlier subsection that the 66-cap in PG(4,5) and the 195-cap in PG(5,5) result from a recursive construction which only uses a conic and its embedding in the plane as ingredients. It is therefore up to discussion if those caps should be considered as atoms. The automorphism group of the 195-cap is isomorphic to $A_5 \times \mathbb{Z}_4 \times \mathbb{Z}_2$.

A 434-cap in PG(5,7)

The Glynn cap makes use of a certain mapping $\gamma : PG(2,q^2) \longrightarrow PG(5,q)$. The image Γ_q of this mapping is a set of $(q^4 - q)/2$ points. In case $q = 4$, this is the Glynn cap. In [21], a computer program produced a subset of $\Gamma_7 \subset PG(5,7)$, which is a 434-cap whose automorphism group has order $672 = 4 \times 168$. This automorphism group is not solvable. It involves the simple group of order 168. It may be possible to find further large caps as subsets of Γ_q. If synthetic constructions can be found, it may be the case that the Glynn cap is the beginning of an infinite family of caps in PG$(5,q)$.

Most of the automorphism groups were calculated using Thomas Feulner's program [29] which is available online, see also the paper [28].

10. An Asymptotic Problem

As in Section 5., let $C_k(q)$ be the maximum size of a cap in AG(k,q). Define

$$\mu(q) = \limsup_{k \longrightarrow \infty} \log_q(C_k(q))/k.$$

Clearly, we could use caps in PG(k,q) instead of AG(k,q) and obtain the same limit. Working with affine caps has the advantage that because of the product construction of Theorem 2, each value $C_k(q)$ in a concrete dimension k yields a lower bound: $\mu(q) \geq \log_q(C_k(q))/k$. In particular, the affine part of the elliptic quadric in PG$(3,q)$ yields $\mu(q) \geq 2/3$. A basic open problem is to show that $\mu(q) < 1$. The best known lower bound for general q seems to follow from an application of Theorem 4 to elliptic quadrics, see [20]. This leads to a cap of size $(q^2+1)^2 - 1 = q^4 + 2q^2$ in PG$(6,q)$. It is easy to see that there is a hyperplane meeting this cap in q^2+1 points. This leads to an (q^4+q^2-1)-cap in AG$(6,q)$ and the lower bound $\mu(q) \geq \log_q(q^4+q^2-1)/6$. For $q = 4$, the affine part of the Glynn-cap yields a better lower bound: $\mu(4) \geq \log_4(120)/5 = 0,6906\ldots$ As is to be expected, the ternary case has been studied most intensively. The recursive constructions of Calderbank-Fishburn [14] based on the Hill cap have been further refined in [17]. Currently, the best known lower bound is $\mu(3) \geq 0,724851\ldots$

11. Additive Codes and Quantum Caps

Additive codes are a far-reaching generalization of linear codes. Here we view the alphabet of size q^m not as a field but rather as a vector space over the subfield $\mathbb{F}_q$ and assume linearity only over $\mathbb{F}_q$. Of particular interest is the quaternary case ($q = m = 2$).

Definition 5. *Let k be such that $2k$ is a positive integer. An additive quaternary $[n,k]_4$-code C (length n, dimension k) is a $2k$-dimensional subspace of $\mathbb{F}_2^{2n}$, where the coordinates come in pairs of two. We view the codewords as n-tuples where the coordinate entries are elements of $\mathbb{F}_2^2$.*

A **generator matrix** *of C is a binary $(2k,2n)$-matrix whose rows form a basis of the binary vector space C.*

One reason to concentrate on the quaternary case is the link with quantum error-correction established in [15]. It may be described equivalently using the symplectic form, which is a basic notion from geometric algebra.

Definition 6. *Let $V = V(2n,q)$ be a $2n$-dimensional vector space over $\mathbb{F}_q$. A* **symplectic form** *on V is a mapping $\langle , \rangle : V \oplus V \longrightarrow \mathbb{F}_q$ which satisfies the following conditions:*

- $\langle x_1 + x_2, y\rangle = \langle x_1, y\rangle + \langle x_2, y\rangle$, $\langle x, y_1 + y_2\rangle = \langle x, y_1\rangle + \langle x, y_2\rangle$ *and* $\langle cx, y\rangle = \langle x, cy\rangle = c\langle x,y\rangle$ *for all* $x, x_i, y, y_i \in V, c \in \mathbb{F}_q$.
- $\langle x,x\rangle = 0$ *for all* $x \in V$.
- *The only vector x satisfying $\langle x,y\rangle = 0$ for all $y \in V$ is $x = 0$.*

If $\langle x,y\rangle = 0$ we also write $x \perp y$ and $y \in x^\perp$. Let $W \subset V$. The dual of W is a subspace defined by

$$W^\perp = \{y | y \in V, \langle w,y\rangle = 0 \text{ for all } w \in W\}.$$

A symplectic space V possesses a **symplectic basis** $\{v_1, \dots, v_n, w_1, \dots, w_n\}$ such that $\langle v_i, v_j\rangle = \langle w_i, w_j\rangle = 0$ for all i,j and $\langle v_i, w_j\rangle = \delta_{i,j}$.

The pertinent notion is the following.

Definition 7. *A pure additive quantum stabilizer $[[n,m,d]]$-code C (short: quantum code) is a quaternary additive code C of length n and dimension $(n-m)/2$ which satisfies*

- $C \subseteq C^\perp$ *where the dual is with respect to the symplectic form.*
- $C^\perp$ *has distance* $\geq d$.

The translation into geometry is as follows, see [11]:

Theorem 6. *The following are equivalent:*

- *A pure $[[n, n-r, t+1]]$ quantum stabilizer code.*
- *A set of n lines, the* **codelines**, *in* $\mathrm{PG}(r-1,2)$ *satisfying:*
  - *any t codelines are in general position and*

– *the quantum condition: for every secundum (subspace* $\mathrm{PG}(r-3,2)$*) S, the number of codelines skew to S is even.*

In particular, pure quantum codes are always described in terms of sets of pairwise skew lines in binary projective space. When $d=3$, the only additional condition to satisfy is the quantum condition. In contrast to the classical theory of linear codes, even case $d=3$ is not trivial. The classification of all parameters n,m such that $[[n,m,3]]$ quantum codes exist is very recent, see [8].

The smallest open case is $d=4$ and the corresponding quantum codes form a natural generalization of the concept of a cap. Under the additional hypothesis that the code be not only additive, but also $\mathbb{F}_4$-linear, the concept of a quantum cap is obtained:

Definition 8. *A* **pre-quantum cap** *is an n-cap $K \subset PG(m-1,4)$ which satisfies the following equivalent conditions:*

- *$K \cap H$ has the same parity as n for every hyperplane H.*
- *The corresponding quaternary linear cap-code $C(K)$ has all weights even.*
- *$C(K)$ is self-orthogonal with respect to the Hermitian form.*

A **quantum cap** *in* $\mathrm{PG}(m-1,4)$ *is a pre-quantum cap which is not contained in a proper subspace.*

Here the Hermitian form on $\mathbb{F}_4^m$ is defined by $B((x_1,x_2,\ldots,x_m),(y_1,y_2,\ldots,y_m)) = \sum_{i=1}^m x_i y_i^2$. A quantum n-cap in $\mathrm{PG}(m-1,4)$ is equivalent to a pure $[[n,n-2m,4]]$ quantum code which is $\mathbb{F}_4$-linear. As an example, consider the elliptic quadric in $\mathrm{PG}(3,4)$. As this cap has 17 points and plane intersections of sizes 1 or 5, the conditions of Definition 8 are satisfied. The corresponding cap-code is a $[17,4,12]_4$-code and it is a $[[17,9,4]]$ quantum code. The smallest quantum cap in $\mathrm{PG}(3,4)$ has 8 points. It may be constructed as the complement of $\mathrm{PG}(2,2)$ in $\mathrm{PG}(3,2)$, where $\mathrm{PG}(3,2)$ is embedded in $\mathrm{PG}(3,4)$. This is a quantum $[[8,0,4]]$-code. The cardinalities of quantum caps in $\mathrm{PG}(3,4)$ are $8,12,14,17$. The cardinalities of quantum caps in $\mathrm{PG}(4,4)$ are a priori between 10 (the obvious theoretical minimum) and 41, the size of the largest cap in $\mathrm{PG}(4,4)$. In fact, one of the two 41-caps in $\mathrm{PG}(4,4)$ is quantum as is the uniquely determined largest cap in $\mathrm{AG}(4,4)$, which has 40 points. Here is a construction of a quantum 10-cap in $\mathrm{PG}(4,4)$: choose two planes Π_1,Π_2 in $\mathrm{PG}(4,4)$ which intersect in a point P. Choose ovals $o_i \subset \Pi_i$ such that P is the nucleus of o_i. Then $o_1 \cup o_2$ is a quantum cap. The most obvious recursive construction is the following

Theorem 7. *Let K_1,K_2 be disjoint pre-quantum caps in* $\mathrm{PG}(m-1,4)$*. If $K_1 \cup K_2$ is a cap, then it is a pre-quantum cap.*

Let $K_1 \subset K_2$ be pre-quantum caps. Then also $K_2 \setminus K_1$ is a pre-quantum cap.

This theorem can be used in two ways. One is to start from a quantum cap K_2 and construct quantum caps $K_1 \subset K_2$. This point of view was adopted by Tonchev [56] who found quantum caps contained in the quantum 41-cap in $\mathrm{PG}(4,4)$ (of sizes $n \in \{10,12,14-27,29,31,33,35\}$) and in the Glynn cap, a 126-cap in $\mathrm{PG}(5,4)$ which is quantum. The question which subcaps of a given quantum cap are pre-quantum can be expressed in terms of a certain binary code.

Definition 9. *Let K be a cap in* PG$(m-1,4)$ *and M a corresponding generator matrix. The* **associated binary code** *A is the binary linear code of length n generated by the supports of the quaternary codewords of the code generated by M.*

Observe that by definition, K is pre-quantum if and only if A is contained in the all-even code. This leads to the following characterization.

Theorem 8. *Let $K \subset PG(m-1,4)$ be pre-quantum and $K_1 \subseteq K$. Then K_1 (and its complement $K \setminus K_1$) is pre-quantum if and only if the characteristic vector of K_1 is contained in the dual $A^{\perp}$ of the binary code A associated to K.*

This is essentially Theorem 7 of [15].

The other way how to use Theorem 7 is to construct quantum caps as a union $K_1 \cup K_2$ of two disjoint pre-quantum caps K_1 and K_2. This often leads to more transparent constructions. For example, a quantum 12-cap in PG$(3,4)$ can be constructed simply as the union of two disjoint hyperovals on two planes. Bartoli [3] describes a quantum 20-cap in PG$(4,4)$ and constructs more quantum caps in PG$(4,4)$ of cardinalities $29,30,32,33,34$ in [5]. Theorem 7 can be generalized.

Theorem 9. *Let Π_1,Π_2 be different hyperplanes of* PG$(m,4)$ *and $K_i \subset \Pi_i$ be pre-quantum caps such that $K_1 \cap \Pi_1 \cap \Pi_2 = K_2 \cap \Pi_1 \cap \Pi_2$. Then the symmetric sum $K_1 + K_2 = (K_1 \setminus K_2) \cup (K_2 \setminus K_1)$ is a pre-quantum cap.*

Theorem 10. *Let Π_1,Π_2 be different $(m-2)$-dimensional subspaces of* PG$(m,4)$ *which together generate* PG$(m,4)$*. Let $K_i \subset \Pi_i$ be pre-quantum caps such that $K_1 \cap \Pi_1 \cap \Pi_2 = K_2 \cap \Pi_1 \cap \Pi_2$. Then the symmetric sum $K_1 + K_2$ is a pre-quantum cap.*

As an application of Theorem 10, choose two planes Π_1,Π_2 in PG$(4,4)$ which meet in a point X. Let $K_i \cup \{X\}$ be a hyperoval in Π_i, for $i = 1,2$. Then the symmetric sum $K_1 \cup K_2$ is a quantum 10-cap in PG$(4,4)$. In [4], we give geometric constructions of quantum 36-caps and of quantum 38-caps in PG$(4,4)$. This yields new quantum codes with parameters $[[36,26,4]]$ and $[[38,28,4]]$.

Tonchev [56] found a quantum 27-cap in PG$(6,4)$ by the action of an automorphism of order 13. It turns out that the dual distance is in fact 5, so this yields a quaternary linear $[[27,13,5]]$-quantum code.

In [11], a quantum 5040-cap in PG$(9,4)$ and a quantum 756-cap in PG$(7,4)$ are constructed. For a long time, the smallest open problem on additive quantum codes concerned the existence of $[[13,5,4]]$-quantum codes. This has been settled in [12]: such a quantum code does not exist.

12. A Problem in Additive Number Theory

Definition 10. *Let A be an abelian group, written additively, and $e = exp(A)$ its exponent, i.e. the lowest common multiple of its element orders. A* **sequence** *over A is a mapping $\sigma : A \longrightarrow \{0,1,2,\ldots\}$. We think of a sequence as a multiset, where each element $a \in A$ occurs with multiplicity $\sigma(a)$. The size of a sequence is $\sum_a \sigma(a)$.*
A sequence $S(A)$ is a sequence over A which does not contain subsequences of size e which sum to 0. Denote by $l(A)$ the largest size of a sequence $S(A)$.

The problem is the determination of $l(A)$. This problem and certain related problems have a long history in additive number theory. Clearly, all multiplicities of elements in a sequence $S(A)$ are bounded by $e-1$.

In the literature mostly the case of homocyclic groups $A=\mathbb{Z}_m^n$ is considered. One reason for this may have been the following observation: $l(\mathbb{Z}_m^n)+1$ is the smallest number N such that each set of N points in the rank n integer lattice $\mathbb{Z}^n$ contains a subset of m points whose centroid is in $\mathbb{Z}^n$. Clearly $exp(\mathbb{Z}_m^n)=exp(\mathbb{Z}_m)=m$. In case $m=3$, there is an obvious link to affine caps. Recall from Section 5. that $C_n(q)$ denotes the largest size of a cap in $\mathrm{AG}(n,q)$.

Proposition 1. $l(\mathbb{Z}_3^n)=2C_n(3)$.

Proof. This follows directly from the fact that a subset of $\mathrm{AG}(n,3)$ is a cap if and only if it does not contain a 3-subset summing to 0. If K is a cap, then the multiset $2K$, where each element of K appears with multiplicity 2, is a sequence $S(\mathbb{Z}_3^n)$. On the other hand, if a sequence $S(\mathbb{Z}_3^n)$ is given, then using each element of multiplicity >0 with multiplicity 2 produces a sequence $S(\mathbb{Z}_3^n)$ which has the form $2K$, where K is a cap. □

Recall that each abelian group can be written as a direct product $A=\mathbb{Z}_{m_1}\times\cdots\times\mathbb{Z}_{m_r}$ where $m_1\mid\cdots\mid m_r$ in a unique way and r is the **rank** of A, the largest rank of its Sylow p-subgroups, where p varies over the prime divisors of $|A|$. For rank ≤ 2, the answer to our problem is known:

Theorem 11. *Let $A=\mathbb{Z}_{m_1}\times\mathbb{Z}_{m_2}$ where $m_1\mid m_2$. Then $l(A)=2m_1+2m_2-4$.*

A proof is in [32]. For rank one, this implies $l(\mathbb{Z}_m)=2m-2$. This implies that each sequence of $2m-1$ integers contains a subsequence of m integers which sum to 0 mod m. This is the Erdös-Ginzburg-Ziv theorem, see Section 2.4 of Nathanson [43].

A Global Approach

Here is a related global problem.

Definition 11. *A subset $U\subset\{0,1,2,\ldots\}^n$ is a sequence $S(n,\mathbb{Z})$ if for each odd integer m, the multiset $(m-1)(U \bmod m)$ is a sequence $S(\mathbb{Z}_m^n)$.*

Here $(m-1)(U \bmod m)$ stands for the following: each element of U is read mod m in each component, the resulting tuple in $\mathbb{Z}_m^n$ is used with multiplicity $m-1$. In particular each sequence $S(n,\mathbb{Z})$ of cardinality u yields a sequence $S(\mathbb{Z}_m^n)$ of cardinality $(m-1)u$, for each odd m. Choosing $m=3$, we see that U mod 3 is a cap in $\mathrm{AG}(n,3)$, consequently $|U|\leq C_n(3)$.

Proposition 2. *Let $U=\{0,1\}^n$. Then S is a sequence $S(n,\mathbb{Z})$ of size 2^n.*

Proof. Assume S is a multisubset of $(m-1)(U \bmod m)$, defined by multiplicities $\mu_v\leq m-1$ for $v\in S$, such that $\sum\mu_v=m$ and $\sum\mu_v v\equiv 0 \pmod m$. The coordinate entries in $\sum\mu_v v$ are 0 or m. Let $v\neq v'$ such that $\mu_v>0,\mu_{v'}>0$. Choose notations such that there exists a coordinate i with $v_i=0,v'_i=1$. Then coordinate i yields a contradiction. There is therefore only one v such that $\mu_v>0$. This yields the contradiction $\mu_v=m$. □

Proposition 2 is due to Harborth [35]. This result implies $l(\mathbb{Z}_m^n) \geq (m-1)2^n$. Sets $S(3,\mathbb{Z})$ and $S(4,\mathbb{Z})$ of maximal sizes $C_3(3) = 9$ and $C_4(3) = 20$, respectively, were constructed in [24, 27]. Here is the sequence $S(3,\mathbb{Z})$ of size 9 as given in [27]. It is in fact contained in $\{0,1,2\}^3$ and consists of the following triples:

$$(0,0,0),(1,0,0),(0,1,0),(0,0,1),(1,0,1),(0,1,1),(1,1,2),(1,2,2),(2,1,2).$$

Let us check the defining property for $m = 3$. This is equivalent with the statement that the set of points $(1 : x_2 : x_3 : x_4) \in AG(3,3)$, where $x = (x_2,x_3,x_4)$ varies over the nine triples above and entries are interpreted in $\mathbb{Z}/3\mathbb{Z}$, form a cap. In fact all those points are on the quadric $x_2^2 + x_3^2 + x_4^2 - x_1x_2 - x_1x_3 - x_1x_4 + x_2x_3 = 0$. This is an elliptic quadric in PG$(3,3)$ whose points therefore form a cap. As a consequence $l(\mathbb{Z}_m^3) \geq 9(m-1)$ for all odd $m \geq 3$. It is conjectured in Gao-Thangadurai [31] that equality always holds. The conjecture has been confirmed for $m = 3^a5^b$, see [30]. An analogous conjecture concerning $\mathbb{Z}_m^4$ is made in [24]: $l(\mathbb{Z}_m^4) = 20(m-1)$ for all odd $m \geq 3$.

In [18], a product construction is used to produce sequences $S(5,\mathbb{Z}), S(6,\mathbb{Z}), S(7,\mathbb{Z})$ of sizes 42, 96, and 192, respectively. As 42 is the size of the second-largest complete cap in AG$(5,3)$ (this fact is proved in [18]), it follows that any sequence $S(5,\mathbb{Z})$ of size > 42 must have the property that its image mod 3 is contained in the affine Hill cap. The existence of such a sequence $S(5,\mathbb{Z})$ remains an open problem. Another open problem concerns the following conjecture: Each sequence $S(A)$ of maximal length $l(A)$ arises from a subset of A by using each element with multiplicity $e-1$.

Acknowledg ments

The research of the second author takes place within the project "Linear codes and cryptography" of the Research Foundation – Flanders (Belgium) (FWO) (Project nr. G.0317.06), and is supported by the Interuniversitary Attraction Poles Programme - Belgian State - Belgian Science Policy: project P6/26-Bcrypt.

References

[1] J. Barát, Y. Edel, R. Hill, and L. Storme, *On complete caps in the projective geometries over $\mathbb{F}_3$. II. New improvements*, J. Combin. Math. Combin. Comput., 49 (2004), pp. 9–31.

[2] A. Barlotti, *Some topics in finite geometrical structures*, Tech. Rep. 439, Institute of Statistics, University of Carolina, Mimeo Series, 1965.

[3] D. Bartoli, *Quantum codes and related geometric properties*, PhD thesis, University of Perugia, 2008.

[4] D. Bartoli, J. Bierbrauer, S. Marcugini, and F. Pambianco, *Geometric constructions of quantum codes*, in Error-Correcting Codes, Cryptography and Finite Geometries, A. A. Bruen and D. L. Wehlau, eds., vol. 523 of Contemp. Math., Amer. Math. Soc., Providence, RI, 2010, pp. 149—154.

[5] D. Bartoli, S. Marcugini, and F. Pambianco, *New quantum caps in* $PG(4,4)$. manuscript.

[6] J. Bierbrauer, *Large caps*, J. Geom., 76 (2003), pp. 16–51. Combinatorics, 2002 (Maratea).

[7] ——, *Introduction to coding theory*, Discrete Mathematics and its Applications (Boca Raton), Chapman & Hall/CRC, Boca Raton, FL, 2005.

[8] ——, *The spectrum of stabilizer quantum codes of distance* 3, IEEE Trans. Inform. Theory, submitted, (2010).

[9] J. Bierbrauer and Y. Edel, *A family of caps in projective* 4*-space in odd characteristic*, Finite Fields Appl., 6 (2000), pp. 283–293.

[10] ——, *Bounds on affine caps*, J. Combin. Des., 10 (2002), pp. 111–115.

[11] J. Bierbrauer, G. Faina, M. Giulietti, S. Marcugini, and F. Pambianco, *The geometry of quantum codes*, Innov. Incidence Geom., 6/7 (2007/08), pp. 53–71.

[12] J. Bierbrauer, R. Fears, S. Marcugini, F. Pambianco, *The non-existence of a* $[[13,5,4]]$ *quantum stabilizer code*, IEEE Trans. Inform. Theory, to appear.

[13] R. C. Bose, *Mathematical theory of the symmetrical factorial design*, Sankhyā, 8 (1947), pp. 107–166.

[14] R. Calderbank and P. C. Fishburn, *Maximal three-independent subsets of* $\{0,1,2\}^n$, Des. Codes Cryptogr., 4 (1994), pp. 203–211.

[15] R. Calderbank, E. M. Rains, P. W. Shor, and N. J. A. Sloane, *Quantum error correction via codes over* $\mathrm{GF}(4)$, IEEE Trans. Inform. Theory, 44 (1998), pp. 1369–1387.

[16] B. L. Davis and D. Maclagan, *The card game SET*, Math. Intelligencer, 25 (2003), pp. 33–40.

[17] Y. Edel, *Extensions of generalized product caps*, Des. Codes Cryptogr., 31 (2004), pp. 5–14.

[18] ——, *Sequences in abelian groups G of odd order without zero-sum subsequences of length* $\exp(G)$, Des. Codes Cryptogr., 47 (2008), pp. 125–134.

[19] Y. Edel and J. Bierbrauer, 41 *is the largest size of a cap in* $\mathrm{PG}(4,4)$, Des. Codes Cryptogr., 16 (1999), pp. 151–160.

[20] ——, *Recursive constructions for large caps*, Bull. Belg. Math. Soc. Simon Stevin, 6 (1999), pp. 249–258.

[21] ——, *Large caps in small spaces*, Des. Codes Cryptogr., 23 (2001), pp. 197–212.

[22] ——, *The largest cap in* $\mathrm{AG}(4,4)$ *and its uniqueness*, Des. Codes Cryptogr., 29 (2003), pp. 99–104.

[23] ——, *Caps of order* $3q^2$ *in affine 4-space in characteristic 2*, Finite Fields Appl., 10 (2004), pp. 168–182.

[24] Y. Edel, C. Elsholtz, A. Geroldinger, S. Kubertin, and L. Rackham, *Zero-sum problems in finite abelian groups and affine caps*, Q. J. Math., 58 (2007), pp. 159–186.

[25] Y. Edel, S. Ferret, I. Landjev, and L. Storme, *The classification of the largest caps in* $\mathrm{AG}(5,3)$, J. Combin. Theory Ser. A, 99 (2002), pp. 95–110.

[26] Y. Edel, L. Storme, and P. Sziklai, *New upper bounds on the sizes of caps in* $\mathrm{PG}(N,5)$ *and* $\mathrm{PG}(N,7)$, J. Combin. Math. Combin. Comput., 60 (2007), pp. 7–32.

[27] C. Elsholtz, *Lower bounds for multidimensional zero sums*, Combinatorica, 24 (2004), pp. 351–358.

[28] T. Feulner, *The automorphism groups of linear codes and canonic representatives of their semilinear isometry classes*, Adv. Math. Commun., 3 (2009), pp. 363–383.

[29] ——. `http://www.algorithm.uni-bayreuth.de/en/research/Coding_Theory/CanonicalForm/index.html`, 2010.

[30] W. D. Gao, Q. H. Hou, W. A. Schmid, and R. Thangadurai, *On short zero-sum subsequences. II*, Integers, 7 (2007), pp. A21, 22 pp. (electronic).

[31] W. D. Gao and R. Thangadurai, *On zero-sum sequences of prescribed length*, Aequationes Math., 72 (2006), pp. 201–212.

[32] A. Geroldinger and F. Halter-Koch, *Non-unique factorizations*, vol. 278 of Pure and Applied Mathematics (Boca Raton), Chapman & Hall/CRC, Boca Raton, FL, 2006. Algebraic, combinatorial and analytic theory.

[33] D. G. Glynn, *A 126-cap of* $\mathrm{PG}(5,4)$ *and its corresponding* $[126,6,88]$*-code*, Util. Math., 55 (1999), pp. 201–210.

[34] V. D. Goppa, *Codes and information*, Uspekhi Mat. Nauk, 39 (1984), pp. 77–120.

[35] H. Harborth, *Ein Extremalproblem für Gitterpunkte*, J. Reine Angew. Math., 262/263 (1973), pp. 356–360. Collection of articles dedicated to Helmut Hasse on his seventy-fifth birthday.

[36] R. Hill, *On the largest size of cap in* $S_{5,3}$, Atti Accad. Naz. Lincei Rend. Cl. Sci. Fis. Mat. Natur. (8), 54 (1973), pp. 378–384 (1974).

[37] ——, *Caps and groups*, in Colloquio Internazionale sulle Teorie Combinatorie (Rome, 1973), Tomo II, Accad. Naz. Lincei, Rome, 1976, pp. 389–394. Atti dei Convegni Lincei, No. 17.

[38] J. W. P. Hirschfeld and L. Storme, *The packing problem in statistics, coding theory and finite projective spaces: update 2001*, in Finite geometries, vol. 3 of Dev. Math., Kluwer Acad. Publ., Dordrecht, 2001, pp. 201–246.

[39] H.-J. Kroll and R. Vincenti, *Antiblocking systems and PD-sets*, Discrete Math., 308 (2008), pp. 401–407.

[40] Y. I. Manin, *What is the maximum number of points on a curve over* $\mathbf{F}_2$*?*, J. Fac. Sci. Univ. Tokyo Sect. IA Math., 28 (1981), pp. 715–720 (1982).

[41] R. Meshulam, *On subsets of finite abelian groups with no* 3*-term arithmetic progressions*, J. Combin. Theory Ser. A, 71 (1995), pp. 168–172.

[42] A. C. Mukhopadhyay, *Lower bounds on* $m_t(r,s)$, J. Combinatorial Theory Ser. A, 25 (1978), pp. 1–13.

[43] M. B. Nathanson, *Additive number theory*, vol. 165 of Graduate Texts in Mathematics, Springer-Verlag, New York, 1996. Inverse problems and the geometry of sumsets.

[44] F. Pambianco and L. Storme, *Small complete caps in spaces of even characteristic*, J. Combin. Theory Ser. A, 75 (1996), pp. 70–84.

[45] G. Pellegrino, *Sul massimo ordine delle calotte in* $S_{4,3}$, Matematiche (Catania), 25 (1970), pp. 149–157 (1971).

[46] A. Potechin, *Maximal caps in* $\mathrm{AG}(6,3)$, Des. Codes Cryptogr., 46 (2008), pp. 243–259.

[47] B. Qvist, *Some remarks concerning curves of the second degree in a finite plane*, Ann. Acad. Sci. Fennicae. Ser. A. I. Math.-Phys., 1952 (1952), p. 27.

[48] R. Schoof and M. van der Vlugt, *Hecke operators and the weight distributions of certain codes*, J. Combin. Theory Ser. A, 57 (1991), pp. 163–186.

[49] B. Segre, *Sulle ovali nei piani lineari finiti*, Atti Accad. Naz. Lincei. Rend. Cl. Sci. Fis. Mat. Nat. (8), 17 (1954), pp. 141–142.

[50] ——, *Ovals in a finite projective plane*, Canad. J. Math., 7 (1955), pp. 414–416.

[51] ——, *Le geometrie di Galois*, Ann. Mat. Pura Appl. (4), 48 (1959), pp. 1–96.

[52] ——, *On complete caps and ovaloids in three-dimensional Galois spaces of characteristic two*, Acta Arith., 5 (1959), pp. 315–332 (1959).

[53] L. Storme, J. A. Thas, and S. K. J. Vereecke, *New upper bounds for the sizes of caps in finite projective spaces*, J. Geom., 73 (2002), pp. 176–193.

[54] G. Tallini, *Calotte complete di* $S_{4,q}$ *contenenti due quadriche ellittiche quali sezioni iperpiane*, Rend. Mat. e Appl. (5), 23 (1964), pp. 108–123.

[55] J. Tits, *Ovoïdes et groupes de Suzuki*, Arch. Math., 13 (1962), pp. 187–198.

[56] V. D. Tonchev, *Quantum codes from caps*, Discrete Math., 308 (2008), pp. 6368–6372.

In: Current Research Topics in Galois Geometry
Editors: J. De Beule and L. Storme
ISBN: 978-1-61209-523-3

Chapter 5

THE POLYNOMIAL METHOD IN GALOIS GEOMETRIES

Simeon Ball*
Departament de Matemàtica Aplicada IV,
Universitat Politècnica de Catalunya, Jordi Girona 1-3, Mòdul C3,
Campus Nord, 08034 Barcelona, Spain

Abstract

The polynomial method refers to the application of polynomials to combinatorial problems. The method is particularly effective for Galois geometries and a number of problems and conjectures have been solved using the polynomial method. In many cases the polynomial approach is the only method which we know of that works. In this article, the various polynomial techniques that have been applied to Galois geometries are detailed and, to demonstrate how to apply these techniques, some of the problems referred to above are resolved.

1. Introduction

In this article we shall introduce the polynomial method that allows us to solve some problems in Galois geometries by considering properties of certain polynomials of $\mathbb{F}_q[X]$. In general the method is the following. Given an object $\mathcal{O}$ in a Galois geometry over $\mathbb{F}_q$ with a regular property, define a polynomial f with coefficients in $\mathbb{F}_q$, or some finite extension of $\mathbb{F}_q$, which translates the geometrical property of $\mathcal{O}$ into an algebraic property of f. Using this algebraic property of f we then try to deduce further algebraic properties which translate back into further geometrical properties of $\mathcal{O}$.

This is best seen by way of an example. Consider a set $\mathcal{S}$ of points of $\mathrm{AG}(n,q)$ with the property that every hyperplane of $\mathrm{AG}(n,q)$ is incident with a point of $\mathcal{S}$. We wish to prove a lower bound on $|\mathcal{S}|$ and construct an example to show that this bound is best possible. A combinatorial counting argument gives a bound of roughly $q+\sqrt{q}$ for $n=2$, whereas by construction the best we can do is $2q-1$. For the construction one can take the points on the union of two intersecting lines.

*E-mail address: simeon@ma4.upc.edu

A hyperplane of $\mathrm{AG}(n,q)$ which does not contain the origin is defined by an equation

$$a_1X_1 + \ldots + a_nX_n + 1 = 0.$$

By assumption there is a point $s \in \mathcal{S}$ which is incident with this hyperplane or, in other words,

$$a_1s_1 + \ldots + a_ns_n + 1 = 0.$$

Let

$$f(X) = \prod_{s \in \mathcal{S}} (s_1X_1 + \ldots + s_nX_n + 1).$$

Assuming that the origin is an element of $\mathcal{S}$ this polynomial has degree $|\mathcal{S}| - 1$. It has the property that

$$f(a) = 0$$

for all $a \in \mathbb{F}_q^n$, provided that $a \neq 0$. Moreover, $f(0) = 1$.

We shall show in the next section that the degree of a polynomial with such properties is at least $n(q-1)$, which will imply that $|\mathcal{S}| \geq n(q-1)+1$. This bound was first proven by Jamison [36] and will be referred to as Jamison's theorem.

Note that the various sections are written in such a way that, apart from Section 3., they stand alone and can be read independently.

2. Combinatorial Nullstellensatz

Let $\mathbb{F}$ be a field, not necessarily finite. Let $f \in \mathbb{F}[X]$ be a polynomial with the property that $f(x) = 0$ for all $x \in \mathcal{S}_1$, where $\mathcal{S}_1$ is some finite subset of $\mathbb{F}$. If we define

$$g_1(X) = \prod_{s \in \mathcal{S}_1} (X - s)$$

then we can write $f = g_1h_1$, for some polynomial h_1 of degree $f^\circ - g_1^\circ$, where f° will denote the degree of a polynomial f.

The Combinatorial Nullstellensatz of Alon extends this observation to polynomials in more indeterminates. The proof is straightforward induction so we shall not include it here, those interested can find a proof in the article by Alon [1].

For $i = 1, \ldots, n$, let $\mathcal{S}_i$ be finite subsets of $\mathbb{F}$ and define

$$g_i(X_i) = \prod_{s \in \mathcal{S}_i} (X_i - s).$$

Theorem 2.1. *If $f \in \mathbb{F}[X_1, \ldots, X_n]$ has the property that $f(s_1, \ldots, s_n) = 0$ for all $(s_1, \ldots, s_n) \in \mathcal{S}_1 \times \ldots \times \mathcal{S}_n$ then*

$$f = \sum_{i=1}^{n} g_ih_i,$$

for some polynomials h_i of degree at most $f^\circ - g_i^\circ$.

Let us return to the polynomial f from the previous section. The following theorem is essentially the proof of Jamison's theorem given by Brouwer and Schrijver in [27].

Theorem 2.2. *If $f \in \mathbb{F}_q[X_1,\ldots,X_n]$ is a polynomial with the property that $f(s)=0$ for all $s \in \mathbb{F}_q^n$, $s \neq 0$ and $f(0)=1$, then $f^\circ \geq n(q-1)$.*

Proof. Let

$$g_i(X_i) = \prod_{s\in\mathbb{F}_q}(X_i - s) = X_i^q - X_i.$$

We can write

$$f = \sum g_i u_i + w$$

for some polynomials u_i in such a way that the polynomial $w = w(X_1,\ldots,X_n) \neq 0$ has degree at most $q-1$ in X_i and $f^\circ \geq w^\circ$.

For every i the polynomial $X_i w$ has the property that $(X_i w)(s_1,\ldots,s_n) = 0$ for all $(s_1,\ldots,s_n) \in \mathbb{F}_q^n$. However, for $j \neq i$ the degree in X_j of $X_i w$ is at most $q-1$. When we apply Theorem 2.1 to $X_i w$ the $g_j h_j$ terms are zero since the X_j^q term in g_j would give terms on the right-hand side that do not appear in $X_i w$. Hence

$$X_i w = g_i h_i,$$

for some polynomial h_i. Thus $g_i(X_i)$ divides $X_i w$ for all i. The g_i are polynomials in different indeterminates and so are pairwise coprime and therefore $\prod g_i(X_i)$ divides $(\prod X_i)w$. Thus the degree of w, and therefore the degree of f, is at least $\sum g_i^\circ - n = nq - n$. □

This can be easily extended to more general sets where $\mathbb{F}_q$ is replaced by arbitrary finite subsets of a field, see [15].

The consequences of Theorem 2.2 for the set of points $\mathcal{S}$ were already mentioned in the previous section. Namely we get Jamison's theorem, which is the following.

Corollary 2.3. *If $\mathcal{S}$ is a set of points of $\mathrm{AG}(n,q)$ with the property that every hyperplane is incident with a point of $\mathcal{S}$ then*

$$|\mathcal{S}| \geq n(q-1)+1.$$

Proof. Since the degree of the polynomial f is bounded below by $n(q-1)$, the set of points $\mathcal{S}$ has size at least $n(q-1)+1$. □

This bound can be obtained by taking the set of points that is the union of n lines which span $\mathrm{AG}(n,q)$ and all concurrent with a point x.

3. Nullstellensätze for Lower Dimensional Subspaces?

In the previous section we proved that a set of points $\mathcal{S}$, with the property that every hyperplane of $\mathrm{AG}(n,q)$ is incident with a point of $\mathcal{S}$, has size at least $n(q-1)+1$. There are various generalisations which we may consider. If we replace the condition "one point" with "t points" then similar techniques to those mentioned before have been used to prove lower bounds on the size of the set, see [3] and [28], although it is doubtful in most cases that these bounds are attainable. We shall not consider them here.

Another possible generalisation would be to replace "hyperplane" with "k-dimensional subspace", where $k \leq n-2$. Here, we can combine a combinatorial counting approach, with the theorem we obtained using the polynomial method in the previous section, to obtain a lower bound on the size of $\mathcal{S}$.

Suppose that $\mathcal{S}$ is a set of points with the property that every k-dimensional subspace of $\mathrm{AG}(n,q)$ is incident with a point of $\mathcal{S}$. Let ρ be a k-dimensional subspace incident with exactly one point x of $\mathcal{S}$. Let σ be a $(k+1)$-dimensional subspace containing ρ. The set $\mathcal{S}\cap\sigma$ has the property that every hyperplane of σ is incident with a point of $\mathcal{S}\cap\sigma$ and so by Jamison's theorem, which we proved in the previous section, it has size at least $(k+1)(q-1)+1$. There are $(q^{n-k}-1)/(q-1)$ subspaces σ containing ρ which all contain the point x but share no other point of $\mathcal{S}$. Thus

$$|\mathcal{S}| \geq (k+1)(q^{n-k}-1)+1.$$

This bound is, more or less, the best known bound. The polynomial method does not seem to allow us to improve on this, although that may be because we simply do not know how to apply it to this more general case.

The known constructions are somewhat crude. For example, let $\mathcal{S}$ be a set of points of $\mathrm{AG}(3,q)$ with the property that every line is incident with a point of $\mathcal{S}$. For q square, the smallest known example has size roughly $2q^2+2q\sqrt{q}$ and is constructed using a double blocking set of $\mathrm{PG}(2,q)$ at infinity and forming a cone with a vertex point of the affine space. However, the lower bound we obtained with $n=3$ and $k=1$ is $2q^2-1$, so we are some way short of the size of the set in the construction.

Let us see where a polynomial approach, similar to that used in the introduction leads to. Define

$$f(X_1,X_2,X_3)=\prod_{s\in\mathcal{S}}(s_1X_1+s_2X_2+s_3X_3+1).$$

We would like to translate the geometric property of $\mathcal{S}$, that every line is incident with a point of $\mathcal{S}$, into an algebraic property of f. An affine line is defined by two equations of the form

$$a_1X_1+a_2X_2+a_3X_3+1=0,\quad b_1X_1+b_2X_2+b_3X_3=0,$$

where a and b are linearly independent. By assumption, for every $a,b\in\mathbb{F}_q^3$, linearly independent, there is a point $s\in\mathcal{S}$ with the property that

$$a_1s_1+a_2s_2+a_3s_3+1=0,\quad b_1s_1+b_2s_2+b_3s_3=0.$$

Therefore

$$f(a_1+b_1X,a_2+b_2X,a_3+b_3X)=\prod_{s\in\mathcal{S}}(s_1a_1+s_2a_2+s_3a_3+1+(s_1b_1+s_2b_2+s_3b_3)X)=0$$

for all $a,b\in\mathbb{F}_q^3$ linearly independent, and $f(0)=1$. It is not clear what lower bound can be proved for the degree of a polynomial with such properties but clearly any lower bound would give a lower bound for the size of $\mathcal{S}$.

There are a number of objects in higher dimensional spaces for which properties can be deduced using planar results but where a direct application of the polynomial method doesn't appear to offer more insight but probably should. The example mentioned above is just one example of these.

4. Lacunary Polynomials

It was Rédei who first worked on lacunary polynomials over finite fields and wrote the book [40]. In Chapter VI, §36, he applies a theorem on lacunary polynomials to functions over a finite field which determine few directions. This may have been the first application of the polynomial method to a geometrical problem. Before considering the geometrical problem, let us prove a generalisation of one of Rédei's result on lacunary polynomials, which is due to Blokhuis [18].

Let (g,h) denote the greatest common divisor of polynomials g and h.

Lemma 4.1. *Suppose that $f(X) = g(X)X^q + h(X)$ is a polynomial in $\mathbb{F}_q[X]$ which factorises completely into linear factors in $\mathbb{F}_q[X]$. If $\max(g^\circ, h^\circ) \leq (q-1)/2$ then $f(X) = g(X)(X^q - X)$ or $f(X) = (g,h)e(X^p)$ for some $e \in \mathbb{F}_q[X]$, where $q = p^h$.*

Proof. We can suppose that g and h have no common factors since removing them does not affect the hypothesis.

The factors of f are factors of $X^q - X$ and so factors of $f - (X^q - X)g = Xg + h$.

The factors of multiplicity $m \geq 2$ are factors of multiplicity at least $m-1$ of

$$f' = \frac{df}{dX} = g'X^q + h',$$

and so are factors of multiplicity at least $m-1$ of $f'g - g'f = h'g - g'h$.

Therefore f is a factor of $(Xg+h)(h'g - g'h)$. This polynomial has degree at most $(q+1)/2 + g^\circ + h^\circ - 1 \leq q - 1 + g^\circ$, whereas $f^\circ = q + g^\circ$. Since f cannot be a factor of a non-zero polynomial of less degree than itself, it follows that $(Xg+h)(h'g-g'h) = 0$.

If $Xg + h = 0$ then $f(X) = g(X)(X^q - X)$.

If $h'g - g'h = 0$ then h divides h' (assuming $(g,h) = 1$) and so $h' = 0$ and $g' = 0$. Thus in this case g and h are in $\mathbb{F}_q[X^p]$ and the lemma is proved. □

This lemma was used by Blokhuis to prove his theorem on blocking sets in $\mathrm{PG}(2,p)$ which we shall see later. Note that the bound on g° and h° is tight for q odd since the polynomial $X^q - X^{(q+1)/2}$ factors into linear factors in $\mathbb{F}_q[X]$.

Consider the graph of a function ϕ over a finite field $\mathbb{F}_q$; in other words the set of q points $\{(x,\phi(x)) \mid x \in \mathbb{F}_q\}$ of $\mathrm{AG}(2,q)$. The set of directions determined by this set is

$$\mathcal{D}_\phi = \{\frac{\phi(y)-\phi(x)}{y-x} \mid y \neq x,\ x,y \in \mathbb{F}_q\}.$$

For a typical function ϕ, the set $\mathcal{D}_\phi$ will be the set of all elements of $\mathbb{F}_q$. However, there are functions for which $\mathcal{D}_\phi$ is not all $\mathbb{F}_q$. The linear functions determine only one direction of course. For a function ϕ, which is linear over a proper subfield $\mathbb{F}_s$ of $\mathbb{F}_q$, $\mathcal{D}_\phi$ satisfies

$$\frac{q}{s} + 1 \leq |\mathcal{D}_\phi| \leq \frac{q-1}{s-1},$$

and there are functions which attain both bounds. If q is odd then, for the function ϕ defined by the monomial $x^{(q+1)/2}$, the set $\mathcal{D}_\phi$ has $(q+3)/2$ elements. Thus, for every q (since $\mathbb{F}_2$ is a subfield of $\mathbb{F}_q$ when q is even), there is some function ϕ for which

$$|\mathcal{D}_\phi| \leq \frac{q+3}{2}.$$

Rédei started the investigation which would eventually lead to proving that the functions ϕ that determine less than $(q+3)/2$ directions are linear over a subfield. Some improvements to Rédei's initial work are included in [21], the classification being all but obtained in [19], and finally obtained in [4]. It can be summarised in the following theorem.

Theorem 4.2. *If, for some function ϕ from $\mathbb{F}_q$ to $\mathbb{F}_q$, the set $\mathcal{D}_\phi$ has less than $(q+3)/2$ elements, then there is a subfield $\mathbb{F}_s$ of $\mathbb{F}_q$ such that*

$$\frac{q}{s}+1 \leq |\mathcal{D}_\phi| \leq \frac{q-1}{s-1},$$

and ϕ is linear over $\mathbb{F}_s$.

We shall prove the classification in the prime case, as Rédei did in his book, and leave the interested reader to consult the references for the non-prime case. We shall in fact prove something stronger, that was first proved by Blokhuis in [18]. He proved that a set $\mathcal{S}$ of at most $(3p+1)/2$ points of PG$(2,p)$, p prime, with the property that every line is incident with a point of $\mathcal{S}$, contains all the points of a line.

Consider a set of q points $\mathcal{S}$ of AG$(2,q)$ and let

$$\mathcal{D} = \{\frac{s_2-t_2}{s_1-t_1} \mid s \neq t,\ s,t \in \mathcal{S}\},$$

be the set of directions determined by the points of $\mathcal{S}$. Let us assume that $\infty \in \mathcal{D}$. Note that if S is the graph of a function then $\infty \notin \mathcal{D}$. However, we can apply an affine transformation to $\mathcal{S}$ so that ∞ is an element of $\mathcal{D}$.

An element $-x \in \mathcal{D}$ if and only if there are elements $s,t \in \mathcal{S}$ with the property that $xs_1+s_2 = xt_1+t_2$. Therefore if $-x \notin \mathcal{D}$ the set $\{xs_1+s_2 \mid s \in \mathcal{S}\} = \mathbb{F}_q$.

Let $\mathcal{E} = (\mathbb{F}_q \cup \{\infty\}) \setminus \mathcal{D}$. We are interested in the case $|\mathcal{D}| \leq (q+1)/2$ or equivalently $|\mathcal{E}| \geq (q+1)/2$.

Let us generalise the situation to a set $\mathcal{S}$ of $q+k$ points where

$$\mathcal{E} = \{-x \in \mathbb{F}_q \mid \{xs_1+s_2 \mid s \in \mathcal{S}\} = \mathbb{F}_q\}$$

has size at least $(q+1)/2+k$. The parallel lines with direction m are defined by equations of the form $X_2 = mX_1+c$. The lines in this set of lines are all incident with a point of $\mathcal{S}$ if and only if $m \in \mathcal{E}$.

We shall prove that in the prime case $\mathcal{S}$ contains all the points of a line.

Firstly we introduce a polynomial f which translates the geometric property of $\mathcal{S}$ into an algebraic property of f. Let

$$f(X_1,X_2) = \prod_{s\in\mathcal{S}}(X_1+s_1X_2+s_2).$$

The polynomial f has the property that the polynomial $X_1^q - X_1$ is a factor of $f(X_1,x)$ if and only if $-x \in \mathcal{E}$.

The proof of Theorem 4.3 follows Blokhuis' approach in [18].

Theorem 4.3. *Let p be a prime and let $f \in \mathbb{F}_p[X_1,X_2]$ be the product of $p+k$ linear polynomials in $\mathbb{F}_p[X_1,X_2]$. If there are at least $(p+1)/2+k \leq p-1$ elements $x \in \mathbb{F}_p$ with the property that $X_1^p - X_1$ is a factor of $f(X_1,x)$ then f has a factor*

$$X_1^p - X_1 - c(X_2+m)^{p-1} + c = \prod_{a_1 \in \mathbb{F}_p} (X_1 + a_1X_2 + ma_1 + c),$$

for some $m, c \in \mathbb{F}_p$.

Proof. Define polynomials $h_j(X_2)$ of degree at most j by writing

$$f(X_1,X_2) = \sum_{j=0}^{p+k} h_j(X_2)X_1^{p+k-j}.$$

Let $\mathcal{E} = \{x \in \mathbb{F}_p \mid X_1^p - X_1 \text{ divides } f(X_1,x)\}$.

If $x \in \mathcal{E}$ then

$$f(X_1,x) = (X_1^p - X_1)g(X_1)$$

for some $g(X_1)$, dependent on x, of degree at most k. Therefore $h_{k+1}(x) = \ldots = h_{p-1}(x) = 0$. Since a non-zero polynomial h has at most h° roots, the polynomials $h_{k+1}(X_2) = \ldots = h_{|\mathcal{E}|-1}(X_2) = 0$.

Therefore

$$f(X_1,X_2) = \sum_{j=0}^{k} h_j(X_2)X_1^{p+k-j} + \sum_{j=|\mathcal{E}|}^{p+k} h_j(X_2)X_1^{p+k-j}.$$

If $y \notin \mathcal{E}$ then

$$f(X_1,y) = X^p g(X_1) + h(X_1),$$

where $\max(g^\circ, h^\circ) = k \leq (p-1)/2$ and $X_1^p - X_1$ is not a factor. By Lemma 4.1,

$$f(X_1,y) = g(X_1)(X_1^p + c)$$

where $c \in \mathbb{F}_p$. Note that here we use the fact that we are working over a prime field.

The polynomial $f(X_1,y)$ has a factor $X_1 + c$ of multiplicity p and so $f(X_1,X_2)$ has p factors $X_1 + a_1X_2 + a_2$ for which $a_1y + a_2 = c$. Defining $m = -y$ proves the theorem. □

Let us return to the set of points $\mathcal{S}$.

Corollary 4.4. *Let $\mathcal{S}$ be a set of points of* AG$(2,p)$. *If there are at least $|\mathcal{S}| - (p-1)/2$ and at most $p-1$ parallel classes for which the lines of these parallel classes are all incident with at least one point of $\mathcal{S}$ then $\mathcal{S}$ contains all the points of a line.*

Proof. Since there are at most $p-1$ parallel classes for which the lines of these parallel classes are all incident with at least one point of $\mathcal{S}$, we can assume that there is an m such that the parallel class of lines defined by equations $X_2 = mX_1 + c$ are not all incident with a point of $\mathcal{S}$.

Define

$$f(X_1,X_2) = \prod_{s \in \mathcal{S}} (X_1 + s_1X_2 + s_2),$$

a product of $|\mathcal{S}| = p+k$ linear polynomials. By hypothesis there are at least $(p+1)/2+k$ elements $x \neq m$ with the property that $X_1^p - X_1$ is a factor of $f(X_1,x)$.

Applying Theorem 4.3, we can conclude that $\mathcal{S}$ contains all the points on the line $X_2 = mX_1 + c$, for some c. □

The corollary above implies Blokhuis' theorem on blocking sets in $\mathrm{PG}(2,p)$.

Corollary 4.5. *Let $\mathcal{B}$ be a set of points in $\mathrm{PG}(2,p)$ with the property that every line is incident with at least one point of $\mathcal{B}$. If $|\mathcal{B}| \leq (3p+1)/2$ then $\mathcal{B}$ contains all the points of a line.*

Proof. Suppose that $|\mathcal{B}| \leq (3p+1)/2$. Let l_∞ be a line which is incident with $n \geq 2$ points of $\mathcal{B}$. Let $\mathcal{S} = \mathcal{B} \setminus l_\infty$. Then $|\mathcal{S}| = |\mathcal{B}| - n$ and there are $p+1-n$ parallel classes for which the lines of these parallel classes are all incident with at least one point of $\mathcal{S}$. Since $p-1 \geq p+1-n \geq |\mathcal{S}| - (p-1)/2$ we can apply Corollary 4.4. Hence $\mathcal{B}$ contains all the points of an affine line. If it does not contain the point where this line meets l_∞ then it must contain a point on the p other lines through this point, which would imply $|\mathcal{B}| \geq 2p$. □

We are, of course, also interested in the case $q = p^h$ non-prime. It is conjectured that a minimal blocking set in $\mathrm{PG}(2,q)$ of size at most $(3q+1)/2$ is of a certain type but this will not be discussed here. It is known that every line is incident with 1 mod p points of a minimal blocking set of size at most $(3q+1)/2$, from the work of Szőnyi [46], see also [43].

There have been some results obtained using lacunary polynomials in several indeterminates, see for example [13], [31]. However, it seems that many of these can also be obtained using field extensions as we shall see in Section 6., so they will not be directly discussed here.

5. Vector Spaces of Polynomials and Functions over $\mathbb{F}_q$

Let $\mathbb{K}$ be a field.

Let $\mathcal{E}$ be a non-empty subset of $\mathbb{K}^n$. The set $\mathcal{E}^{\mathbb{F}_q}$ of functions from $\mathcal{E}$ to $\mathbb{F}_q$ is a vector space over $\mathbb{F}_q$ of dimension $|\mathcal{E}|$. A basis for this vector space is

$$\{f_y \mid y \in \mathcal{E}\},$$

where $f_y(x) = 1$ if $y = x$ and $f_y(x) = 0$ if $y \neq x$.

The set $\mathbb{K}_d[X_1,\ldots,X_n]$ of polynomials of degree at most d with coefficients from $\mathbb{K}$ is a vector space over $\mathbb{F}_q$ of dimension $\binom{n+d}{d}$. It has a basis

$$\{X_1^{d_1}\cdots X_n^{d_n} \mid d_1+\ldots+d_n \leq d\}.$$

The set $\mathbb{K}_{[d]}[X_1,\ldots,X_n]$ of polynomials of degree at most d in each variable, and with coefficients from $\mathbb{K}$, is a vector space over $\mathbb{F}_q$ of dimension $(d+1)^n$. It has a basis

$$\{X_1^{d_1}\cdots X_n^{d_n} \mid d_i \leq d\}.$$

Lemma 5.1. *For every function $\phi \in (\mathbb{F}_q^n)^{\mathbb{F}_q}$ there is a unique polynomial $f \in (\mathbb{F}_q)_{[q-1]}[X_1,\ldots,X_n]$ with the property that $\phi(x) = f(x)$ for all $x \in \mathbb{F}_q^n$.*

Proof. By Alon's Nullstellensatz, Theorem 2.1, a polynomial $f(X_1,\ldots,X_n)$, with the property that $f(x)=0$ for all $x \in \mathbb{F}_q^n$, is an element of the ideal

$$I = \langle X_1^q - X_1, \ldots, X_n^q - X_n \rangle.$$

If f and g are polynomials in n variables whose evaluations define the same function from $\mathbb{F}_q^n$ to $\mathbb{F}_q$ then $f-g \in I$. If they are both of degree at most $q-1$ in each variable then $f=g$.

The vector space of functions from $\mathbb{F}_q^n$ to $\mathbb{F}_q$ has dimension q^n and this set of polynomials in n variables of degree at most $q-1$ in each variable also has dimension q^n. Thus, each function $\phi \in (\mathbb{F}_q^n)^{\mathbb{F}_q}$ is uniquely represented by a polynomial f, of degree at most $q-1$ in each variable, where $\phi(x_1,\ldots,x_n) = f(x_1,\ldots,x_n)$. □

We shall now use this observation to obtain shorter proofs of Theorem 2.2. The following proof is due to Blokhuis, Brouwer and Szőnyi [20].

Proof. The polynomials $f_b(X_1,\ldots,X_n) = f(X_1-b_1,\ldots,X_n-b_n)$ have the property that $f_b(b)=1$ and $f_b(a)=0$ for $a \neq b$. Thus, their evaluations form a basis for the vector space $(\mathbb{F}_q^n)^{\mathbb{F}_q}$ and therefore a basis for the set of polynomials in n variables of degree at most $q-1$ in each variable. This set contains the monomial $X_1^{q-1}\cdots X_n^{q-1}$, so f has must have degree at least $n(q-1)$. □

The following proof, which was noted by Pepe [39], is similar to that of Bruen [29, Theorem 1.8, Proof 1] and Wilson [29, Theorem 1.8, Proof 3].

Proof. Let $f_0 = f \bmod I$, where the degree of f_0 in each indeterminate is at most $q-1$. The function defined by evaluating f is the same as the function defined by evaluating the polynomial

$$g(X) = \prod_{i=1}^{n}(1-X_i^{q-1}),$$

which is the same as the function defined by evaluating f_0. However, the degree of f_0 in each indeterminate is at most $q-1$ and so $f_0 = g$. Hence, the degree of f is at least $n(q-1)$. □

We will apply the following lemma to a distinct geometrical problem. In the following we are interested in the degree of the polynomial and not the degree in each variable.

Lemma 5.2. *Let E be a subset of $\mathbb{F}_q^n$. If*

$$|E| < \binom{d+n}{d}$$

then there is a non-zero polynomial f of degree at most d with the property that $f(x)=0$ for all $x \in E$.

Proof. The dimension of the vector space of functions from E to $\mathbb{F}_q$ is $|E|$. The dimension of the vector space of polynomials in n variables of degree at most d is $\binom{d+n}{d}$. Since $|E| < \binom{d+n}{d}$ there are distinct polynomials g and h which agree on E. Let $f=g-h$. □

Suppose that a subset $\mathcal{E}$ of $\mathrm{AG}(n,q)$ has the property that for any direction m, there is a line ℓ with direction m contained in $\mathcal{E}$. In other words, all the points of ℓ are points of $\mathcal{E}$. A set $\mathcal{E}$ with such a property is sometimes called a Besikovitch set and is related to the Kakeya problem which concerns the real space analogue. We wish to prove a lower bound for $|\mathcal{E}|$.

The following Lemma 5.3 and Theorem 5.4 are due to Dvir [32].

Lemma 5.3. *Let $\mathcal{E}$ be a set of points of* $\mathrm{AG}(n,q)$ *which contains a line in every direction. A non-zero polynomial f, which is zero at all elements of $\mathcal{E}$, has degree at least q.*

Proof. The geometrical property that $\mathcal{E}$ contains a line in every direction translates to the following algebraic property of f. Namely, for all $y \in \mathbb{F}_q^n$, $y \neq 0$, there is an x with the property that $f(x+\lambda y) = 0$ for all $\lambda \in \mathbb{F}_q$.

Suppose that $f^\circ = d \leq q-1$. Write

$$f(X+\lambda Y) = \sum_{i=1}^{d} g_i(X,Y)\lambda^i.$$

The polynomial g_d is non-zero, of degree d and depends only on Y, so we can write $g_d(X,Y) = g_d(Y)$.

Since $d \leq q-1$ and f vanishes for all $\lambda \in \mathbb{F}_q$, for each i, the coefficient of λ^i is zero. Specifically $g_d(y) = 0$. By Alon's Nullstellensatz, Theorem 2.1, $g_d \in \langle Y_1^q - Y_1, \ldots, Y_n^q - Y_n \rangle$. However, the polynomial g has degree $d \leq q-1$ and so $g_d = 0$, which is a contradiction. Therefore $f^\circ \geq q$. □

Theorem 5.4. *A set of points $\mathcal{E}$ of* $\mathrm{AG}(n,q)$ *which contains a line in every direction contains at least $\binom{n+q-1}{n}$ points.*

Proof. If not then by Lemma 5.2, there is a non-zero polynomial f of degree at most $q-1$ which is zero on $\mathcal{E}$, which contradicts Lemma 5.3. □

There are many questions which arise as a result of Dvir's Theorem, the most obvious being to ask how good the bound is. For $n=2$, it is tight for q even and can be improved to $q(q+1)/2 + (q-1)/2$ for q odd. Blokhuis and Mazzocca [23] classified all sets which meet this bound for q odd. For $n=3$, it is not clear if a lower bound of approximately $q^3/6$ is near to being the true value. For n large and q small, there are probably better bounds to be found.

There are some obvious generalisations to be considered. If we replace lines by k-dimensional subspaces, for example, or if we replace one line in every direction with t lines in every direction. For the moment, this has yet to be done.

The approach of Dvir should be applicable to more geometrical problems, as should the following idea, which was developed by Gács.

Consider again $\mathcal{S}$, the graph of a function ϕ from $\mathbb{F}_q$ to $\mathbb{F}_q$. The set $\mathcal{S}$ contains q points and is a subset of points of $\mathrm{AG}(2,q)$ (and therefore of $\mathbb{F}_q^2$). Applying Lemma 5.2, we have that there is a non-zero polynomial $f(X_1,X_2)$ of degree less than $\sqrt{2q}$ with the property that $f(x_1,x_2) = 0$ for all $x \in \mathcal{S}$. In other words, $\mathcal{S}$ lies on an algebraic curve of degree at most $\sqrt{2q}$.

Gács was interested in the case that the function ϕ is a map from $\mathbb{F}_p$ to $\mathbb{F}_p$, where p is a prime. By Corollary 4.4, if the number of directions $M(\phi)$, not determined by ϕ, is at least $(p+1)/2$ then ϕ is linear. Lovász and Schrijver [38] proved that if $M(\phi)=(p-1)/2$ then the graph of the function ϕ is affinely equivalent to the graph of the function $x \mapsto x^{(p+1)/2}$. Megyesi [40] provided examples of functions where $M(\phi)=(p-1)/d$, whenever $p=1$ mod d, using the multiplicative subgroup of $\mathbb{F}_p$ of index d. In Megyesi's examples the graph of the function is contained in the union of two lines and so in an algebraic plane curve of degree two. Gács wanted to prove that there were no examples of functions ϕ for which $(p-1)/3 < M(\phi) < (p-1)/2$, which he almost succeeded in doing. In [34], he proved the following.

Theorem 5.5. *If $M(\phi) > (p+2)/3$ then the graph of ϕ is contained in the union of two lines.*

This allowed him to apply the following theorem of Szőnyi [44]. Note that the generalised examples of Megyesi mentioned in the following also have $M(\phi)=(p-1)/d$, for some d dividing $p-1$.

Theorem 5.6. *If $M(\phi) \geq 2$ and the graph of ϕ is contained in the union of two lines then f is affinely equivalent to a generalised example of Megyesi.*

In [10], his approach, which we shall summarise below, led to the following theorem and conjecture. Let $\varepsilon = 0$ if $M(\phi)$ is even and $\varepsilon = 1$ if $M(\phi)$ is odd.

Theorem 5.7. *If $M(\phi) > (p-1-2\varepsilon)/t + t - 3 + \varepsilon$ for some integer $t \geq 2$, then every line of $\mathrm{AG}(2,p)$ is incident with at least $M(\phi)+4-t$ points of the graph of ϕ or at most $t-1$ points of the graph of ϕ.*

Conjecture 5.8. *If $M(\phi) > (p-1-2\varepsilon)/t + t - 3 + \varepsilon$ for some integer $t \geq 2$, then the graph of ϕ is contained in an algebraic curve of degree $t-1$.*

The Gács approach starts in the same way as that of Lovász and Schrijver [38]. If $-c$ is a direction not determined by ϕ then the map

$$x \mapsto \phi(x) + cx$$

is a permutation. By [37, Lemma 7.3], c is a zero of the polynomials

$$h_k(Y) = \sum_{x \in \mathbb{F}_p} (\phi(x) + xY)^k = \sum_{i+j=k} \sum_{x \in \mathbb{F}_p} \binom{k}{i} x^j \phi(x)^i Y^j,$$

for $p-2 \geq k \geq 1$. The degree of these polynomials h_k is at most $k-1$ and so for $1 \leq k \leq M(\phi)-1$ the polynomials h_k are zero. Since $k < p$ the binomial coefficient $\binom{k}{i} \neq 0$ and so we conclude that

$$\sum_{x \in \mathbb{F}_p} x^j \phi(x)^i = 0,$$

for all $1 \leq i+j \leq M(\phi)-1$.

For any polynomial $g(X) = \sum_{i=0}^{p-1} g_i X^i$, of degree at most $p-1$, the sum

$$\sum_{x \in \mathbb{F}_p} g(x) = -g_{p-1}.$$

Therefore the above implies that the polynomial that represents the function $\phi(x)^i$ has degree at most $p - M(\phi) + i - 1$ for $i = 1, \ldots, M(\phi) - 1$.

The dimension of a subspace of polynomials is equal to the number of distinct degrees of polynomials occuring in the subspace.

We wish to combine this fact with our observation that the polynomials

$$\sum_{i=1}^{M(\phi)-1} F_i \phi^i,$$

where $F_i^{\circ} \leq M(\phi) - i - 1$, are of degree at most $p-2$.

In [10], linear maps ψ from $\{(F_1, \ldots, F_s) \mid F_i^{\circ} \leq s - i\}$ to $\mathbb{F}_p[X]$, defined by

$$\psi(F_1, \ldots, F_s) = F_1 \phi + \ldots + F_s \phi^s$$

are considered. If $s < M(\phi)/2$ then for all polynomials $g, h \in \mathrm{Im}(\psi)$ the product gh does not have degree $p-1$. Since it can be written as a sum of the type

$$\sum_{i=1}^{M(\phi)-1} G_i \phi^i,$$

where $G_i^{\circ} \leq M(\phi) - i - 1$, it has degree at most $p-2$.

Therefore, only half the degrees can occur amongst the polynomials in $\mathrm{Im}(\psi)$ and so its dimension is bounded by roughly $p/2$. This then gives a lower bound for the dimension of the kernel of ψ. For an element $(F_1, \ldots, F_s)$ in the kernel of ψ and x, not a zero of ϕ,

$$-F_1 = F_2 \phi + \ldots + F_s \phi^{s-1}.$$

If the number of zeros of ϕ is limited then this equation is valid for sufficiently many elements that it is a polynomial identity. The condition that ϕ has few zeros is equivalent to saying that the line, defined by the second coordinate is zero, contains few points of the graph of ϕ. This line can be chosen arbitrarily so, under the assumption that some line is incident with a bounded number of points of the graph of ϕ, we can consider further iterative linear maps reducing s by one each time. Note that Conjecture 5.8 is true if and only if the map ψ has a non-trivial kernel when $s = t - 1$.

This approach should extend to other combinatorial objects. One could hope to obtain further properties for any object that can be parameterised by a function ϕ (which may be in several indeterminates) and whose combinatorial property implies that the powers of ϕ are represented by polynomials which do not have certain degrees.

6. Field Extensions as Vector Spaces

The field $\mathbb{F}_{q^h}$ is a vector space of dimension h over $\mathbb{F}_q$. Since $\mathrm{PG}(n-1,q)$ and $\mathrm{AG}(n,q)$ are constructed from the n-dimensional vector space over $\mathbb{F}_q$, one can also construct them from $\mathbb{F}_{q^n}$, or more generally from

$$\prod_{i=1}^{k} \mathbb{F}_{q^{r_i}},$$

where $\sum_{i=1}^{k} r_i = n$ and $r_1|r_2|\ldots|r_k$. Note that the last condition implies that all the fields $\mathbb{F}_{q^{r_i}}$ are subfields of $\mathbb{F}_{q^{r_k}}$.

Up until now we have only considered the case $r_1 = \ldots = r_n = 1$. Let us consider the other extremal case $r_1 = n$.

The hyperplanes of the vector space $\mathbb{F}_{q^n}$ are defined by equations of the form $Tr(ax) = 0$, where a is a non-zero element of $\mathbb{F}_{q^n}$ and

$$Tr(X) = X + X^q + \ldots + X^{q^{n-1}}.$$

The k-dimensional subspaces are defined by equations of the form $f(x) = 0$, where

$$f(X) = X^{q^k} + b_{k-1}X^{q^{k-1}} + \ldots + b_1 X^q + b_0 X,$$

and the b_i satisfy relations, which are determined by the divisibility $f(X)$ divides $X^{q^n} - X$.

In the case of the 1-dimensional subspaces, $X^q - aX$ divides $X^{q^n} - X$ if and only if $a^{(q^n-1)/(q-1)} = 1$.

In $\mathrm{AG}(n,q)$ the lines are cosets of the one-dimensional subspaces of $\mathbb{F}_q^n$ and so are defined by equations of the form

$$x^q - ax = b.$$

The points are cosets of the zero dimensional subspace and so are simply the elements of $\mathbb{F}_{q^n}$. For the line joining the points x and y, $a = (x-y)^{q-1}$, which corresponds to the direction of the line. The point z is on this line if and only if $(x-z)^{q-1} = (x-y)^{q-1}$, so we can interpret this condition as a collinearity condition for three points x, y and z.

Let $\mathcal{S}$ be a subset of points of $\mathrm{AG}(n,q)$ in this model, so $\mathcal{S}$ is a subset of $\mathbb{F}_{q^n}$. Consider the polynomial

$$f(T,X) = \prod_{s\in\mathcal{S}} (T - (X-s)^{q-1}).$$

Two factors of $f(T,x)$ are the same if and only if there are two elements $s,t \in \mathcal{S}$, for which $(x-s)^{q-1} = (x-t)^{q-1}$; or in other words, if x, s and t are collinear. Thus, the polynomial f can be used for any set of points $\mathcal{S}$ of $\mathrm{AG}(n,q)$ which has some regular property with respect to lines. Note that the linear factors of $f(T,x)$ are factors of the polynomial $T^{(q^n-1)/(q-1)} - 1$.

Blokhuis [17] used this model with $n = 2$ to prove Theorem 6.1. An *external nucleus* to a set of points $\mathcal{S}$ is a point x with the property that every line incident with x is incident with at least one point of $\mathcal{S}$.

Theorem 6.1. *A set $\mathcal{S}$ of $q+k$ points of* $\mathrm{AG}(2,q)$ *has at most $k(q-1)$ external nuclei.*

Proof. We can assume that $k \leq q-1$ otherwise there is nothing to prove. For any external nucleus x, the polynomial $f(T,x)$ has every factor of $T^{q+1}-1$ amongst its linear factors. Therefore, there is a polynomial $g(T,x)$, of degree $k-1$, for which

$$f(T,x) = (T^{q+1}-1)g(T,x).$$

The coefficient of T^q in f is a polynomial $\sigma(X)$, which by definition is

$$(-1)^k \sum (X-s_1)^{q-1} \dots (X-s_k)^{q-1},$$

where the sum is taken over all k-subsets of $\mathcal{S}$. There are $\binom{|\mathcal{S}|}{k} = \binom{q+k}{k}$ such subsets and so the leading term has degree $k(q-1)$ and coefficient $(-1)^k\binom{q+k}{k} = (-1)^k$.

For any external nucleus x we have seen that $\sigma(x) = 0$ and since σ is a polynomial of degree $k(q-1)$, there are at most $k(q-1)$ external nuclei. □

The bound in Theorem 6.1 is attainable by taking, for example, $\mathcal{S} = \ell \cup \{x_1,\dots,x_k\}$, where ℓ is a line and the points $x_1,\dots,x_k$ are points on distinct lines parallel to ℓ.

Blokhuis extended Theorem 6.1 to t-fold nuclei [16], where a *t-fold external nucleus* to a set of points $\mathcal{S}$ is a point x with the property that every line incident with x is incident with at least t points of $\mathcal{S}$.

Let us consider another application of the polynomial f. Suppose $\mathcal{S}$ is a set of points with the property that every line of $\mathrm{AG}(2,q)$ is incident with a multiple of r points, for some fixed r. It is trivial to prove that $|\mathcal{S}| \geq (r-1)q+r$ and that r divides q. We shall sketch a proof that the lower bound can be improved to $|\mathcal{S}| \geq (r-1)q+(p-1)r$, where $q=p^h$. This implies that, for q odd, there are no non-trivial sets of points with intersection number 0 or r (so-called *maximal arcs*), which was the main result of [9] and [7]. The result stated here is from [8], although the sketched proof is that used to prove the main result of [7].

Theorem 6.2. *If $\mathcal{S}$ is a set of points of $\mathrm{AG}(2,q)$ with the property that every line is incident with a multiple of r points of $\mathcal{S}$ then $|S| \geq (r-1)q+(p-1)r$.*

Proof. (sketch) Suppose that $|\mathcal{S}| = (r-1)q+kr$, where $k < p-1$.

The geometrical property translates to the following algebraic properties for the polynomial f. Namely, if $x \in \mathcal{S}$ then

$$f(T,x) = T(T^{q+1}-1)^{r-1}g(T),$$

where g is a polynomial of degree $(k-1)r$. If $x \notin \mathcal{S}$ then $f(T,x)$ has factors repeated a multiple of r times and so is an r-th power. As in the proof of Theorem 6.1, we focus on one particular coefficient of f, in this case the coefficient $\sigma(X)$ of $T^{|\mathcal{S}|-kr}$. As in the proof of Theorem 6.1, we can deduce that it has a leading term of degree $kr(q-1)$. If $x \in \mathcal{S}$ then the fact that $g(T)$ has degree $(k-1)r$ implies $\sigma(x) = 0$. Thus the polynomial

$$a(X) = \prod_{s \in \mathcal{S}} (X-s)$$

divides $\sigma(X)$. One then exploits the divisibility

$$f(T,x) \text{ divides } (T^{q+1}-1)\frac{\partial f}{\partial X}(T,x)$$

to prove that $a(X)^{p-1}$ divides $\sigma(X)$. This implies that $(p-1)|\mathcal{S}| \leq kr(q-1)$ which gives $k \geq p-1$. □

Let us consider how to use another representation of $\mathbb{F}_q^n$, specifically

$$\prod_{i=1}^{k} \mathbb{F}_{q^{r_i}},$$

where $r_1 = 1$ and $r_2 = n-1$.

The points of the affine space $\mathrm{AG}(n,q)$ are elements of

$$\mathbb{F}_q \times \mathbb{F}_{q^{n-1}}.$$

Suppose that $s = (s_1, s_2)$ and $t = (t_1, t_2)$ are points of $\mathrm{AG}(n,q)$. The direction of the line joining s and t is given by the projective point $\langle (s_1 - t_1, s_2 - t_2) \rangle$, which if $s_1 \neq t_1$ is the point $\langle (1, x) \rangle$ where

$$x = \frac{s_2 - t_2}{s_1 - t_1}.$$

We shall look at two related problems with this representation and use slightly differing polynomials. In the first problem we wish to obtain further geometrical properties for the higher dimensional analogue of the graph of a function which determines few directions and in the second problem we shall prove a stability result for such graphs.

Let ϕ be a function from $\mathbb{F}_q^{n-1}$ to $\mathbb{F}_q$. The graph of the function ϕ is the set of points

$$\{(\phi(s_2), s_2) \mid s_2 \in \mathbb{F}_{q^{n-1}}\}.$$

Let $M(\phi)$ be the number of directions not determined by ϕ. Let $\mathcal{S}$ be a set of q^{n-1} points of $\mathrm{AG}(n,q)$, affinely equivalent to the graph of ϕ, for which the number of points of $\mathcal{S}$ on the hyperplane defined by the first coordinate being zero, is not q^{n-2}. This condition implies that the directions not determined by the set of points $\mathcal{S}$ are not of the form $\langle (0,x) \rangle$ for any $x \in \mathbb{F}_{q^{n-1}}$.

Consider the polynomial

$$h(T,X) = \prod_{s \in \mathcal{S}} (T - (s_1 X - s_2)^{q-1}).$$

If $\langle (1,x) \rangle$ is a direction not determined by $\mathcal{S}$ then $s_1 x - s_2 \neq t_1 x - t_2$ and so $s_1 x - s_2$ are distinct values of $\mathbb{F}_{q^{n-1}}$. Therefore

$$h(T,x) = \prod_{\lambda \in \mathbb{F}_{q^{n-1}}} (T - \lambda^{q-1}) = T(T^{(q^{n-1}-1)/(q-1)} - 1)^{q-1}. \tag{1}$$

The polynomial h allows us to prove the following theorem from [6], which improves on a similar result in [14], where the representation $r_1 = \ldots = r_n = 1$ was used.

Let $q = p^h$, p prime.

Theorem 6.3. *If, for some non-negative integer $e \le (n-2)h-1$, there are more than $p^e(q-1)$ directions not determined by a set $\mathcal{S}$ of q^{n-1} points of* AG(n,q) *then every hyperplane is incident with a multiple of p^{e+1} points of $\mathcal{S}$.*

Proof. The coefficient of $T^{q^{n-1}-p^e}$ in $h(T,X)$ is a polynomial $\sigma(X)$ which, by definition, is

$$(-1)^{p^e}\sum(s_1X-s_2)^{q-1}\dots(t_1X-t_2)^{q-1},$$

where the sum is taken over all p^e-subsets of $\mathcal{S}$. The coefficient of $X^{p^e(q-1)}$ is

$$\sum s_1^{q-1}\dots t_1^{q-1}$$

where the sum is taken over all p^e-subsets of $\mathcal{S}$. Note that $s_1 \in \mathbb{F}_q$, so the terms in this sum are 1 for every p^e subset of $\mathcal{S}$ in which all the points in the subset have first-coordinate non-zero. Let N be the number of points in $\mathcal{S}$ with first coordinate zero. Then the coefficient of $X^{p^e(q-1)}$ in $\sigma(X)$ is $\binom{|\mathcal{S}|-N}{p^e}=\binom{-N}{p^e}$.

On the other hand, if $\langle(1,x)\rangle$ is a direction not determined by $\mathcal{S}$ then (1) implies that $\sigma_{p^e}(x)=0$. By assumption, there are more than $p^e(q-1)$ directions not determined by $\mathcal{S}$ and the degree of σ is at most $p^e(q-1)$, so we conclude that it is identically zero. Therefore $N=0$ modulo p^{e+1} and so the number of points of $\mathcal{S}$, on the hyperplane of points with first coordinate zero, is 0 mod p^{e+1}. Since this hyperplane was chosen arbitrarily, the theorem is proved. □

Theorem 6.3 has an immediate corollary for ovoids of the parabolic quadric Q$(4,q)$. Indeed, using the Tits representation of Q$(4,q)$ as T$_2(\mathcal{O})$, where $\mathcal{O}$ is a conic, one obtains the following result, which first appeared in [5] and for $p=2$ in [2].

Corollary 6.4. *An ovoid of $Q(4,q)$ and an elliptic quadric $Q^-(3,q)$ embedded in $Q(4,q)$ intersect in* 1 *modulo p points, where $q=p^h$.*

With some combinatorial counting this leads to the following theorem from [12].

Theorem 6.5. *An ovoid of $Q(4,p)$, where p is prime, is an elliptic quadric.*

Corollary 6.4 can be improved in the q even case. The following is from [30].

Theorem 6.6. *An ovoid of $Q(4,q)$ and an elliptic quadric $Q^-(3,q)$ embedded in $Q(4,q)$ intersect in* 1 *modulo* 4 *points, where $q=2^h$.*

Theorem 6.3 implies that if $\mathcal{S}$ is a set of p^2 points in AG$(3,p)$, which does not determine at least p directions, then every plane is incident with a multiple of p points of $\mathcal{S}$. The only examples which we are aware of which have this property are the cylinders, i.e. the points on the union of p parallel lines. This leads to the following conjecture, which is called the *strong cylinder conjecture*.

Conjecture 6.7. *If $\mathcal{S}$ is a set of p^2 points in* AG$(3,p)$ *with the property that every plane is incident with a multiple of p points of $\mathcal{S}$ then $\mathcal{S}$ is a cylinder.*

We shall use the same representation of $\mathrm{AG}(n,q)$ to prove a stability result for sets of points that do not determine all directions. Here we shall use a slightly different (although somewhat familiar) polynomial

$$e(X_1,X_2)=\prod_{s\in\mathcal{S}}(X_1+s_1X_2+s_2).$$

As for the polynomial $f(X_1,x)$, we conclude that if $\mathcal{S}$ is a set of q^{n-1} points and $\langle(1,-x)\rangle$ is a direction not determined by $\mathcal{S}$ then

$$e(X_1,x)=X_1^{q^{n-1}}-X_1.$$

Now consider a set $\mathcal{S}$ of $q^{n-1}-2$ points and suppose that $\mathcal{D}$ is the set of directions not determined by $\mathcal{S}$. We wish to show that if $\mathcal{D}$ is large enough then $\mathcal{S}$ can be extended to a set of q^{n-1} points which do not determine the directions $\mathcal{D}$. This type of result is called a stability theorem.

More precisely we prove the following, which was proved in [31] using the representation $r_1=\ldots=r_n=1$.

Theorem 6.8. *A set of $q^{n-1}-2$ points of $\mathrm{AG}(n,q)$, $q=p^h$ odd, which does not determine a set $\mathcal{D}$, of at least $p+2$ directions, can be extended to a set of q^{n-1} points not determining the set of directions $\mathcal{D}$.*

Proof. Writing $e(X_1,X_2)$ as a polynomial in X_1, we define polynomials $\sigma_j(X_2)$ of degree at most j by

$$e(X_1,X_2)=\sum_{j=0}^{|\mathcal{S}|}\sigma_j(X_2)X_1^{|\mathcal{S}|-j}.$$

The polynomial

$$\sigma_1(X_2)=\sum_{s\in\mathcal{S}}(s_1X_2+s_2).$$

By making the translation $(s_1,s_2)\mapsto(s_1+\lambda_1,s_2+\lambda_2)$, where $\lambda_1=(\sum_{s\in\mathcal{S}}s_1)/2$ and $\lambda_2=(\sum_{s\in\mathcal{S}}s_2)/2$, then we can assume $\sigma_1(X_2)=0$. Here we use the assumption q is odd.

Suppose that $\langle(1,-x)\rangle$ is a direction not determined by $\mathcal{S}$. Then, by the discussion preceding the theorem,

$$e(X_1,x)(X_1^2-\sigma_2(x))=X_1^{q^{n-1}}-X_1.$$

This implies that $\sigma_{2k}(x)=\sigma_2(x)^k$ for all $k<q^{n-1}/2$.

Let $\pi_k(X_2)=\sum_{s\in\mathcal{S}}(s_1X_2+s_2)^k$. The Newton identities relate the symmetric functions σ_k and the power sums π_k by the equations

$$k\sigma_k=\sum_{j=1}^{k}(-1)^{j-1}\pi_j\sigma_{k-j}.$$

Solving these equation recursively implies $\pi_{2k}=-2\sigma_2^k$. Thus, for $2k=p+1$, we have

$$(-2\sigma_2)(x)^{(p+1)/2}=\sum_{s\in\mathcal{S}}(s_1x+s_2)^{p+1}=c_{p+1}x^{p+1}+c_px^p+c_1x+c_0,$$

for some $c_i \in \mathbb{F}_{q^{n-1}}$.

Write $-2\sigma_2(X_2) = d_2X_2^2 + d_1X_2 + d_0$. We have shown that the polynomial

$$(d_2X_2^2 + d_1X_2 + d_0)^{(p+1)/2} - (c_{p+1}X_2^{p+1} + c_pX_2^p + c_1X_2 + c_0)$$

is zero for every direction not determined by $\mathcal{S}$. Since, by assumption, there are at least $p+2$ of these, this polynomial is zero. Thus, either $d_0 = d_1 = c_0 = c_1 = c_p = 0$ and $d_2^{(p+1)/2} = c_{p+1}$, or $d_2 = d_1 = c_{p+1} = c_p = c_1 = 0$ and $d_0^{(p+1)/2} = c_0$.

In the first case $\sigma_2(X_2) = d_2X_2^2$. When $\langle(1,-x)\rangle$ is not a direction determined by $\mathcal{S}$

$$T^2 - \sigma_2(x) = (T - d_2^{1/2}x)(T + d_2^{1/2}x),$$

so d_2 is a square. We can then extend $\mathcal{S}$ with the points $(-d^{1/2},0)$ and $(d^{1/2},0)$ without determining any of the directions not determined by $\mathcal{S}$. The other case is similar. □

Again, using the Tits representation of $\mathrm{Q}(4,q)$ as $\mathrm{T}_2(\mathcal{O})$, where $\mathcal{O}$ is a conic, Theorem 6.8 has the following consequences for partial ovoids of $\mathrm{Q}(4,q)$.

Corollary 6.9. *A partial ovoid of $Q(4,q)$, q odd and not a prime, of size q^2-1 can be extended to an ovoid.*

Curiously, for $q = 5,7$ and 11, there are examples of partial ovoids of size q^2-1 which cannot be extended to an ovoid.

7. Algebraic Curves over Finite Fields

In this section, we shall give an example of how to apply bounds on the number of points on an algebraic curve defined over $\mathbb{F}_q$ to a geometrical problem of the type discussed before.

The following is from Szőnyi [45].

Lemma 7.1. *Suppose $f \in \mathbb{F}_q[X_1,X_2]$ is a polynomial of degree d. If f has no linear factor in $\mathbb{F}_q[X_1,X_2]$ and $2 \le d \le \sqrt{q}/2$ then f has at most $d(q+1)/2$ zeros in $\mathbb{F}_q^2$.*

Proof. Let N be the number of zeros of f in $\mathbb{F}_q^2$.

If f is absolutely irreducible then Weil's theorem [35, Corollary 2.29] implies

$$N \le q+1+(d-1)(d-2)\sqrt{q} \le d(q+1)/2.$$

If not then f factorises into irreducible factors $f = f_1 \dots f_k$ over the algebraic closure of $\mathbb{F}_q$. Let N_i be the number of zeros of f_i in $\mathbb{F}_q[X_1,X_2]$ and let d_i be the degree of f_i.

If $f_i \in \mathbb{F}_q[X_1,X_2]$ then Weil's theorem implies $N_i \le d_i(q+1)/2$.

If $f_i \notin \mathbb{F}_q[X_1,X_2]$ then by [35, Lemma 2.24] $N_i \le d_i^2 < d_i(q+1)/2$. Thus,

$$N \le \sum_{i=1}^{k} N_i \le (\sum_{i=1}^{k} d_i)(q+1)/2 = d(q+1)/2.$$ □

Using Lemma 7.1, we shall prove the following stability result for the graphs of functions from $\mathbb{F}_q$ to $\mathbb{F}_q$. Again this is from Szőnyi [45]. Compare this to Theorem 6.8.

Theorem 7.2. *A set of $q-k > q-\sqrt{q}/2$ points of* AG$(2,q)$ *which does not determine a set* $\mathcal{D}$*, of more than* $(q+1)/2$ *directions, can be extended to a set of q points not determining the set of directions* $\mathcal{D}$.

Proof. For any polynomial f of degree n one can construct a polynomial g of degree m with the property that $fg = X^{n+m} + h$, where the degree of h is at most $n-1$, by choosing the coefficient of X^{m-j} in g, for $j = 1,\ldots,m$, in such a way that the coefficient of X^{n+m-j} on the right-hand side is zero.

Apply this observation to the polynomial

$$f(X_1,X_2) = \prod_{s \in \mathcal{S}} (X_1 + s_1 X_2 + s_2),$$

with $m = k$, by considering this polynomial as a polynomial in X_1 with coefficients that are polynomials in X_2. The polynomial $g(X_1,X_2)$ obtained has overall degree at most k, and

$$f(X_1,X_2)g(X_1,X_2) = X_1^q + h(X_1,X_2),$$

where the degree of h in X_1 is at most $q-k-1$.

If $-x \in \mathcal{D}$, a direction not determined by $\mathcal{S}$, then $f(X_1,x)$ divides $X_1^q - X_1$. The quotient of this division is of degree k and so is $g(X_1,x)$. Therefore, $g(X_1,x)$ is the product of distinct linear factors over $\mathbb{F}_q$ and has k zeros in $\mathbb{F}_q$. Hence, $g(X_1,X_2)$ has $kM \geq k(q+1)/2$ zeros, where M is the number of directions not determined by $\mathcal{S}$. By Lemma 7.1, $g(X_1,X_2)$ has a linear factor $X_1 + t_1 X_2 + t_2$ in $\mathbb{F}_q[X_1,X_2]$.

The set $\mathcal{S} \cup \{(t_1,t_2)\}$ does not determine a direction $-x$, not determined by $\mathcal{S}$, since $X_1 + t_1 x + t_2$ is a factor of $g(X_1,x)$, whose factors are different to the factors of $f(X_1,x)$. In other words, for all $s \in \mathcal{S}$, $t_1 x + t_2 \neq s_1 x + s_2$. Thus, $\mathcal{S}$ can be extended, and repeating the above, can be extended to a set of q points, which does not determine any of the directions in $\mathcal{D}$. □

Further applications of bounds on the number of points on algebraic curves over finite fields from both Weil's lemma, those deduced from Stöhr-Voloch [42], and the number of points in the intersection of two curves deduced from Bezout's theorem, can be found in articles such as [11], [33] and [24].

8. Resultant of Polynomials in Two Variables

In [47] Szőnyi showed that a generalisation of the resultant of two polynomials could be applied to finite geometrical problems. This was further developed by Weiner [50] and together with Szőnyi in [48].

Suppose that f and g are polynomials of degree n and at most $n-1$ respectively. Let $b = \sum_{i=0}^{m-1} b_i X^i + X^m$ and $a = \sum_{i=0}^{m-1} a_i X^i$ be polynomials of degree m and at most $m-1$ respectively, with the property that

$$af + bg = 0.$$

Considering the coefficients of $X^{n-m-1},\dots,X^{n+m-1}$ gives $2m$ linear equations which can be written in matrix form

$$(a_1,\dots,a_{m-1},b_0,\dots,b_{m-1})R_m = (g_0,\dots,g_{2m-1}),$$

where the entries in the $2m\times 2m$ matrix R_m are the suitable coefficients of f and g.

Suppose that $h=(f,g)$ has degree $n-k$.

If $m\geq k+1$ then there are multiple solution to the above equation, choosing b to be a non-constant multiple of f/h and $a=-bg/h$. Hence, the system of linear equations has multiple solutions and therefore $\det R_m = 0$.

If $m=k$ then there is a unique solution to the above equation $b=\gamma f/h$ and $a=-bg/h$, where γ is chosen so that b is monic. Thus, $\det R_k\neq 0$.

Now suppose that $f=f(X_1,X_2)$ and $g=g(X_1,X_2)$ are polynomials in two variables. By writing the polynomials as polynomials in X_1, with coefficients which are polynomials in X_2, the determinant $\det R_m$ becomes a polynomial in X_2.

Lemma 8.1. *Suppose that there is an element $x_2\in\mathbb{F}_q$ for which*

$$\deg(f(X_1,x_2),g(X_1,x_2)) = n-k.$$

If there are n_h elements $y\in\mathbb{F}_q$ for which

$$\deg(f(X_1,y),g(X_1,y)) = n-(k-h)$$

then

$$\sum_{h=1}^{k-1} hn_h \leq \deg(\det R_k).$$

Proof. (sketch) The determinant of the matrix R_k is a polynomial in X_2 and $(\det R_k)(x_2)\neq 0$ by the above discussion. If, for $y\in\mathbb{F}_q$, the degree of $(f(X_1,y),g(X_1,y))$ is $n-(k-h)$ then it can be shown that y is a zero of $\det R_k$ (of multiplicity h). The discussion preceding the lemma implies that y is a zero of $\det R_k$, if $h\geq 1$. □

This lemma has been applied to a variety of problems, see for example [47] and [50]. The following is from [48].

Theorem 8.2. *Let $\mathcal{S}$ be a set of points of* AG$(2,q)$ *and suppose $|\mathcal{S}|\neq q$. Let n_h be the number of directions d for which exactly h of the lines with direction d are incident with $\mathcal{S}$. If $n_k\neq 0$ then*

$$\sum_{h=k+1}^{q} hn_h \leq (|\mathcal{S}|-k)(q-k).$$

Proof. Let

$$f(X_1,X_2) = \prod_{(s_1,s_2)\in\mathcal{S}} (X_1+s_1X_2+s_2) = \sum_{j=0}^{|\mathcal{S}|} f_j(X_2)X_1^{|\mathcal{S}|-j},$$

where the degree of $f_j(X_2)$ is at most j.

Consider the matrix R_k for $f(X_1,X_2)$ and $g(X_1,X_2)=X_1^q-X_1$. One should check that the determinant $\det R_k$ is a polynomial in X_2 of degree at most $(|\mathcal{S}|-k)(q-k)$.

If there are exactly t lines with direction m which are incident with $\mathcal{S}$ then $\deg(f(X_1,m),X_1^q - X_1) = t$. Since $n_k \neq 0$ we have that $\det R_k \neq 0$. Applying Lemma 8.1, the theorem follows. □

This theorem has Metsch's conjecture as a corollary.

Corollary 8.3. *Let $\mathcal{S}$ be a point set of* PG$(2,q)$. *Let x be a point not in $\mathcal{S}$. If there are exactly r lines incident with x that are incident with $\mathcal{S}$, then the total number of lines incident with $\mathcal{S}$ is at most*

$$1 + rq + (|\mathcal{S}| - r)(q + 1 - r).$$

Note that when $|S| = 2q-2$ and $r = q$ then the above implies that the total number of lines incident with $\mathcal{S}$ is at most $q^2 + q - 1$, which gives yet another proof of Jamison's theorem, Corollary 2.3, in the plane.

9. Open Problems

In this section I have listed some problems which we would like to see resolved. Most are stated in the form that implies a conjecture. For example, "Prove that" implies that the statement is thought more likely to hold than the contrary.

Section 3.

1. Let $1 \leq k \leq n-2$ and let $\mathcal{S}$ be a set of points of AG(n,q) with the property that every k-dimensional subspace is incident with a point of $\mathcal{S}$. It should be possibly to prove that there are examples for which $|\mathcal{S}|/(k+1)q^{n-k} \to 0$ as $q \to \infty$. It would be interesting to know the order of magnitude of $|\mathcal{S}| - (k+1)q^{n-k}$. In the smallest case $n = 3$ and $k = 1$ we only have that $c < |\mathcal{S}| - 2q^2 < 2q^{\frac{3}{2}}$ for some constant c.

2. Let $\mathcal{S}$ be a set of points of AG(n,q) with the property that every hyperplane is incident with at least t points of $\mathcal{S}$. Prove a lower bound for $|\mathcal{S}|$ of about $(t+n-1)q-n$ for most t. See [3] for a proof in the case $t \leq q-1$.

Section 4.

1. The projective plane PG$(2,q)$ consists of points and lines which are the one and two dimensional subspaces of $\mathbb{F}_q^3$. If $q = p^t$, where p is prime, then these subspaces are respectively rank t and rank $2t$ subspaces of $\mathbb{F}_p^{3t}$. Here, rank refers to vector space dimension. Let $\mathcal{U}$ be a rank $(t+1)$ subspace of $\mathbb{F}_p^{3t}$. The set of points of PG$(2,q)$ whose corrsponding rank t subspace has a non-trivial intersection with $\mathcal{U}$ is denoted $B(\mathcal{U})$, the *bubble* of $\mathcal{U}$. Prove that if $\mathcal{S}$ is a set of less than $3(q+1)/2$ points of PG$(2,q)$ with the property that every line is incident with a point of $\mathcal{S}$ then $\mathcal{S} = B(\mathcal{U})$, for some rank $(t+1)$ subspace $\mathcal{U}$ of $\mathbb{F}_p^{3t}$.

Section 5.

1. Prove a lower bound for the size of a set $\mathcal{E}$ of points of $\mathrm{AG}(n,q)$ with the property that for every direction (slope, gradient) d there are at least t lines of direction d contained in $\mathcal{E}$.

2. Let π be a hyperplane of $\mathrm{PG}(n,q)$ and consider the affine space $\mathrm{PG}(n,q) \setminus \pi$. We say that two affine subspaces $U_1 \setminus \pi$ and $U_2 \setminus \pi$ have the same direction if $U_1 \cap \pi = U_2 \cap \pi$. For a fixed k, prove a lower bound for the size of a set $\mathcal{E}$ of points of $\mathrm{AG}(n,q)$ with the property that for every direction d, there is at least one k-dimensional subspace of direction d contained in $\mathcal{E}$.

3. Prove Conjecture 5.8.

4. Apply Gács' approach to other geometrical objects that can be defined by one or more polynomials $f \in \mathbb{F}_p[X_1,\ldots,X_n]$, whose geometrical property implies that the power sums
$$\sum_{x \in \mathbb{F}_p^n} f(x)^i = 0,$$
for some i's.

Section 6.

1. Prove that an ovoid of $\mathrm{Q}(4,q)$ and an elliptic quadric $\mathrm{Q}^-(3,q)$ embedded in $\mathrm{Q}(4,q)$ intersect in 1 mod p^r points, for some $2 \leq r < h/2$, where $q = p^h$ for some prime p.

2. Prove the cylinder conjecture, Conjecture 6.7. If not, prove a weaker form of this conjecture in which one assume that there are at least p directions not determined by $\mathcal{S}$.

3. Prove a version of Theorem 6.8 in which a larger set $\mathcal{D}$ implies more stability. For example, if $|\mathcal{D}| > p^2$ then the $q^{n-1} - 2$ can be replaced by $q^{n-1} - f(q)$, for some function of q.

Section 7.

1. In [24] some stability is proven for sets of $q+k$ points in $\mathrm{AG}(2,q)$. Comparing this with Theorem 6.8 and Theorem 7.2, one may be able to extend this stability to sets of points in higher dimensional spaces.

2. In [11] the Stöhr-Voloch bound is used to prove that a set of p points in $\mathrm{AG}(3,p)$, which does not determine approximately $p^2/3$ line directions (see Problem 2. of Section 5. for the definition of a line direction) is contained in a plane. Prove that this can be extended to p^2/d, for a larger $d \in \mathbb{N}$, with few exceptions. It may be possible using Gács' approach from Section 5..

Section 8.

1. Applications of Lemma 8.1 have centered on the cases $g(X_1,X_2) = X_1^q - X_1$ and $g(X_1,X_2) = \frac{\partial f}{\partial X_1}$. There is a huge scope for further applications here, using other polynomials for g, introducing more indeterminates, or even more polynomials.

10. Final Comments

There are also many results obtained using Menelaus theorem, an approach introduced by Segre in [41]. This is not elaborated here but some examples are included in [22], [26], [25] and [49].

I would like to thank Andras Gács, Péter Sziklai and Zsuzsa Weiner for their suggestions and corrections to an earlier version of this manuscript.

Acknowledg nents

The author acknowledges the support of the project MTM2008-06620-C03-01 of the Spanish Ministry of Science and Education and the project 2009-SGR-01387 of the Catalan Research Council.

References

[1] N. Alon, *Combinatorial Nullstellensatz*, Combin. Probab. Comput., 8 (1999), pp. 7–29. Recent trends in combinatorics (Mátraháza, 1995).

[2] B. Bagchi and N. S. Narasimha Sastry, *Even order inversive planes, generalized quadrangles and codes*, Geom. Dedicata, 22 (1987), pp. 137–147.

[3] S. Ball, *On intersection sets in Desarguesian affine spaces*, European J. Combin., 21 (2000), pp. 441–446.

[4] ——, *The number of directions determined by a function over a finite field*, J. Combin. Theory Ser. A, 104 (2003), pp. 341–350.

[5] ——, *On ovoids of* $\mathrm{O}(5,q)$, Adv. Geom., 4 (2004), pp. 1–7.

[6] ——, *On the graph of a function in many variables over a finite field*, Des. Codes Cryptogr., 47 (2008), pp. 159–164.

[7] S. Ball and A. Blokhuis, *An easier proof of the maximal arcs conjecture*, Proc. Amer. Math. Soc., 126 (1998), pp. 3377–3380.

[8] S. Ball, A. Blokhuis, A. Gács, P. Sziklai, and Zs. Weiner, *On linear codes whose weights and length have a common divisor*, Adv. Math., 211 (2007), pp. 94–104.

[9] S. Ball, A. Blokhuis, and F. Mazzocca, *Maximal arcs in Desarguesian planes of odd order do not exist*, Combinatorica, 17 (1997), pp. 31–41.

[10] S. Ball and A. Gács, *On the graph of a function over a prime field whose small powers have bounded degree*, European J. Combin., 30 (2009), pp. 1575–1584.

[11] S. Ball, A. Gács, and P. Sziklai, *On the number of directions determined by a pair of functions over a prime field*, J. Combin. Theory Ser. A, 115 (2008), pp. 505–516.

[12] S. Ball, P. Govaerts, and L. Storme, *On ovoids of parabolic quadrics*, Des. Codes Cryptogr., 38 (2006), pp. 131–145.

[13] S. Ball and M. Lavrauw, *How to use Rédei polynomials in higher dimensional spaces*, Matematiche (Catania), 59 (2004), pp. 39–52 (2006).

[14] ——, *On the graph of a function in two variables over a finite field*, J. Algebraic Combin., 23 (2006), pp. 243–253.

[15] S. Ball and O. Serra, *Punctured combinatorial Nullstellensätze*, Combinatorica, 29 (2009), pp. 511–522.

[16] A. Blokhuis, *On multiple nuclei and a conjecture of Lunelli and Sce*, Bull. Belg. Math. Soc. Simon Stevin, 1 (1994), pp. 349–353. A tribute to J. A. Thas (Gent, 1994).

[17] ——, *On nuclei and affine blocking sets*, J. Combin. Theory Ser. A, 67 (1994), pp. 273–275.

[18] ——, *On the size of a blocking set in* $\mathrm{PG}(2,p)$, Combinatorica, 14 (1994), pp. 111–114.

[19] A. Blokhuis, S. Ball, A. E. Brouwer, L. Storme, and T. Szőnyi, *On the number of slopes of the graph of a function defined on a finite field*, J. Combin. Theory Ser. A, 86 (1999), pp. 187–196.

[20] A. Blokhuis, A. E. Brouwer, and T. Szőnyi, *Covering all points except one*, J. Algebraic Combin., 32 (2010), pp. 59–66.

[21] ——, *The number of directions determined by a function* f *on a finite field*, J. Combin. Theory Ser. A, 70 (1995), pp. 349–353.

[22] A. Blokhuis, A. A. Bruen, and J. A. Thas, *Arcs in* $\mathrm{PG}(n,q)$*, MDS-codes and three fundamental problems of B. Segre—some extensions*, Geom. Dedicata, 35 (1990), pp. 1–11.

[23] A. Blokhuis and F. Mazzocca, *The finite field Kakeya problem*, in Building bridges, vol. 19 of Bolyai Soc. Math. Stud., Springer, Berlin, 2008, pp. 205–218.

[24] A. Blokhuis, R. Pellikaan, and T. Szőnyi, *Blocking sets of almost Rédei type*, J. Combin. Theory Ser. A, 78 (1997), pp. 141–150.

[25] A. Blokhuis, Á. Seress, and H. A. Wilbrink, *On sets of points in* $\mathrm{PG}(2,q)$ *without tangents*, Mitt. Math. Sem. Giessen, (1991), pp. 39–44.

[26] A. Blokhuis, T. Szőnyi, and Zs. Weiner, *On sets without tangents in Galois planes of even order*, Des. Codes Cryptogr., 29 (2003), pp. 91–98.

[27] A. E. Brouwer and A. Schrijver, *The blocking number of an affine space*, J. Combinatorial Theory Ser. A, 24 (1978), pp. 251–253.

[28] A. A. Bruen, *Polynomial multiplicities over finite fields and intersection sets*, J. Combin. Theory Ser. A, 60 (1992), pp. 19–33.

[29] ——, *Applications of finite fields to combinatorics and finite geometries*, Acta Appl. Math., 93 (2006), pp. 179–196.

[30] D. K. Butler, *On the intersection of ovoids sharing a polarity*, Geom. Dedicata, 135 (2008), pp. 157–165.

[31] J. De Beule and A. Gács, *Complete arcs on the parabolic quadric* $\mathrm{Q}(4,q)$, Finite Fields Appl., 14 (2008), pp. 14–21.

[32] Z. Dvir, *On the size of Kakeya sets in finite fields*, J. Amer. Math. Soc., 22 (2009), pp. 1093–1097.

[33] S. Ferret, L. Storme, P. Sziklai, and Zs. Weiner, *A* $t \pmod{p}$ *result on weighted multiple* $(n-k)$*-blocking sets in* $\mathrm{PG}(n,q)$, Innov. Incidence Geom., 6/7 (2007/08), pp. 169–188.

[34] A. Gács, *On a generalization of Rédei's theorem*, Combinatorica, 23 (2003), pp. 585–598.

[35] J. W. P. Hirschfeld, *Projective geometries over finite fields*, Oxford Mathematical Monographs, The Clarendon Press Oxford University Press, New York, second ed., 1998.

[36] R. E. Jamison, *Covering finite fields with cosets of subspaces*, J. Combinatorial Theory Ser. A, 22 (1977), pp. 253–266.

[37] R. Lidl and H. Niederreiter, *Finite fields*, vol. 20 of Encyclopedia of Mathematics and its Applications, Cambridge University Press, Cambridge, second ed., 1997. With a foreword by P. M. Cohn.

[38] L. Lovász and A. Schrijver, *Remarks on a theorem of Rédei*, Studia Sci. Math. Hungar., 16 (1983), pp. 449–454.

[39] V. Pepe, *Personal communication.*

[40] L. Rédei, *Lacunary polynomials over finite fields*, North-Holland Publishing Co., Amsterdam, 1973. Translated from the German by I. Földes.

[41] B. Segre, *Curve razionali normali e k-archi negli spazi finiti*, Ann. Mat. Pura Appl. (4), 39 (1955), pp. 357–379.

[42] K.-O. Stöhr and J. F. Voloch, *Weierstrass points and curves over finite fields*, Proc. London Math. Soc. (3), 52 (1986), pp. 1–19.

[43] P. Sziklai, *On small blocking sets and their linearity*, J. Combin. Theory Ser. A, 115 (2008), pp. 1167–1182.

[44] T. Szőnyi, *Combinatorial problems for abelian groups arising from geometry*, in Proceedings of the Second International Mathematical Miniconference, Part II (Budapest, 1988), vol. 19, 1991, pp. 91–100.

[45] ——, *On the number of directions determined by a set of points in an affine Galois plane*, J. Combin. Theory Ser. A, 74 (1996), pp. 141–146.

[46] ——, *Blocking sets in Desarguesian affine and projective planes*, Finite Fields Appl., 3 (1997), pp. 187–202.

[47] ——, *On the embedding of (k, p)-arcs in maximal arcs*, Des. Codes Cryptogr., 18 (1999), pp. 235–246. Designs and codes—a memorial tribute to Ed Assmus.

[48] T. Szőnyi and Zs. Weiner, *On stability theorems in finite geometry.* Unpublished manuscript, `http://www.cs.elte.hu/~weiner/stab.pdf`

[49] J. F. Voloch, *Arcs in projective planes over prime fields*, J. Geom., 38 (1990), pp. 198–200.

[50] Zs. Weiner, *On (k, p^e)-arcs in Desarguesian planes*, Finite Fields Appl., 10 (2004), pp. 390–404.

In: Current Research Topics in Galois Geometry
Editors: J. De Beule and L. Storme
ISBN: 978-1-61209-523-3

Chapter 6

FINITE SEMIFIELDS

***Michel Lavrauw*[1,*] *and Olga Polverino*[2,†]**
[1] Università degli Studi di Padova,
Dipartimento di Tecnica e Gestione dei Sistemi Industriali,
Stradella S. Nicola, 3, I-36100 Vicenza, Italy
[2] Dipartimento di Matematica, Seconda Università degli Studi di Napoli,
I–81100 Caserta, Italy

1. Introduction and Preliminaries

In this article, we concentrate on the links between Galois geometry and a particular kind of non-associative algebras of finite dimension over a finite field $\mathbb{F}$, called *finite semifields*. Although in the earlier literature (predating 1965) the term semifields was not used, the study of these algebras was initiated about a century ago by Dickson [31], shortly after the classification of finite fields, taking a purely algebraic point of view. Nowadays it is common to use the term *semifields* introduced by Knuth [58] in 1965 with the following motivation:

"We are concerned with a certain type of algebraic system, called a semifield. *Such a system has several names in the literature, where it is called, for example, a "nonassociative division ring" or a "distributive quasifield". Since these terms are rather lengthy, and since we make frequent reference to such systems in this paper, the more convenient name semifield will be used."*

By now, the theory of semifields has become of considerable interest in many different areas of mathematics. Besides the numerous links with finite geometry, most of which considered here, semifields arise in the context of difference sets, coding theory, cryptography, and group theory.

To conclude this prelude we would like to emphasize that this article should not be considered as a general survey on finite semifields, but rather an approach to the subject with the focus on its connections with Galois geometry. There are many other interesting

*E-mail address: michel.lavrauw@unipd.it; This author is a Postdoctoral Research Fellow of the Research Foundation – Flanders (Belgium) (FWO).
†E-mail address: olga.polverino@unina2.it

properties and constructions of finite semifields (and links with other subjects) that are not addressed here.

1.1. Definition and First Properties

A *finite semifield* $\mathbb{S}$ is an algebra with at least two elements, and two binary operations $+$ and $\circ$, satisfying the following axioms.

(S1) $(\mathbb{S},+)$ is a group with identity element 0.

(S2) $x\circ(y+z)=x\circ y+x\circ z$ and $(x+y)\circ z=x\circ z+y\circ z$, for all $x,y,z\in\mathbb{S}$.

(S3) $x\circ y=0$ implies $x=0$ or $y=0$.

(S4) $\exists 1\in\mathbb{S}$ such that $1\circ x=x\circ 1=x$, for all $x\in\mathbb{S}$.

An algebra satisfying all of the axioms of a semifield except (S4) is called a *pre-semifield*. By what is sometimes called Kaplansky's trick, a semifield with identity $u\circ u$ is obtained from a pre-semifield by defining a new multiplication $\hat{\circ}$ as follows

$$(x\circ u)\hat{\circ}(u\circ y)=x\circ y. \tag{1}$$

A finite field is of course a trivial example of a semifield. The first non-trivial examples of semifields were constructed by Dickson in [31]: a semifield $(\mathbb{F}^2_{q^k},+,\circ)$ of order q^{2k} with addition and multiplication defined by

$$\begin{cases} (x,y)+(u,v) &= (x+u,y+v) \\ (x,y)\circ(u,v) &= (xu+\alpha y^q v^q, xv+yu) \end{cases} \tag{2}$$

where q is an odd prime power and α is a non-square in $\mathbb{F}_{q^k}$.

One easily shows that the additive group of a semifield is elementary abelian, and the additive order of the elements of $\mathbb{S}$ is called the *characteristic* of $\mathbb{S}$. Contained in a semifield are the following important substructures, all of which are isomorphic to a finite field. The *left nucleus* $\mathbb{N}_l(\mathbb{S})$, *the middle nucleus* $\mathbb{N}_m(\mathbb{S})$, and the *right nucleus* $\mathbb{N}_r(\mathbb{S})$ are defined as follows:

$$\mathbb{N}_l(\mathbb{S}) := \{x\ :\ x\in\mathbb{S}\mid x\circ(y\circ z)=(x\circ y)\circ z,\ \forall y,z\in\mathbb{S}\}, \tag{3}$$

$$\mathbb{N}_m(\mathbb{S}) := \{y\ :\ y\in\mathbb{S}\mid x\circ(y\circ z)=(x\circ y)\circ z,\ \forall x,z\in\mathbb{S}\}, \tag{4}$$

$$\mathbb{N}_r(\mathbb{S}) := \{z\ :\ z\in\mathbb{S}\mid x\circ(y\circ z)=(x\circ y)\circ z,\ \forall x,y\in\mathbb{S}\}. \tag{5}$$

The intersection of the *associative center* $\mathbb{N}(\mathbb{S})$ (the intersection of the three nuclei) and the commutative center is called the *center* of $\mathbb{S}$ and denoted by $C(\mathbb{S})$. Apart from the usual representation of a semifield as a finite-dimensional algebra over its center, a semifield can also be viewed as a left vector space $V_l(\mathbb{S})$ over its left nucleus, as a left vector space $V_{lm}(\mathbb{S})$ and right vector space $V_{rm}(\mathbb{S})$ over its middle nucleus, and as a right vector space $V_r(\mathbb{S})$ over its right nucleus. Left (resp. right) multiplication in $\mathbb{S}$ by an element x is denoted by L_x (resp. R_x), i.e. $y^{L_x}=x\circ y$ (resp. $y^{R_x}=y\circ x$). It follows that L_x is an endomorphism of $V_r(\mathbb{S})$, while R_x is an endomorphism of $V_l(\mathbb{S})$.

If $\mathbb{S}$ is an n-dimensional algebra over the field $\mathbb{F}$, and $\{e_1,\ldots,e_n\}$ is an $\mathbb{F}$-basis for $\mathbb{S}$, then the multiplication can be written in terms of the multiplication of the e_i, i.e., if $x = x_1e_1 + \cdots + x_ne_n$ and $y = y_1e_1 + \cdots + y_ne_n$, with $x_i, y_i \in \mathbb{F}$, then

$$x \circ y = \sum_{i,j=1}^{n} x_iy_j(e_i \circ e_j) = \sum_{i,j=1}^{n} x_iy_j\left(\sum_{k=1}^{n} a_{ijk}e_k\right) \tag{6}$$

for certain $a_{ijk} \in \mathbb{F}$, called the *structure constants* of $\mathbb{S}$ with respect to the basis $\{e_1,\ldots,e_n\}$. This approach was used by Dickson in 1906 to prove the following characterisation of finite fields.

Theorem 1 ([31]). *A two-dimensional finite semifield is a finite field.*

In [58] Knuth noted that the action, of the symmetric group S_3, on the indices of the structure constants gives rise to another five semifields starting from one semifield $\mathbb{S}$. This set of at most six semifields is called the *S_3-orbit* of $\mathbb{S}$, and consists of the semifields $\{\mathbb{S}, \mathbb{S}^{(12)}, \mathbb{S}^{(13)}, \mathbb{S}^{(23)}, \mathbb{S}^{(123)}, \mathbb{S}^{(132)}\}$.

1.2. Projective Planes and Isotopism

As mentioned before, the study of semifields originated around 1900, and the link with projective planes through the coordinatisation method inspired by Hilbert's *Grundlagen der Geometrie* (1999), and generalised by Hall [39] in 1943, was a further stimulation for the development of the theory of finite semifields. Everything which is contained in this section concerning projective planes and the connections with semifields can be found with more details in [28], [44], [47], and [73]. It is in this context that the notion of isotopism is of the essence.

Two semifields $\mathbb{S}$ and $\hat{\mathbb{S}}$ are called *isotopic* if there exists a triple (F,G,H) of non-singular linear transformations from $\mathbb{S}$ to $\hat{\mathbb{S}}$ such that $x^F \hat{\circ} y^G = (x \circ y)^H$, for all $x,y,z \in \mathbb{S}$. The triple (F,G,H) is called an *isotopism*. An isotopism where H is the identity is called a *principal isotopism*. The set of semifields isotopic to a semifield $\mathbb{S}$ is called the *isotopism class of* $\mathbb{S}$ and is denoted by $[\mathbb{S}]$. Note that the size of the center as well as the size of the nuclei of a semifield are invariants of its isotopism class, and since the nuclei are finite fields, it is allowed to talk about the nuclei of an isotopism class $[\mathbb{S}]$.

A *projective plane* is a geometry consisting of a set $\mathcal{P}$ of *points* and a set $\mathcal{L}$ of subsets of $\mathcal{P}$, called *lines*, satisfying the following three axioms

(PP1) Each two different points are contained in exactly one line.

(PP2) Each two different lines intersect in exactly one point.

(PP3) There exist four points, no three of which are contained in a line.

Two projective planes π and π' are *isomorphic* if there exists a one-to-one correspondence between the points of π and the points of π' preserving collinearity, i.e., a line of π is mapped onto a line of π'. A projective plane is called *Desarguesian* if it is isomorphic to PG$(2,\mathbb{F})$, for some (skew) field $\mathbb{F}$. An isomorphism of a projective plane π is usually called a *collineation* and a (P,ℓ)*-perspectivity* of π is a collineation of π that fixes every line on P and every point on ℓ. Because of the self-dual property of the set of axioms {(PP1),(PP2),(PP3)}, interchanging points and lines of a projective plane π, one obtains another projective plane, called the *dual plane*, which we denote by π^d. If there exists a line ℓ in a projective plane π, such that for each point P on ℓ the group of (P,ℓ)-perspectivities acts transitively on the points of the affine plane $\pi \setminus \ell$, then π is called a *translation plane*, and ℓ is called a translation line of π. If both π and π^d are translation planes, then π is called a *semifield plane*. The point of a semifield plane corresponding to the translation line of the dual plane is called the *shears point*. It can be shown that, unless the plane is Desarguesian, the translation line (shears point) of a translation plane (dual translation plane) is unique, and the shears point of a semifield plane π lies on the translation line of π. The importance of the notion of isotopism arises from the equivalence between the isomorphism classes of projective planes and the isotopism classes of finite semifields, as shown by A. A. Albert in 1960.

Theorem 2 ([1]). *Two semifield planes are isomorphic if and only if the corresponding semifields are isotopic.*

The connection between semifield planes and the notion of semifields as we introduced them (as an algebra) is given by the coordinatisation method of projective planes. Without full details here, let us give an overview using homogeneous coordinates, following Knuth [58]. Let π be a projective plane, and let (R,T) be a ternary ring coordinatising π, with respect to a frame $\mathcal{G}$ in π. The points of π are represented by $(1,a,b)$, $(0,1,a)$, or $(0,0,1)$, where $a,b \in R$ and the lines are represented by $[1,c,d]$, $[0,1,c]$, $[0,0,1]$, with $c,d \in R$, where the frame $\mathcal{G} = \{(1,0,0),(0,1,0),(0,0,1),(1,1,1)\}$. The point (a,b,c) lies on the line $[d,e,f]$ if and only if

$$dc = T(b,e,af). \tag{7}$$

Since d and a must be either 0 or 1, it is clear what dc and af means.

It follows that T satisfies certain properties, and in fact one can list the necessary and sufficient properties that a ternary ring has to satisfy in order to be a ternary ring obtained by coordinatising a projective plane (by "inverse coordinatisation", i.e. constructing the plane starting from the ternary ring). In this case (R,T) is called a *planar ternary ring* , usually abbreviated to PTR. Now define two operations $a \circ b := T(a,b,0)$, and $a+b := T(a,1,b)$, and consider the structure $(R,\circ,+)$. This turns $(R,+)$ and $(R,\circ)$ into loops, with respective identities 0 and 1. With this setup, one is able to connect the algebraic properties of the PTR with the geometric properties of the plane, or more specifically, with the properties of the automorphism group of the plane π, using the following standard terminology.

A PTR is called *linear* if $T(a,b,c) = a \circ b + c, \forall a,b,c \in R$. A *cartesian group* is a linear PTR with associative addition; a *(left) quasifield* is a cartesian group in which the left distributive law holds; and a *semifield* is a quasifield in which both distributive laws hold, consistent with (S1)-(S4). These algebraic properties correspond to the following geometric properties. A linear PTR is a cartesian group if and only if π is $((0,0,1),[0,0,1])$-transitive. A cartesian group is a quasifield if and only if π is $((0,1,0),[0,0,1])$-transitive, and in this case π is a translation plane with translation line $[0,0,1]$, and $(R,+)$ is abelian. A semifield plane was defined as a translation plane which is also a dual translation plane, and we leave it to the reader to check the consistency of this definition.

1.3. Spreads and Linear Sets

An elegant way to construct a translation plane is by using so-called *spreads* of projective spaces. This construction is often called the *André-Bruck-Bose* construction.

Let $\mathcal{S}$ be a set of $(t-1)$-dimensional subspaces of $\mathrm{PG}(n-1,q)$. Then $\mathcal{S}$ is called a $(t-1)$-*spread* of $\mathrm{PG}(n-1,q)$ if every point of $\mathrm{PG}(n-1,q)$ is contained in exactly one element of $\mathcal{S}$. If $\mathcal{S}$ is a set of subspaces of $V(n,q)$ of rank t, then $\mathcal{S}$ is called a t-*spread* of $V(n,q)$ if every vector of $V(n,q)\setminus\{0\}$ is contained in exactly one element of $\mathcal{S}$.

Theorem 3 ([88]). *There exists a* $(t-1)$-*spread in* $\mathrm{PG}(n-1,q)$ *if and only if* t *divides* n.

Suppose t divides n, and put $n=rt$. The $(t-1)$-spread of $\mathrm{PG}(rt-1,q)$ obtained by considering the points of $\mathrm{PG}(r-1,q^t)$ as $(t-1)$-dimensional subspaces over $\mathbb{F}_q$ is called a *Desarguesian spread.* This correspondence between the points of $\mathrm{PG}(r-1,q^t)$ and the elements of a Desarguesian $(t-1)$-spread will often be used in this article, and if the context is clear, we will identify the elements of the Desarguesian $(t-1)$-spread of $\mathrm{PG}(rt-1,q)$ with the points of $\mathrm{PG}(r-1,q^t)$. If T is any subset of $\mathrm{PG}(rt-1,q)$ endowed with a Desarguesian spread $\mathcal{D}$, then by $B_{\mathcal{D}}(T)$ (or $B(T)$ if there is no confusion) we denote the set of elements of $\mathcal{D}$ that intersect T non-trivially.

A set L of points in $\mathrm{PG}(r-1,q_0)$ is called a *linear set* if there exists a subspace U in $\mathrm{PG}(rt-1,q)$, for some $t\geq 1$, $q^t=q_0$, such that L is the set of points corresponding to the elements of a Desarguesian $(t-1)$-spread of $\mathrm{PG}(rt-1,q)$ intersecting U, i.e. $L=B(U)$. If we want to specify the field $\mathbb{F}_q$ over which L is linear, we call L an $\mathbb{F}_q$-*linear set.* If U has dimension d in $\mathrm{PG}(rt-1,q)$, then the linear set $B(U)$ is called a linear set *of rank* $d+1$.

The same notation and terminology is used when U is a subspace of the vector space $V(rt,q)$ instead of a projective subspace. For an overview of the use of linear sets in various other areas of Galois geometries, we refer to [59], [65], and [85].

Let $\mathcal{S}$ be a $(t-1)$-spread in $\mathrm{PG}(2t-1,q)$. Consider $\mathrm{PG}(2t-1,q)$ as a hyperplane of $\mathrm{PG}(2t,q)$. We define an incidence structure $(\mathcal{P},\mathcal{L},I)$ as follows. The pointset $\mathcal{P}$ consists of all points of $\mathrm{PG}(2t,q)\setminus\mathrm{PG}(2t-1,q)$ and the lineset $\mathcal{L}$ consists of all t-spaces of $\mathrm{PG}(2t,q)$ intersecting $\mathrm{PG}(2t-1,q)$ in an element of $\mathcal{S}$. The incidence relation I is containment.

Theorem 4 ([3], [17], [18]). *The incidence structure* $(\mathcal{P},\mathcal{L},I)$ *is an affine plane and its projective completion is a translation plane of order* q^t. *Moreover every translation plane can be constructed in this way.*

For this reason, a $(t-1)$-spread in $\mathrm{PG}(2t-1,q)$ is sometimes called a *planar spread.* Two spreads are said to be *isomorphic* if there exists a collineation of the projective space mapping one spread onto the other.

Theorem 5 ([3], [17], [18]). *Two translation planes are isomorphic if and only if the corresponding spreads are isomorphic.*

These theorems are of fundamental importance in Galois geometry; they imply a one-to-one correspondence between translation planes and planar spreads. The construction of a translation plane from a planar spread is called the *André-Bruck-Bose construction* . If the translation plane obtained is a semifield plane, then the spread is called a *semifield spread* . It follows from the fact that a semifield plane π is a dual translation plane, that a semifield spread $\mathcal{S}$ contains a special element S_∞ (corresponding to the shears point) such that the stabiliser of $\mathcal{S}$ fixes S_∞ pointwise and acts transitively on the elements of $\mathcal{S} \setminus \{S_\infty\}$, and moreover, this property characterises a semifield spread. The next theorem motivates the choice of the term Desarguesian spread.

Theorem 6 ([88]). *A $(t-1)$-spread of* $\mathrm{PG}(2t-1,q)$ *is Desarguesian if and only if the corresponding translation plane is Desarguesian, i.e. isomorphic to* $\mathrm{PG}(2,q^t)$.

By a method called *derivation* , it is possible to construct a non-Desarguesian translation plane from a Desarguesian plane. This construction can in fact be applied to any translation plane corresponding to a spread that contains a regulus. A *regulus* in $\mathrm{PG}(3,q)$ is a set of $q+1$ lines that intersect a given set of three two by two disjoint lines (see [28]). Replacing a regulus by its opposite regulus one obtains another spread, and the corresponding new translation plane is called the *derived* plane.

The spread corresponding to a translation plane π can also be constructed algebraically from the coordinatising quasifield, see e.g. [44]. In order to avoid unnecessary generality, we restrict ourselves to the case where π is a semifield plane. In this case there are essentially two approaches one can take, by considering either the endomorphisms L_x or R_x. In the literature it is common to use the endomorphism R_x. We define the following subspaces of $\mathbb{S} \times \mathbb{S}$. For each $x \in \mathbb{S}$, consider the set of vectors $S_x := \{(y, y^{R_x}) : y \in \mathbb{S}\}$, and put $S_\infty := \{(0,y) : y \in \mathbb{S}\}$. It is an easy exercise to show that $\mathcal{S} := \{S_x : x \in \mathbb{S}\} \cup \{S_\infty\}$ is a spread of $\mathbb{S} \times \mathbb{S}$. The set of endomorphisms

$$S := \{R_x \ : \ x \in \mathbb{S}\} \subset \mathrm{End}(V_l(\mathbb{S}))$$

is called the *semifield spread set* corresponding to $\mathbb{S}$. Note that by (S2) the spread set S is closed under addition and, by (S3), the non-zero elements of S are invertible.

More generally, if $\mathcal{S}$ is a t-spread of $\mathbb{F}_q^t \times \mathbb{F}_q^t$, containing $S_0 = \{(y,0) : y \in \mathbb{F}_q^t\}$, and $S_\infty = \{(0,y) : y \in \mathbb{F}_q^t\}$, then we can label the elements of $\mathcal{S}$ different from S_∞ as $S_x := \{(y, y^{\phi_x}) : y \in \mathbb{F}_q^t\}$, with $\phi_x \in \mathrm{End}(\mathbb{F}_q^t)$. The set $S := \{\phi_x : x \in \mathbb{F}_q^t\} \subset \mathrm{End}(\mathbb{F}_q^t)$ of endomorphisms is called a *spread set* associated with $\mathcal{S}$. A spread set S is a semifield spread set if it forms an additive subgroup of $\mathrm{End}(\mathbb{F}_q^t)$.

Two spread sets are called *equivalent* if the corresponding spreads are isomorphic. The following theorem is well known, and should probably be credited to Maduram [74]. By lack of a reference containing the exact same statement, we include a short proof.

Theorem 7. *Two semifield spread sets S, $S' \subset \mathrm{End}(\mathbb{F}^t)$ are equivalent if and only if there exist invertible elements $\omega, \psi \in \mathrm{End}(\mathbb{F}^t)$ and $\sigma \in \mathrm{Aut}(\mathbb{F})$ such that $S' = \{\omega R_x^\sigma \psi \ : \ R_x \in S\}$.*

Proof. Using the properties of a semifield spread, we may assume that an equivalence between two semifield spread sets S and S' is induced by an isomorphism β between the corresponding spreads $\mathcal{S}$ and $\mathcal{S}'$ which fixes $S_\infty = S'_\infty$ and $S_0 = S'_0$ with the notation from above. It follows that β is of the form $(x,y) \mapsto (Ax^\sigma, By^\sigma)$, where A, B are elements of $GL(n,q)$ and $\sigma \in \mathrm{Aut}(\mathbb{F}_q)$. Calculating the effect on the elements of the spread set concludes the proof (see e.g. [63, page 908]). □

1.4. Dual and Transpose of a Semifield, the Knuth Orbit

Knuth proved that the action of S_3, defined above, on the indices of the structure constants of a semifield $\mathbb{S}$ is well-defined with respect to the isotopism classes of $\mathbb{S}$, and by the *Knuth orbit of* $\mathbb{S}$ (notation $\mathcal{K}(\mathbb{S})$), we mean the set of isotopism classes corresponding to the S_3-orbit of $\mathbb{S}$, i.e.,

$$\mathcal{K}(\mathbb{S}) = \{[\mathbb{S}], [\mathbb{S}^{(12)}], [\mathbb{S}^{(13)}], [\mathbb{S}^{(23)}], [\mathbb{S}^{(123)}], [\mathbb{S}^{(132)}]\}. \tag{8}$$

The advantage of using Knuth's approach to the coordinatisation with homogeneous coordinates, is that we immediately notice the duality. The semifield corresponding to the dual plane $\pi(\mathbb{S})^d$ of a semifield plane $\pi(\mathbb{S})$ is the plane $\pi(\mathbb{S}^{opp})$, where $\mathbb{S}^{opp}$ is the *opposite algebra* of $\mathbb{S}$ obtained by reversing the multiplication $\circ$, or in other words, the semifield corresponding to the dual plane is $\mathbb{S}^{(12)}$, which we also denote by $\mathbb{S}^d$, i.e.,

$$\mathbb{S}^d = \mathbb{S}^{(12)} = \mathbb{S}^{opp}. \tag{9}$$

Similarly, it is easy to see that the semifield $\mathbb{S}^{(23)}$ can be obtained by transposing the matrices corresponding to the transformations L_{e_i}, $e_i \in \mathbb{S}$, with respect to some basis $\{e_1, e_2, \ldots, e_n\}$ of $V_r(\mathbb{S})$, and for this reason $\mathbb{S}^{(23)}$ is also denoted by $\mathbb{S}^t$, called the *transpose of* $\mathbb{S}$. With this notation, the Knuth orbit becomes

$$\mathcal{K}(\mathbb{S}) = \{[\mathbb{S}], [\mathbb{S}^d], [\mathbb{S}^t], [\mathbb{S}^{dt}], [\mathbb{S}^{td}], [\mathbb{S}^{dtd}]\}. \tag{10}$$

Taking the transpose of a semifield can also be interpreted geometrically as dualising the semifield spread (Maduram [74]). The resulting action on the set of nuclei of the isotopism class $\mathbb{S}$ is as follows. The permutation (12) fixes the middle nucleus and interchanges the left and right nuclei; the permutation (23) fixes the left nucleus and interchanges the middle and right nuclei. Summarising, the action of the dual and transpose generate a series of at most six isotopism classes of semifields, with nuclei according to Figure 1.

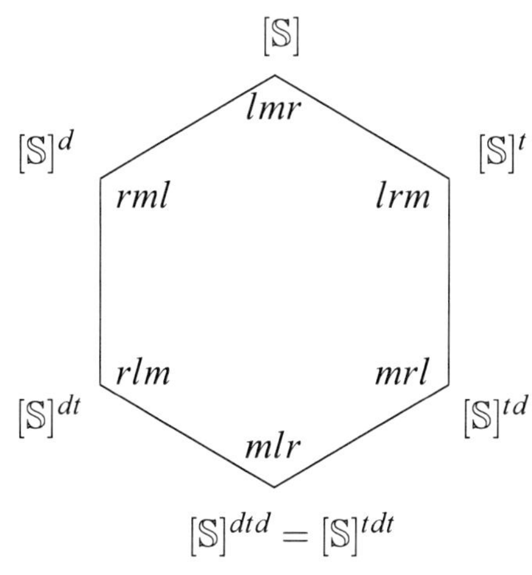

Figure 1. The Knuth orbit of a semifield $\mathbb{S}$ with nuclei *lmr*.

2. Semifields: A Geometric Approach

In this section, we explain a geometric approach to finite semifields, which has been very fruitful in recent years. In what follows, we consider the set of endomorphisms corresponding to right multiplication in the semifield, and by doing so it is natural to consider the semifield as a left vector space over (a subfield of) its left nucleus. It should be clear to the reader that this is just a matter of choice and the same geometric approach can be taken by considering the set of endomorphisms corresponding to left multiplication in the semifield. The left nucleus should then be replaced by the right nucleus in what follows.

Let $\mathbb{S} = (\mathbb{S}, +, \circ)$ be a finite semifield and let S be the semifield spread set associated with $\mathbb{S}$. Clearly, for any subfield $\mathbb{F} \subset \mathbb{N}_l(\mathbb{S})$, $\mathbb{S}$ is a left vector space over $\mathbb{F}$, and S is also an additive subgroup of $\mathrm{End}(\mathbb{F}^n)$ (if $|\mathbb{S}| = |\mathbb{F}|^n$) by considering R_x as elements of $\mathrm{End}(\mathbb{F}^n)$ instead of $\mathrm{End}(V_l(\mathbb{S}))$. Conversely, any subgroup S of the additive group of $\mathrm{End}(\mathbb{F}^n)$ whose non-zero elements are invertible defines a semifield $\mathbb{S}$ whose left nucleus contains the field $\mathbb{F}$. If S does not contain the identity map, then S defines a pre–semifield.

This means that semifields, n–dimensional over a subfield $\mathbb{F}_q$ of their left nucleus, can be investigated via the semifield spread sets of $\mathbb{F}_q$–linear maps of $\mathbb{F}_{q^n}$, regarded as a vector space over $\mathbb{F}_q$. An element φ of $\mathrm{End}_{\mathbb{F}_q}(\mathbb{F}_{q^n})$ can be represented in a unique way as a q–polynomial over $\mathbb{F}_{q^n}$, that is a polynomial of the form

$$\sum_{i=0}^{n-1} a_i X^{q^i} \in \mathbb{F}_{q^n}[X],$$

and φ is invertible if and only if $\det(A) \neq 0$, where

$$A = \begin{pmatrix} a_0 & a_{n-1}^q & a_{n-2}^{q^2} & \cdots & a_1^{q^{n-1}} \\ a_1 & a_0^q & a_{n-1}^{q^2} & \cdots & a_2^{q^{n-1}} \\ a_2 & a_1^q & a_0^{q^2} & \cdots & a_3^{q^{n-1}} \\ \vdots & \vdots & \vdots & & \vdots \\ a_{n-1} & a_{n-2}^q & a_{n-3}^{q^2} & \cdots & a_0^{q^{n-1}} \end{pmatrix}$$

(see e.g. [67, page 362]).

Hence, any spread set S of linear maps defining a semifield of order q^n can be seen as a set of q^n linearized polynomials, closed with respect to the addition, containing the zero map and satisfying the above mentioned non-singularity condition.

2.1. Linear Sets and the Segre Variety

Let $\mathbb{M}(n,q)$ denote the n^2-dimensional vector space of all $(n \times n)$-matrices over $\mathbb{F}_q$. The Segre variety $\mathcal{S}_{n,n}$ of the projective space $\mathrm{PG}(\mathbb{M}(n,q), \mathbb{F}_q) = \mathrm{PG}(n^2-1, q)$ is an algebraic variety corresponding to the matrices of $\mathbb{M}(n,q)$ of rank one and the $(n-2)$–th secant variety $\Omega(\mathcal{S}_{n,n})$ of $\mathcal{S}_{n,n}$ is the hypersurface corresponding to the non-invertible matrices of $\mathbb{M}(n,q)$ (also called a *determinantal hypersurface*). There are two systems $\mathcal{R}_1$ and $\mathcal{R}_2$ of maximal subspaces contained in $\mathcal{S}_{n,n}$ and each element of $\mathcal{R}_i$ has dimension $n-1$. If $n = 2$, then $\mathcal{S}_{2,2}$ is a hyperbolic quadric $Q^+(3,q)$ of a 3-dimensional projective space and $\mathcal{R}_1$ and $\mathcal{R}_2$ are the reguli of $Q^+(3,q)$.

By the well-known isomorphism between the vector spaces $\mathbb{M}(n,q)$ and $\mathbb{V} = \mathrm{End}_{\mathbb{F}_q}(\mathbb{F}_{q^n})$, we have that the elements of $\mathbb{V}$ with kernel of rank $n-1$ correspond to a Segre variety $\mathcal{S}_{n,n}$ of the projective space $\mathrm{PG}(\mathbb{V}) = \mathrm{PG}(n^2-1,q)$ and the non–invertible elements of $\mathbb{V}$ correspond to the $(n-2)$–th secant variety $\Omega(\mathcal{S}_{n,n})$ of $\mathcal{S}_{n,n}$.

Also, the collineations of $\mathrm{PG}(\mathbb{V})$ induced by the semilinear maps

$$\Gamma_{\psi\sigma\omega} : \varphi \mapsto \psi\varphi^{\sigma}\omega, \tag{11}$$

(where ω and ψ are invertible elements of $\mathbb{V}$ and $\sigma \in \mathrm{Aut}(\mathbb{F}_q)$) form the automorphism group $\mathcal{H}(\mathcal{S}_{n,n})$ of $\mathcal{S}_{n,n}$ preserving the systems $\mathcal{R}_1$ and $\mathcal{R}_2$ of $\mathcal{S}_{n,n}$ (see [41]). The group $\mathcal{H}(\mathcal{S}_{n,n})$ has index two in the stabiliser $\mathcal{G}(\mathcal{S}_{n,n})$ of $\mathcal{S}_{n,n}$ inside $\mathrm{P\Gamma L}(n^2,q)$.

Now, let $\mathbb{S}$ be a semifield and let S be its semifield spread set consisting of $\mathbb{F}_q$–linear maps of $\mathbb{F}_{q^n}$. Since S is an additive subgroup of $\mathbb{V}$, it is an $\mathbb{F}_s$–subspace of $\mathbb{V}$, for some subfield $\mathbb{F}_s$ of $\mathbb{F}_q$ (say $q = s^t$), of dimension nt. This implies that (using the terminology of linear sets from above) $\mathbb{S}$ defines an $\mathbb{F}_s$–linear set $L(\mathbb{S}) := B(S)$ in $\mathrm{PG}(n^2-1,q)$ of rank nt. Note that $\mathbb{F}_s$ is contained in the center of $\mathbb{S}$. Since each non-zero element of S is invertible, the linear set $L(\mathbb{S})$ is disjoint from the variety $\Omega(\mathcal{S}_{n,n})$ of $\mathrm{PG}(\mathbb{V})$. Conversely, if L is an $\mathbb{F}_s$–linear set of $\mathrm{PG}(\mathbb{V}) = \mathrm{PG}(n^2-1,s^t)$ of rank nt disjoint from $\Omega(\mathcal{S}_{n,n})$, then the set S of $\mathbb{F}_q$–linear maps underlying L satisfies the properties of a semifield spread set except, possibly, the existence of the identity map and hence L defines a pre–semifield of order $q^n = s^{nt}$, whose associated semifield has left nucleus containing $\mathbb{F}_q$ and center containing $\mathbb{F}_s$. So we have the following theorem.

Theorem 8 ([64]). *To any semifield $\mathbb{S}$ of order q^n $(q = s^t)$, with left (right) nucleus containing $\mathbb{F}_q$ and center containing $\mathbb{F}_s$, there corresponds an $\mathbb{F}_s$–linear set $L(\mathbb{S})$ of the projective space* $\mathrm{PG}(n^2-1,q)$ *of rank nt disjoint from the $(n-2)$–th secant variety $\Omega(\mathcal{S}_{n,n})$ of a Segre variety, and conversely.*

Note that, if $\mathbb{F}_q$ is a subfield of the center of the semifield (i.e., if $t = 1$), then the corresponding linear set is simply an $(n-1)$–dimensional subspace of $\mathrm{PG}(n^2-1,q)$. Now, rephrasing Theorem 7, using (11), in the projective space $\mathrm{PG}(\mathbb{V})$ we have the following theorem.

Theorem 9 ([64]). *Two semifields $\mathbb{S}_1$ and $\mathbb{S}_2$ with corresponding $\mathbb{F}_s$–linear sets $L(\mathbb{S}_1)$ and $L(\mathbb{S}_2)$ in* $\mathrm{PG}(n^2-1,q)$ *are isotopic if and only if there exists a collineation $\Phi \in \mathcal{H}(\mathcal{S}_{n,n})$ such that $L(\mathbb{S}_2) = L(\mathbb{S}_1)^{\Phi}$.*

By the previous arguments, it is clear that linear sets $L(\mathbb{S}_1)$ and $L(\mathbb{S}_2)$ having a different geometric structure with respect to the collineation group $\mathcal{H}(\mathcal{S}_{n,n})$, determine non–isotopic semifields $\mathbb{S}_1$ and $\mathbb{S}_2$, and hence non–equivalent semifield spread sets S_1 and S_2, and non–isomorphic semifield spreads $\mathcal{S}(\mathbb{S}_1)$ and $\mathcal{S}(\mathbb{S}_2)$. Theorems 8 and 9 can be found in [64]; they generalize previous results obtained in [63], and in [69] and [21] where rank two semifields are studied.

Using the geometric approach, the transpose operation $\mathbb{S} \mapsto \mathbb{S}^t$ can be read in the following way. If τ is any polarity of the projective space $\mathrm{PG}(\mathbb{S}\times\mathbb{S},\mathbb{F}_q)$, then $\mathcal{S}(\mathbb{S})^{\tau}$ is a semifield spread as well and the corresponding semifield is isotopic to the transpose semifield $\mathbb{S}^t$ of $\mathbb{S}$.

It can be shown that any polarity of $\mathrm{PG}(\mathbb{S}\times\mathbb{S},\mathbb{F}_q)$ fixing the subspaces S_∞ and S_0 induces in $\mathrm{PG}(n^2-1,q)$ a collineation of $\mathcal{G}(\mathcal{S}_{n,n})$ interchanging the systems of $\mathcal{S}_{n,n}$ (see [72]). Hence, since $\mathcal{H}(\mathcal{S}_{n,n})$ has index two in $\mathcal{G}(\mathcal{S}_{n,n})$, by Theorem 9 we have the following.

Theorem 10 ([72]). *If Φ is a collineation of $\mathcal{G}(\mathcal{S}_{n,n})$ not belonging to $\mathcal{H}(\mathcal{S}_{n,n})$ then the linear set $L(\mathbb{S})^{\Phi}$ corresponds to the isotopism class of the transpose semifield $\mathbb{S}^t$ of $\mathbb{S}$.*

2.2. BEL-construction

In this section we concentrate on a geometric construction of finite semifield spreads. The construction we give here is taken from [64], but the main idea is the slightly less general construction given in [7] (where L is a subspace, i.e. $t = 1$).

We define a *BEL-configuration* as a triple $(\mathcal{D}, U, W)$, where $\mathcal{D}$ a Desarguesian $(n-1)$-spread of $\Sigma_1 := \mathrm{PG}(rn-1, s^t)$, $t \geq 1$, $r \geq 2$; U is an nt-dimensional subspace of $\mathbb{F}_s^{rnt}$ such that $L = B(U)$ is an $\mathbb{F}_s$-linear set of Σ_1 of rank nt; and W is a subspace of Σ_1 of dimension $rn-n-1$, such that no element of $\mathcal{D}$ intersects both L and W. From a BEL-configuration one can construct a semifield spread as follows.

- Embed Σ_1 in $\Lambda_1 \cong \mathrm{PG}(rn+n-1, s^t)$ and extend $\mathcal{D}$ to a Desarguesian spread $\mathcal{D}_1$ of Λ_1.
- Let $L' = B(U')$, $U \subset U'$ be an $\mathbb{F}_s$-linear set of Λ_1 of rank $nt+1$ which intersects Σ_1 in L.
- Let $\mathcal{S}(\mathcal{D}, U, W)$ be the set of subspaces defined by L' in the quotient geometry $\Lambda_1/W \cong \mathrm{PG}(2n-1, s^t)$ of W, i.e.,

$$\mathcal{S}(\mathcal{D}, U, W) = \{\langle R, W\rangle / W \ : \ R \in \mathcal{D}_1, R \cap L' \neq \emptyset\}.$$

Theorem 11 ([64]). *The set $\mathcal{S}(\mathcal{D}, U, W)$ is a semifield spread of $\mathrm{PG}(2n-1, s^t)$. Conversely, for every finite semifield spread $\mathcal{S}$, there exists a BEL-configuration $(\mathcal{D}, U, W)$, such that $\mathcal{S}(\mathcal{D}, U, W) \cong \mathcal{S}$.*

The pre-semifield corresponding to $\mathcal{S}(\mathcal{D}, U, W)$ is denoted by $\mathbb{S}(\mathcal{D}, U, W)$. Using this BEL-construction it is not difficult to prove the following characterisation of the linear sets corresponding to a finite field.

Theorem 12 ([63]). *The linear set $L(\mathbb{S})$ of $\mathrm{PG}(n^2-1, q)$ disjoint from $\Omega(\mathcal{S}_{n,n})$ corresponds to a pre–semifield isotopic to a field if and only if there exists a Desarguesian $(n-1)$–spread of $\mathrm{PG}(n^2-1, q)$ containing $L(\mathbb{S})$ and a system of $\mathcal{S}_{n,n}$.*

If $r = 2$ and $s = 1$, then we can use the symmetry in the definition of a BEL-configuration to construct two semifields, namely $\mathbb{S}(\mathcal{D}, U, W)$ and $\mathbb{S}(\mathcal{D}, W, U)$, and in this way we can extend the Knuth orbit by considering the operation

$$\kappa := \mathbb{S}(\mathcal{D}, U, W) \mapsto \mathbb{S}(\mathcal{D}, W, U). \tag{12}$$

Except in the case where the semifield is a rank two semifield, in which case κ becomes the translation dual (see Section 4.), it is not known whether κ is well defined on the set of isotopism classes (see [7], [54]).

3. Rank Two Semifields

Semifields of dimension two over (a subfield of) their left nucleus (*rank two semifields*) correspond to semifield spreads of 3–dimensional projective spaces, as explained in Section 1.. In the last years, the connection between semifields and linear sets described in Section 2 has been intensively used to construct and characterize families of rank two semifields.

If $\mathbb{S} = (\mathbb{F}_{q^2}, +, \circ)$ is a semifield with left nucleus containing $\mathbb{F}_q$ and center containing $\mathbb{F}_s$, $q = s^t$, then by Theorem 8 its semifield spread set S defines an $\mathbb{F}_s$-linear set $L(S)$ of rank $2t$ in the 3-dimensional projective space $\Sigma = PG(\mathbb{V}, \mathbb{F}_q) = PG(3, q)$, where $\mathbb{V} = \mathrm{End}_{\mathbb{F}_q}(\mathbb{F}_{q^2})$, disjoint from the hyperbolic quadric $Q^+(3,q)$ of Σ defined by the non–invertible elements of $\mathbb{V}$, and conversely. Also, by Theorem 9 the study up to isotopy of semifields of order q^2 with left nucleus containing $\mathbb{F}_q$ and center containing $\mathbb{F}_s$ corresponds to the study of $\mathbb{F}_s$-linear sets of rank $2t$ of Σ with respect to the action of the collineation group of Σ fixing the reguli of the hyperbolic quadric $Q^+(3,q)$.

In this case the Knuth orbit of $\mathbb{S}$ can be extended in the following way. If $b(X,Y)$ is the bilinear form associated with $Q^+(3,q)$, then by field reduction we can use the bilinear form

$$b_s(X,Y) := Tr_{q/s}(b(X,Y)),$$

where $Tr_{q/s}$ is the trace function from $\mathbb{F}_q$ to $\mathbb{F}_s$, to obtain another linear set $L(\mathbb{S})^{\perp}$ disjoint from $Q^+(3,q)$ induced by the semifield spread set

$$S^{\perp} := \{x \in \mathbb{V} \ : \ b_s(x,y) = 0, \forall y \in \mathbb{V}\}.$$

Theorem 13. *The set $S^{\perp}$ is a semifield spread set of $\mathbb{F}_q$–linear maps of $\mathbb{F}_{q^2}$.*

The pre–semifield arising from the semifield spread set $S^{\perp}$ is called the *translation dual* $\mathbb{S}^{\perp}$ of the semifield $\mathbb{S}$. The translation dual of a rank two semifield has been introduced in [69] in terms of translation ovoids of $Q^+(5,q)$ generalizing the relationship between semifield flocks and translation ovoids of $Q(4,q)$ that will be detailed in Section 5. In [61], it was shown that this operation links the two sets of three semifields associated with a semifield flock from [6], and that this operation is a special case of the semifield operation κ (see (12)) from [7] (see (12) at the end of Section 2.). The translation dual operation is well defined on the set of isotopism classes and leaves invariant the sizes of the nuclei of a semifield $\mathbb{S}$, as proven in [71, Theorem 5.3]. This implies that in general $[\mathbb{S}^{\perp}]$ is not contained in the Knuth orbit $\mathcal{K}(\mathbb{S})$ and hence in the 2–dimensional case we have a chain of possibly twelve isotopism classes $\mathcal{K}(\mathbb{S}) \cup \mathcal{K}(\mathbb{S}^{\perp})$, with nuclei as illustrated by Figure 2.

To our knowledge, the known examples of semifields $\mathbb{S}$ for which $\mathbb{S}^{\perp}$ is not isotopic to $\mathbb{S}$ are: the symplectic semifield of order $q = 3^{2t}$ ($t > 2$) from Cohen-Ganley [23], and Thas-Payne [92], the symplectic semifield of order 3^{10} from Penttila–Williams [84], the HMO–semifields of order q^4 (for $q = p^k$, k odd, $k \geq 3$ and p prime with $p \equiv 1 \pmod 4$) exhibited in [54, Example 5.8] and their translation duals. But, in all of these cases, the size of $\mathcal{K}(\mathbb{S}) \cup \mathcal{K}(\mathbb{S}^{\perp})$ is six, since these are self-transpose semifields, i.e. $[\mathbb{S}] = [\mathbb{S}^t]$.

Using the geometric approach from Section 2., in [21], the authors classify all semifields of order q^4 with left nucleus of order q^2 and center of order q (see Theorem 26).

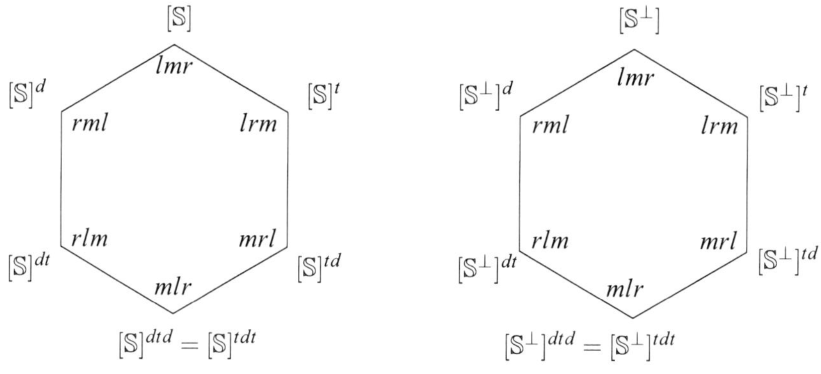

Figure 2. The isotopism classes $\mathcal{K}(\mathbb{S}) \cup \mathcal{K}(\mathbb{S}^\perp)$ of a rank two semifield $\mathbb{S}$ with nuclei *lmr*.

In [75], [48] and [34], semifields of order q^6, with left nucleus of order q^3 and center of order q, are studied using the same geometric approach, giving the following result.

Theorem 14 ([75], [48]). *Let $\mathbb{S}$ be a semifield of order q^6 with left nucleus of order q^3 and center of order q. Then there are eight possible geometric configurations for the corresponding linear set $L(\mathbb{S})$ in $\mathrm{PG}(3,q^3)$. The corresponding classes of semifields are partitioned into eight non-isotopic families, labeled $\mathcal{F}_0$, $\mathcal{F}_1$, $\mathcal{F}_2$, $\mathcal{F}_3$, $\mathcal{F}_4^{(a)}$, $\mathcal{F}_4^{(b)}$, $\mathcal{F}_4^{(c)}$ and $\mathcal{F}_5$.*

The families $\mathcal{F}_i$, $i = 0,1,2$, are completely characterized: the family $\mathcal{F}_0$ contains only Generalized Dickson/Knuth semifields with the given parameters; the family $\mathcal{F}_1$ contains only the symplectic semifield associated with the Payne–Thas ovoid of $Q^+(4,3^3)$; the family $\mathcal{F}_2$ contains only the semifield associated with the Ganley flock of the quadratic cone of $\mathrm{PG}(3,3^3)$.

So far, only few examples of semifields belonging to $\mathcal{F}_3$ and $\mathcal{F}_4^{(b)}$ are known for small values of q. These were obtained by using a computer algebra software package.

A further investigation of families $\mathcal{F}_4^{(a)}$ and $\mathcal{F}_4^{(c)}$ led to the construction of new infinite families of semifields (Section 6., EMPT2 semifields).

Moreover, all semifields of order q^6 with left nucleus of order q^3, right and middle nuclei of order q^2, and center of order q fall in family $\mathcal{F}_4^{(c)}$ and they are completely classified (see Section 6, Theorem 27).

Finally, semifields belonging to the family $\mathcal{F}_5$ are called *scattered semifields* , because their associated linear sets are of maximum size $q^5+q^4+\cdots+q+1$, i.e., are scattered following [14]. In [75], it has been proved that to any semifield $\mathbb{S}$ belonging to $\mathcal{F}_5$ is associated an *$\mathbb{F}_q$–pseudoregulus* $\mathcal{L}(\mathbb{S})$ of $\mathrm{PG}(3,q^3)$, which is a set of q^3+1 pairwise disjoint lines with exactly two transversal lines. An $\mathbb{F}_q$-pseudoregulus of $\mathrm{PG}(3,q^3)$ defines a *derivation set* in a similar way as the pseudoregulus of $\mathrm{PG}(3,q^2)$ defined by Freeman [37]. The known examples of semifields belonging to the family $\mathcal{F}_5$ are the Generalized twisted fields and the two families of Knuth semifields of type III and IV with the involved parameters. In [75], they are also characterized in terms of the associated $\mathbb{F}_q$–pseudoreguli. Precisely, in the case of Knuth semifields the transversal lines of the associated pseudoregulus are contained in a regulus of $Q^+(3,q)$; whereas in the case of Generalized twisted fields the transversal lines of the associated pseudoregulus are pairwise polar external lines of $Q^+(3,q)$ and the set of lines of the pseudoregulus is preserved by the polarity $\perp$ induced by $Q^+(3,q)$.

Recent results obtained in [66] have shown that various other possible geometric configurations of the transversal lines of a pseudoregulus of $\mathrm{PG}(3,q^3)$ can produce new semifields in family $\mathcal{F}_5$.

The results obtained in the case q^6 inspired a more general construction method that led to the discovery of new infinite families of rank two semifields of size q^{2t} for arbitrary values of q and t (see Section 6, EMPT1 semifields).

Some other existence and classification results for rank two semifields obtained by the geometric approach of linear sets can be found in [49], [76] and [77].

4. Symplectic Semifields and Commutative Semifields

A semifield spread $\mathcal{S}$ of the projective space $\mathrm{PG}(2n-1,q)$ is *symplectic* when all subspaces of $\mathcal{S}$ are totally isotropic with respect to a symplectic polarity of $\mathrm{PG}(2n-1,q)$.

Starting from a semifield $\mathbb{S}$, we can construct a family of semifield spreads; precisely, we can associate to $\mathbb{S}$ a semifield spread $\mathcal{S}_{\mathbb{F}}$ for any subfield $\mathbb{F}$ of its left nucleus (see Section 1.3.). By [53] and [70], if $\mathcal{S}_{\mathbb{F}}$ is a symplectic semifield spread then any other semifield spread arising from $\mathbb{S}$ is symplectic. Hence, it makes sense to define a *symplectic semifield* as a semifield whose associated semifield spread is symplectic.

In terms of the associated linear set, a symplectic semifield can be characterized in the following way.

Theorem 15 ([72]). *The semifield $\mathbb{S}$ with corresponding linear set $L(\mathbb{S})$ in $\mathrm{PG}(\mathrm{End}_{\mathbb{F}_q}(\mathbb{F}_{q^n}))$ is symplectic if and only if there is a subspace Γ of $\mathrm{PG}(\mathrm{End}_{\mathbb{F}_q}(\mathbb{F}_{q^n}))$ of dimension $\frac{n(n+1)}{2}-1$ such that $\Gamma \cap \mathcal{S}_{n,n}$ is a quadric Veronesean and $L(\mathbb{S}) \subset \Gamma$.*

Symplectic semifields and commutative semifields are related via the S_3-action in the following way.

Theorem 16 ([52]). *A pre–semifield $\mathbb{S}$ is isotopic to a commutative semifield if and only if the pre–semifield $\mathbb{S}^{td}$ is symplectic.*

It follows from the above that the Knuth orbit $\mathcal{K}(\mathbb{S})$ of a symplectic semifield consists of the isotopism classes $\{[\mathbb{S}] = [\mathbb{S}^t], [\mathbb{S}^d] = [\mathbb{S}^{td}], [\mathbb{S}^{dt}] = [\mathbb{S}^{tdt}]\}$ (see Figure 3).

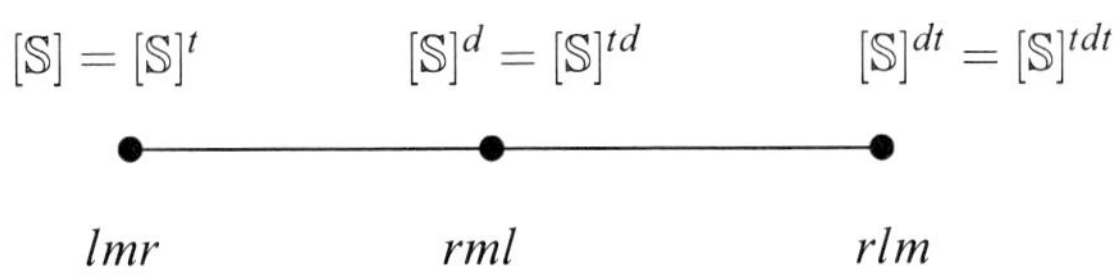

Figure 3. The Knuth orbit $\mathcal{K}(\mathbb{S})$ of a symplectic semifield $\mathbb{S}$ with nuclei *lmr*

Using this connection, in [52], a large number of commutative semifields of even order are constructed starting from the symplectic *semifield scions of the Desarguesian spreads*. These spreads were introduced and investigated in [56]. There the study of symplectic semifield spreads in characteristic 2 having odd dimension over $\mathbb{F}_2$ was motivated by their connections with extremal $\mathbb{Z}_4$-linear codes and extremal line sets in Euclidean spaces (see [20]).

In odd characteristic, commutative pre–semifields are related to the notion of planar DO polynomial. A *Dembowski-Ostrom (DO) polynomial* $f \in \mathbb{F}_q[x]$ ($q = p^e$) is a polynomial of the shape

$$f(x) = \sum_{i,j=0}^{k} a_{ij} x^{p^i + p^j};$$

whereas a polynomial $f \in \mathbb{F}_q[x]$ is *planar* or *perfect nonlinear (PN for short)* if the difference polynomial $f(x+a) - f(x) - f(a)$ is a permutation polynomial for each $a \in \mathbb{F}_q^*$. If $f(x) \in \mathbb{F}_q[x]$, q odd, is a planar DO polynomial, then $\mathbb{S}_f = (\mathbb{F}_q, +, \circ)$ is a commutative pre–semifield with multiplication $\circ$ defined by $a \circ b = f(a+b) - f(a) - f(b)$. Conversely, if $\mathbb{S} = (\mathbb{F}_q, +, \circ)$ is a commutative pre–semifield of odd order, then the polynomial given by $f(x) = \frac{1}{2}(x \circ x)$ is a planar DO polynomial and $\mathbb{S} = \mathbb{S}_f$ (see [25], and [27]).

Perfect nonlinear functions are differentially 1-uniform functions and they are of special interest in differential cryptanalysis (see [12], [80]).

For the known examples of symplectic or commutative (pre)semifields, see semifields of type D, A, K, G, CG/TP, CM-DY, PW/BLP, KW/K, CHK, BH, ZKW, Bi and LMPT listed in Section 6..

5. Rank Two Commutative Semifields

In this section, we turn our attention to commutative semifields that are of rank at most two over their middle nucleus, which we will call *rank two commutative semifields* or *RTCS* for short. Note that with this definition, finite fields are examples of RTCS. These semifields deserve special attention because of their importance in Galois geometry. They are connected to many of the central objects in the field, such as flocks of a quadratic cone, translation generalized quadrangles, ovoids, eggs, ... see e.g. [8].

As seen in the previous section, commutative semifields are linked with symplectic semifields, and the study of RTCS is equivalent to the study of symplectic semifields that are of rank two over their left nucleus. Figure 4 diplays the six isotopism classes corresponding to a RTCS, consisting of two Knuth orbits (see [6] and [61] for more details).

Rewriting the example (2) from Dickson [31], we have the following construction of an RTCS. Let σ be an automorphism of $\mathbb{F}_q$, q odd, and define the following multiplication on $\mathbb{F}_q^2$:

$$(x,y) \circ (u,v) = (xv + yu, yv + m x^\sigma u^\sigma), \tag{13}$$

where m is a non-square in $\mathbb{F}_q$. Cohen and Ganley made significant progress in the investigation of RTCS. They put Dickson's construction in the following more general setting. Let $\mathbb{S}$ be an RTCS of order q^2 with middle nucleus $\mathbb{F}_q$, and let $\alpha \in \mathbb{S} \setminus \mathbb{F}_q$ be such that $\{1, \alpha\}$

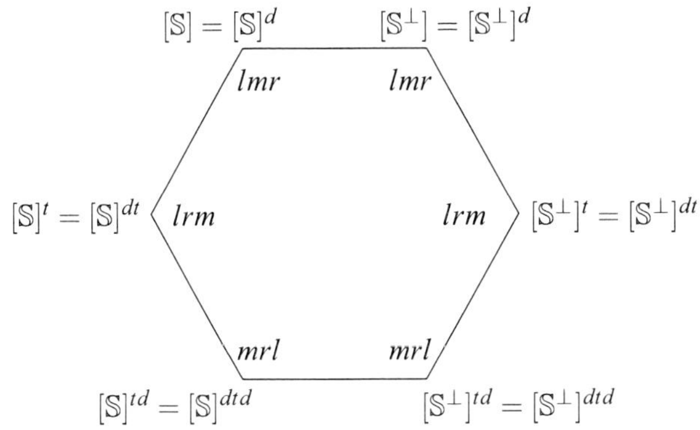

Figure 4. The isotopism classes $\mathcal{K}(\mathbb{S}) \cup \mathcal{K}(\mathbb{S}^\perp)$ corresponding to a RTCS $\mathbb{S}$ with nuclei *lmr*.

is a basis for $\mathbb{S}$. Addition in $\mathbb{S}$ is component-wise and multiplication is defined as

$$(x,y) \circ (u,v) = (xv + yu + g(xu), yv + f(xu)), \tag{14}$$

where f and g are additive functions from $\mathbb{F}_q$ to $\mathbb{F}_q$, such that $x\alpha^2 = g(x)\alpha + f(x)$. We denote this semifield by $\mathbb{S}(f,g)$. Verifying that this multiplication has no zero divisors leads to the following theorem which comes from [23].

Theorem 17. *Let $\mathbb{S}$ be a RTCS of order q^2 and characteristic p. Then there exist $\mathbb{F}_p$-linear functions f and g such that $\mathbb{S} = \mathbb{S}(f,g)$, with multiplication as in (14) and such that $zw^2 + g(z)w - f(z) = 0$ has no solutions for all w, $z \in \mathbb{F}_q$, and $z \neq 0$.*

For q even, Cohen and Ganley obtained the following remarkable theorem proving the non-existence of proper RTCS in even characteristic. To our knowledge, there is no obvious geometric reason why this should be so.

Theorem 18 ([23]). *For q even the only RTCS of order q^2 is the finite field $\mathbb{F}_{q^2}$.*

If q is odd, then the quadratic $zw^2 + g(z)w - f(z) = 0$ in w will have no solutions in $\mathbb{F}_q$ if and only if $g(z)^2 + 4zf(z)$ is a non-square for all $z \in \mathbb{F}_q^*$. In [23], Cohen and Ganley prove that in odd characteristic, in addition to the example with multiplication (13) by Dickson, there is just one other infinite family of proper RTCS, namely of order 3^{2r}, with multiplication given by:

$$(x,y) \circ (u,v) = (xv + yu + x^3u^3, yv + \eta x^9u^9 + \eta^{-1}xu), \tag{15}$$

with η a non-square in $\mathbb{F}_{3^r}$ ($r \geq 2$).

Theorem 19 ([23]). *Suppose that f and g are linear polynomials of degree less than q over $\mathbb{F}_q$, q odd, such that for infinitely many extensions $\mathbb{F}_{q^e}$ of $\mathbb{F}_q$, the functions*

$$f^* : \mathbb{F}_{q^e} \to \mathbb{F}_{q^e} \; : \; x \mapsto f(x), \text{ and}$$

$$g^* : \mathbb{F}_{q^e} \to \mathbb{F}_{q^e} \; : \; x \mapsto g(x),$$

define an RTCS $\mathbb{S}(f^*, g^*)$ of order q^{2e}. Then $\mathbb{S}(f,g)$ is a semifield with multiplication given by (13) or (15), or $\mathbb{S}(f,g)$ is a field.

The only other example of an RTCS was constructed from a translation ovoid of $Q(4,3^5)$, first found by computer in 1999 by Penttila and Williams ([84]). The associated semifield has order 3^{10} and multiplication

$$(x,y)\circ(u,v) = (xv+yu+x^{27}u^{27}, yv+x^9u^9). \tag{16}$$

Summarising, the only known examples of RTCS which are not fields are of *Dickson type* (13), of *Cohen-Ganley type* (15), or of *Penttila-Williams type* (16).

The existence of RTCS was further examined in [15] and [62] obtaining the following theorems which show that there is little room for further examples.

Theorem 20 ([62]). *Let $\mathbb{S}$ be an RTCS of order p^{2n}, p an odd prime. If $p > 2n^2-(4-2\sqrt{3})n+(3-2\sqrt{3})$, then $\mathbb{S}$ is either a field or a RTCS of Dickson type.*

Theorem 21 ([15]). *Let $\mathbb{S}$ be an RTCS of order q^{2n}, q an odd prime power, with center $\mathbb{F}_q$. If $q \geq 4n^2-8n+2$, then $\mathbb{S}$ is either a field or a RTCS of Dickson type.*

In combination with a computational result by Bloemen, Thas, and Van Maldeghem [13], the above implies a complete classification of RTCS of order q^6, with centre of order q.

Theorem 22 ([15]). *Let $\mathbb{S}$ be an RTCS of order q^6 with centre of order q, then either $\mathbb{S}$ is a field, or q is odd and $\mathbb{S}$ is of Dickson type.*

We end this section with the connections between RTCS and some interesting objects in Galois geometry.

5.1. Translation Generalized Quadrangles and Eggs

Let $\mathbb{S}(f,g)$ be an RTCS of order q^{2n} such that f and g are $\mathbb{F}_q$-linear, and for $(a,b)\in\mathbb{F}_{q^n}^2$ define

$$g_t(a,b) := a^2t+g(t)ab-f(t)b^2. \tag{17}$$

Then the set $\mathcal{E}(f,g) := \{E(a,b) \ : \ a,b\in\mathbb{F}_{q^n}\}\cup\{E(\infty)\}$, with

$$E(a,b) := \{\langle(t,-g_t(a,b),-2at-bg(t),ag(t)-2bf(t))\rangle \ : \ t\in\mathbb{F}_{q^n}^*\} \tag{18}$$

$$\text{and } E(\infty) := \{\langle(0,t,0,0)\rangle \ : \ t\in\mathbb{F}_{q^n}^*\}, \tag{19}$$

is a set of $q^{2n}+1$ $(n-1)$-dimensional subspaces of $\mathrm{PG}(4n-1,q)$ satisfying the following properties:

(E1) each three different elements of $\mathcal{E}(f,g)$ span a $(3n-1)$-dimensional subspace of $\mathrm{PG}(4n-1,q)$;

(E2) each element of $\mathcal{E}(f,g)$ is contained in a $(3n-1)$-dimensional subspace of $\mathrm{PG}(4n-1,q)$ that is disjoint from the other elements of $\mathcal{E}(f,g)$.

Such a set of $q^{2n}+1$ $(n-1)$-dimensional subspaces in $\mathrm{PG}(4n-1,q)$, satisfying (E1) and (E2) is called a *pseudo-ovoid, generalized ovoid*, or *egg* of $\mathrm{PG}(4n-1,q)$. These notions can be defined in more generality and were first studied in [89]. A more recent reference containing the general definition is [59]. Analogously to the relationship between planar spreads and translation planes, there is a one-to-one correspondence between eggs and translation generalized quadrangles (TGQ) (see [83]). It is far beyond the scope of this article to give a complete overview of the theory of eggs and TGQ here, and we refer the reader to [59], [83], or [93] for more details. However, we do want to mention the remarkable fact that all known examples of eggs (and hence of TGQ) are either obtained by field reduction from an ovoid or an oval, or they arise from an RTCS, i.e., they correspond to an egg $\mathcal{E}(f,g)$ (or its dual) constructed from an RTCS $\mathbb{S}(f,g)$ as above (see [59, Section 3.8] for more details).

5.2. Semifield Flocks and Translation Ovoids

A *flock of a quadratic cone* $\mathcal{K}$ of $\mathrm{PG}(3,q)$ with vertex v is a partition of $\mathcal{K}\setminus\{v\}$ into irreducible conics. The planes containing the conics of the flock are called the *planes of the flock*. In [90], Thas shows that a flock of a quadratic cone coexists with a set of upper triangular two by two matrices (sometimes called a *q-clan*) for which the difference of any two matrices is anisotropic, i.e. $v(A-B)v^t=0$ implies $v=0$ for $A\neq B$. Previous work, by Kantor [51] and Payne [81] [82], shows that such a set of two by two matrices gives rise to a generalized quadrangle of order (q^2,q).

If $\mathcal{K}$ is the quadratic cone in $\mathrm{PG}(3,q^n)$, q odd, with vertex $v=(0,0,0,1)$ and base the conic C with equation $X_0X_1-X_2^2=0$ in the plane $X_3=0$, then the planes of a flock of $\mathcal{K}$ may be written as

$$\pi_t : tX_0-f(t)X_1+g(t)X_2+X_3=0,\ t\in\mathbb{F}_{q^n}, \tag{20}$$

for some $f,g:\mathbb{F}_{q^n}\to\mathbb{F}_{q^n}$. We denote this flock by $\mathcal{F}(f,g)$. The associated set of two by two matrices consists of the matrices

$$\begin{pmatrix} t & g(t)\\ 0 & -f(t)\end{pmatrix},\ t\in\mathbb{F}_{q^n}. \tag{21}$$

If f and g are linear over a subfield $\mathbb{F}_q$ of $\mathbb{F}_{q^n}$, then $\mathcal{F}(f,g)$ is called a *semifield flock*. Using Theorem 17 and the above, one may conclude that $\mathcal{F}(f,g)$ is a semifield flock if and only if $\mathbb{S}(f,g)$ is an RTCS.

Another well studied object in Galois geometry connected to RTCS are translation ovoids of the generalized quadrangle $Q(4,q)$, consisting of points and lines that are contained in the projective algebraic variety $V(X_0X_1-X_2^2+X_3X_4)$ in $\mathrm{PG}(4,q)$. An *ovoid* of $Q(4,q)$ is a set Ω of points such that each line of $Q(4,q)$ contains exactly one point of Ω. An ovoid Ω is called a *translation ovoid* if there exists a group H of automorphisms of $Q(4,q)$, fixing Ω, a point $x\in\Omega$ and every line through x, acting transitively on the points of Ω not collinear with x. The correspondence between semifield flocks and translation ovoids of $Q(4,q)$ was first explained by Thas in [91], and later by Lunardon [68] with more details. The explicit calculations of what follows can be found in [60, Section 3]. Let $\mathbb{S}(f,g)$ be an RTCS of order q^{2n}, with f and g $\mathbb{F}_q$-linear, i.e. there exist $b_i,c_i\in\mathbb{F}_{q^n}$ such that $g(t)=\sum b_it^{q^i}$, and $f(t)=\sum c_it^{q^i}$. The corresponding ovoid $\Omega(f,g)$ of $Q(4,q^n)$ is then given by the set of points

$$\{(u,F(u,v),v,1,v^2-uF(u,v))\ :\ (u,v)\in\mathbb{F}_{q^n}^2\}\cup\{(0,0,0,0,1)\},$$

with

$$F(u,v) = \sum_{i=0}^{n-1} (c_i u + b_i v)^{1/q^i}.$$

6. Known Examples and Classification Results

In this section, we list the known finite non-associative semifields and some of the known classification results.

In the sequel, p and q will denote a prime and a prime power, respectively. Also, we will say that a semifield $\mathbb{S}'$ is a *Knuth derivative* of a semifield $\mathbb{S}$ if the isotopism class $[\mathbb{S}']$ of $\mathbb{S}'$ belongs to the Knuth orbit of $\mathbb{S}$. Recall that, by Theorem 16, a pre-semifield $\mathbb{S}$ is isotopic to a commutative semifield if and only if its Knuth derivative $\mathbb{S}^{td}$ is symplectic.

1. **(D)** *Dickson commutative semifields* of order p^{2e} with p odd and $e > 1$ [32].
2. **(HK)** *Hughes–Kleinfeld semifields* of order p^{2e} with $e > 1$ [43].
3. **(A)** *Albert Generalized twisted fields* of order q^n with center of order q ($q > 2$ and $n > 2$) [2]. For q odd some Generalized twisted fields are symplectic [4] and their Knuth commutative derivatives are Generalized twisted fields as well. Indeed, the family of the Generalized twisted fields is closed under the Knuth operations (see [52]).
4. **(S)** *Sandler semifields* of order q^{mn} with center of order q and $1 < n \leq m$ [87].
5. **(K)** In [58], Knuth generalizes the Dickson commutative semifields ([58, (7.16)] *Generalized Dickson semifields*) and constructs four types of semifields: *families I, II, III, and IV* of order p^{2e}, with $e > 1$ [58, (7.17)]; semifields of type II are Hughes-Kleinfeld semifields and families II, III and IV belong to the same Knuth orbit (see [9]). Some Generalized Dickson semifields are symplectic semifields (see [50]) and their commutative Knuth derivatives are Dickson semifields (see [52]). In the same paper Knuth also provides a family of commutative semifields of order 2^{mn}, n odd and $mn > 3$: the *Knuth binary semifields* [58, (6.10)].
6. **(G)** *Ganley commutative semifields* [38] of order 3^{2r}, with $r \geq 3$ odd, and their symplectic Knuth derivatives [52, (5.14)].
7. **(CG/TP)** *Cohen–Ganley commutative semifields* of order 3^{2r}, $r \geq 2$ [23, Example 3], their symplectic derivatives (*Thas–Payne symplectic semifields*) and the corresponding semifields associated with a flock [92].
8. **(BL)** *Boerner-Lantz semifield* of order 81 [16].
9. **(JJ)** *Jha–Johnson cyclic semifields of type* (q,m,n), of order q^l where $l = \mathrm{lcm}(n,m)$, $m,n > 1$ and $l > \max\{m,n\}$ [45, Theorem 2]. Jha–Johnson cyclic semifields generalize the Sandler semifields.
10. **(HJ)** *Huang–Johnson semifields*: 7 non-isotopic semifields of order 8^2 (classes II, III, ...,VIII) [42].
11. **(CM − DY)** *Coulter–Matthews/Ding–Yuang commutative pre–semifields* of order 3^n, $n > 1$ odd [27], [33], and their symplectic Knuth derivatives (see [52]).
12. **(PW/BLP)** *Penttila–Williams symplectic semifield* of order 3^{10} [84], its commutative Knuth derivative and the related *Bader-Lunardon-Pinneri semifield* associated with a flock [5].

13. **(KW/K)** *Kantor–Williams symplectic pre-semifields* of order q^m, for q even and $m > 1$ odd [56], and their commutative Knuth derivatives (*Kantor commutative pre-semifields*) [52, (4.2)]. Kantor commutative pre-semifields generalize the Knuth binary semifields.
14. **(CHK)** *Coulter–Henderson–Kosick commutative pre-semifield* of order 3^8 [26].
15. **(CF)** *Cordero–Figueroa semifield* of order 3^6 [47, 37.10].
16. **(De)** *Dempwolff semifields* of order 3^4 [30]. The author in [30] completes the classification of semifields of order 81 and determines 4 Knuth orbits of semifields not previously known (classes I, II, III and V). He also discusses the embedding of semifields of type III and V in an infinite series.
17. **(BH)** *Budaghyan–Helleseth commutative pre-semifields* $\mathbb{B}_{s,k}$ of order p^{2k}, p odd, constructed from PN DO–polynomials of type (i*) with s and k integers such that $0 < s < 2k$, $\gcd(p^s+1, p^k+1) \neq \gcd(p^s+1,(p^k+1)/2)$ and $\gcd(k+s,2k) = \gcd(k+s,k)$; and of type (i**) with s and k integers such that $0 < s < 2k$ and $\gcd(k+s,2k) = \gcd(k+s,k)$ [19].
18. **(MPT)** *Marino–Polverino–Trombetti semifields*: 4 non-isotopic semifields of order 2^{14} [76, Theorem 5.3].
19. **(JMPT)** *Johnson–Marino–Polverino–Trombetti semifields* of order q^{2n} with $n > 1$ odd [49, Theorem 1]. JMPT semifields generalize the Jha–Johnson cyclic semifields of type $(q,2,n)$, n odd. Also, the Huang–Johnson semifield of class VI belongs to this family.
20. **(JMPT($4^5, 16^5$))** *Johnson–Marino–Polverino–Trombetti non-cyclic semifields* of order 4^5 and order 16^5 [49, Theorem 7].
21. **(ZKW)** *Zha–Kyureghyan–Wang commutative pre-semifields* $\mathbb{Z}_{s,k}$ of order p^{3k}, p odd where s and k are integers such that $\gcd(3,k) = 1$, $0 < s < 3k$, $k \equiv s (\mathrm{mod}\ 3)$, $k \neq s$ and $\frac{3k}{\gcd(s,3k)}$ odd, constructed in [95] from PN DO–polynomials.
22. **(EMPT2)** *Ebert–Marino–Polverino–Trombetti semifields* of order q^6 for q odd [36, Theorems 2.7, 2.8].
23. **(EMPT1)** *Ebert–Marino–Polverino–Trombetti semifields* of order q^{2n} with either $n \geq 3$ odd, or $n > 2$ even and q odd [35, Theorem 1.1]. The Huang–Johnson semifields of type VII and VIII belong to this family.
24. **(MT)** *Marino–Trombetti semifield* of order 2^{10} [77].
25. **(Bi)** *Bierbrauer commutative pre-semifields* from PN DO–polynomials [11] and [10].
26. **(RCR)** *Rúa–Combarro–Ranilla semifields* of order 2^6 [86]. The authors in [86] classify all semifields of order 64 and determine 67 Knuth orbits of semifields with 64 elements not previously known.
27. **(LaMPT)** *Lavrauw–Marino–Polverino–Trombetti rank two scattered semifields* of order q^6 for q odd, $q \equiv 1 (\mathrm{mod}\ 3)$ and for $q = 2^{2h}$, $h \equiv 1 (\mathrm{mod}\ 3)$ from [66]. These semifields belong to family $\mathcal{F}_5$.
28. **(LuMPT)** *Lunardon–Marino–Polverino–Trombetti* symplectic semifields of order q^6 for q odd, and their commutative Knuth derivatives [72].

Apart from the Knuth cubical array (see Section 1), the translation dual construction and the BEL geometric model (see Section 2), some other "construction processes" are known to produce semifields starting from a given one: the *lifting construction* (or *HMO construction*) and the *symplectic dual construction*.

The lifting construction produces semifields of order q^4 with left nucleus of order q^2 starting from rank two semifields of order q^2. Note that this process may be iterated producing semifields of order q^{2^i} for any integer $i \geq 2$. Also, the lifting construction is not closed under the isotopy relation, indeed isotopic semifields can produce non-isotopic lifted semifields. This construction method has been introduced by Hiramine, Matsumoto and Oyama in [40], for q odd, and then generalized by Johnson in [46] for any value of q. Semifields lifted from a field are completely determined (see [16], [24] and [21]). For further details on lifting see e.g. [47, Chapter 93] and [54].

The symplectic dual construction has been recently introduced in [72] and produces a symplectic semifield of order q^3 (q odd) with left nucleus containing $\mathbb{F}_q$ starting from a symplectic semifield $\mathbb{S}$ with the same data. As the translation dual construction, the symplectic dual construction is an involutary operation, (i.e., if $\mathbb{S}^\tau$ denotes the symplectic dual of the semifield $\mathbb{S}$, then $(\mathbb{S}^\tau)^\tau = \mathbb{S}$). Indeed the symplectic dual of a semifield is obtained by dualizing the associated linear set with respect to a suitable polarity.

6.1. Classification Results for Any q

We have already seen that all two-dimensional finite semifields are fields. In 1977, G. Menichetti classified all three-dimensional finite semifields proving the following result.

Theorem 23 ([78]). *A semifield of order q^3 with center containing $\mathbb{F}_q$ either is a field or is isotopic to a Generalized twisted field.*

Later on Menichetti generalized the previous result proving the following theorem.

Theorem 24 ([79]). *Let $\mathbb{S}$ be a semifield of prime dimension n over the center $\mathbb{F}_q$. Then there exists an integer $\nu(n)$ depending only on n, such that if $q > \nu(n)$ then $\mathbb{S}$ is isotopic to a Generalized twisted field.*

As a corollary we have that *a semifield of order p^3 is a field or a Generalized twisted field* and that *a semifield of order p^n, n prime, if p is "large enough", is a field or a Generalized twisted field.*

All the other classification results for semifields of given order involve conditions on one or more of their nuclei. In fact, all of them deal with rank two semifields.

The first result in this direction is the following theorem that can be found in [43] (case (a)) and in [58, Theorem 7.4.1].

Theorem 25. *Let $\mathbb{S}$ be a semifield which is not a field and which is a 2-dimensional vector space over a finite field $\mathbb{F}$. Then*
(a) $\mathbb{F} = \mathbb{N}_r = \mathbb{N}_m$ *if and only if $\mathbb{S}$ is a Knuth semifield of type II.*
(b) $\mathbb{F} = \mathbb{N}_l = \mathbb{N}_m$ *if and only if $\mathbb{S}$ is a Knuth semifield of type III.*
(c) $\mathbb{F} = \mathbb{N}_l = \mathbb{N}_r$ *if and only if $\mathbb{S}$ is a Knuth semifield of type IV.*

More recently, using the geometric approach of the linear sets the following results for rank two semifields of order q^4 and q^6 have been obtained in [21] and [49], respectively.

Theorem 26 ([21]). *A semifield $\mathbb{S}$ of order q^4 with left nucleus $\mathbb{F}_{q^2}$ and center $\mathbb{F}_q$ is isotopic to one of the following semifields: Generalized Dickson/Knuth semifields (q odd), Hughes-Kleinfeld semifields, semifields lifted from Desarguesian planes or Generalized twisted fields.*

Theorem 27 ([49]). *Each semifield $\mathbb{S}$ of order q^6, with left nucleus of order q^3 and middle and right nuclei of order q^2 and center of order q is isotopic to a JMPT semifield, precisely $\mathbb{S}$ is isotopic to a semifield $(\mathbb{F}_{q^6}, +, \circ)$ with multiplication given by*

$$x \circ y = (\alpha + \beta u)x + b\gamma x^{q^3}, \quad \textit{where } y = \alpha + \beta u + \gamma b \ (\alpha, \beta, \gamma \in \mathbb{F}_{q^2}),$$

with u a fixed element of $\mathbb{F}_{q^3} \setminus \mathbb{F}_q$ and b an element of $\mathbb{F}_{q^6}$ such that $b^{q^3+1} = u$.

Finally, for classification results concerning rank two commutative semifields (RTCS) we refer to Section 5..

6.2. Classification Results for Small Values of q

All semifields of order $q \leq 125$ are classified. By [58, Theorem 6.1], a non-associative semifield has order p^n, where $n \geq 3$ and $p^n \geq 16$, and by Menichetti's classification result (Theorem 23), semifields of order 27 and 125 are fields or Generalized twisted fields.

Semifields of order 16 and order 32 have been classified in the sixties; those of order 16 form three isotopism classes (see [57]) and those of order 32 form six isotopism classes (see [94]).

Recently in [30] with the aid of the computer algebra software package GAP, Dempwolff has completed the classification of semifields of order 81 proving that there are 27 non-isotopic semifields with 81 elements, partitioned into 12 Knuth orbits.

Finally, in [86], Rúa, Combarro and Ranilla have obtained a computer assisted classification of all semifields of order 64. They have determined 332 non-isotopic semifields with 64 elements, partitioned into 80 Knuth orbits.

7. Open Problems

We conclude this overview of finite semifields with some open problems, at least one problem from every section.

In this article we have encountered a number of different invariants of the isotopism classes of finite semifields, such as the size of semifield, and the size of its nuclei, or the characteristic. Of course, since the isotopism classes for semifields correspond to the isomorphism classes of the corresponding semifield planes, each invariant of the isomorphism classes of projective planes (e.g. the *fingerprint*, *Kennzahl*, *Leitzahl* defined in [22] and [29]) serves as an invariant of the isotopism class of semifields. However, these invariants can sometimes only be computed for semifields of small order, and it often remains very difficult to determine whether a semifield is "new" or not, where "new" means not isotopic to a semifield that was already known before. Moreover, these invariants are perhaps too general, as they apply to general translation planes and not just to semifield planes. As we saw in this article, the geometric approach can sometimes be used in order to distinguishing between isotopism classes of semifield, but there is still no guarantee that different isotopism classes are represented by linear sets that are distinguishable by their geometric properties. This leads us to the following problem.

Problem 1 (Section 1.) Find new invariants of isotopism classes of finite semifields, or even better: find a unique representative for each isotopism class.

The following two problems are related to the geometric construction for semifields (from [7]) explained in Section 2..

Problem 2 (Section 2.) Find examples of semifields $\mathbb{S}$ that are not 2-dimensional over their left nucleus, having $r = 2$ (r is the integer in the BEL-construction), and such that the semifield $\mathbb{S}^{\kappa}$ is new.

Problem 3 (Section 2.) Does the operation κ that interchanges U and W extend to an operation on the isotopism classes, and if so, how many isotopism classes of semifields does this operation produce in conjunction with the Knuth orbit?

The following problem is also related to the Knuth orbit. As pointed out in Section 3., all known examples of rank two semifields $\mathbb{S}$ for which $\mathbb{S}^{\perp}$ is not isotopic to $\mathbb{S}$ have the property that the size of $\mathcal{K}(\mathbb{S}) \cup \mathcal{K}(\mathbb{S}^{\perp})$ is six.

Problem 4 (Section 3.) Find examples of rank two semifields $\mathbb{S}$ for which the set of isotopism classes $\mathcal{K}(\mathbb{S}) \cup \mathcal{K}(\mathbb{S}^{\perp})$ has size twelve.

Theorem 15 gives a characterisation of symplectic semifields, which in combination with Theorem 16 gives an indirect characterisation of commutative semifields. Can we find a more direct characterisation without using the S_3-action?

Problem 5 (Section 4.) Find a geometric characterisation of linear sets associated with a commutative semifield without using Theorem 16.

A longstanding open problem is the classification of RTCS. This would have many interesting corollaries in Galois geometry, for instance in the theory of semifield flocks, translation ovoids, eggs and translation generalized quadrangles.

Problem 6 (Section 5.) Improve on the bounds from [15] and [62], or classify RTCS up to isotopism.

In Section 6., we have listed many examples of finite semifields. Some are contained in infinite families, others are standalone examples. Here is a list of examples that might be embeddable in an infinite family.

Problem 7 (Section 6.) Find infinite families (if they exist) of semifields containing the sporadic examples listed in Section 6 (BL, HJ, PW/BLP, CHK, CF, De, MPT, JMPT($4^5, 16^5$), MT, RCR).

During the last decade a lot of data has been produced including a lot of infinite families of finite semifields. In order to make any progress in the classification of finite semifields, it is important to have strong characterisations for the known families.

Problem 8 (Section 6.) Find characterisations of known families of semifields.

Another classification problem for which progress has already been made concerns rank two semifields of order q^6 that are 6-dimensional over their center (see Theorem 27).

Problem 9 (Section 6.) Complete the classification of semifields of order q^6, 2-dimensional over the left nucleus and 6-dimensional over the center.

Acknowledg ments

The second author acknowledges the support of the Research Project of MIUR (Italian Office for University and Research) *Geometrie su campi di Galois, piani di traslazione e geometrie d'incidenza.*

References

[1] A. A. Albert, *Finite division algebras and finite planes*, in Proc. Sympos. Appl. Math., Vol. 10, American Mathematical Society, Providence, R.I., 1960, pp. 53–70.

[2] ——, *Generalized twisted fields*, Pacific J. Math., 11 (1961), pp. 1–8.

[3] J. André, *Über nicht-Desarguessche Ebenen mit transitiver Translationsgruppe*, Math. Z., 60 (1954), pp. 156–186.

[4] L. Bader, W. M. Kantor, and G. Lunardon, *Symplectic spreads from twisted fields*, Boll. Un. Mat. Ital. A (7), 8 (1994), pp. 383–389.

[5] L. Bader, G. Lunardon, and I. Pinneri, *A new semifield flock*, J. Combin. Theory Ser. A, 86 (1999), pp. 49–62.

[6] S. Ball and M. R. Brown, *The six semifield planes associated with a semifield flock*, Adv. Math., 189 (2004), pp. 68–87.

[7] S. Ball, G. L. Ebert, and M. Lavrauw, *A geometric construction of finite semifields*, J. Algebra, 311 (2007), pp. 117–129.

[8] S. Ball and M. Lavrauw, *Commutative semifields of rank 2 over their middle nucleus*, in Finite fields with applications to coding theory, cryptography and related areas (Oaxaca, 2001), Springer, Berlin, 2002, pp. 1–21.

[9] ——, *On the Hughes-Kleinfeld and Knuth's semifields two-dimensional over a weak nucleus*, Des. Codes Cryptogr., 44 (2007), pp. 63–67.

[10] J. Bierbrauer, *New commutative semifields and their nuclei.* manuscript, 2010.

[11] ——, *New semifields, PN and APN functions*, Des. Codes Cryptogr., 54 (2010), pp. 189–200.

[12] E. Biham and A. Shamir, *Differential cryptanalysis of DES-like cryptosystems*, J. Cryptology, 4 (1991), pp. 3–72.

[13] I. Bloemen, J. A. Thas, and H. Van Maldeghem, *Translation ovoids of generalized quadrangles and hexagons*, Geom. Dedicata, 72 (1998), pp. 19–62.

[14] A. Blokhuis and M. Lavrauw, *Scattered spaces with respect to a spread in* $\mathrm{PG}(n,q)$, Geom. Dedicata, 81 (2000), pp. 231–243.

[15] A. Blokhuis, M. Lavrauw, and S. Ball, *On the classification of semifield flocks*, Adv. Math., 180 (2003), pp. 104–111.

[16] V. Boerner-Lantz, *A class of semifields of order* q^4, J. Geom., 27 (1986), pp. 112–118.

[17] R. H. Bruck and R. C. Bose, *The construction of translation planes from projective spaces*, J. Algebra, 1 (1964), pp. 85–102.

[18] ——, *Linear representations of projective planes in projective spaces*, J. Algebra, 4 (1966), pp. 117–172.

[19] L. Budaghyan and T. Helleseth, *New perfect nonlinear multinomials over* $\mathbb{F}_{p^{2k}}$ *for any odd prime p.* Golomb, Solomon W. (ed.) et al., Sequences and their applications – SETA 2008. 5th international conference, Lexington, KY, USA, September 14–18, 2008 Proceedings. Springer. Lect. Notes Comput. Sci. 5203, 403-414 (2008).

[20] R. Calderbank, P. J. Cameron, W. M. Kantor, and J. J. Seidel, *Z_4-Kerdock codes, orthogonal spreads, and extremal Euclidean line-sets*, Proc. London Math. Soc. (3), 75 (1997), pp. 436–480.

[21] I. Cardinali, O. Polverino, and R. Trombetti, *Semifield planes of order q^4 with kernel F_{q^2} and center F_q*, European J. Combin., 27 (2006), pp. 940–961.

[22] C. Charnes, *Quadratic matrices and the translation planes of order* 5^2, in Coding theory, design theory, group theory (Burlington, VT, 1990), Wiley-Intersci. Publ., Wiley, New York, 1993, pp. 155–161.

[23] S. D. Cohen and M. J. Ganley, *Commutative semifields, two-dimensional over their middle nuclei*, J. Algebra, 75 (1982), pp. 373–385.

[24] M. Cordero and R. Figueroa, *On some new classes of semifield planes*, Osaka J. Math., 30 (1993), pp. 171–178.

[25] R. S. Coulter and M. Henderson, *Commutative presemifields and semifields*, Adv. Math., 217 (2008), pp. 282–304.

[26] R. S. Coulter, M. Henderson, and P. Kosick, *Planar polynomials for commutative semifields with specified nuclei*, Des. Codes Cryptogr., 44 (2007), pp. 275–286.

[27] R. S. Coulter and R. W. Matthews, *Planar functions and planes of Lenz-Barlotti class II*, Des. Codes Cryptogr., 10 (1997), pp. 167–184.

[28] P. Dembowski, *Finite geometries*, Ergebnisse der Mathematik und ihrer Grenzgebiete, Band 44, Springer-Verlag, Berlin, 1968.

[29] U. Dempwolff, *Translation planes of order* 27, Des. Codes Cryptogr., 4 (1994), pp. 105–121.

[30] ——, *Semifield planes of order 81*, J. Geom., 89 (2008), pp. 1–16.

[31] L. E. Dickson, *On commutative linear algebras in which division is always uniquely possible*, Trans. Amer. Math. Soc., 7 (1906), pp. 514–522.

[32] ——, *Linear algebras with associativity not assumed*, Duke Math. J., 1 (1935), pp. 113–125.

[33] C. Ding and J. Yuan, *A family of skew Hadamard difference sets*, J. Combin. Theory Ser. A, 113 (2006), pp. 1526–1535.

[34] G. L. Ebert, G. Marino, O. Polverino, and R. Trombetti, *On the multiplication of some semifields of order* q^6, Finite Fields Appl., 15 (2009), pp. 160–173.

[35] ——, *Infinite families of new semifields*, Combinatorica, 29 (2009), pp. 637–663.

[36] G. L. Ebert, O. Polverino, G. Marino, and R. Trombetti, *Semifields in class* $\mathcal{F}_4^{(a)}$, Electron. J. Combin., 16 (2009). Research Paper 53, 20 pp.

[37] J. W. Freeman, *Reguli and pseudoreguli in* $\mathrm{PG}(3, s^2)$, Geom. Dedicata, 9 (1980), pp. 267–280.

[38] M. J. Ganley, *Central weak nucleus semifields*, European J. Combin., 2 (1981), pp. 339–347.

[39] M. Hall, *Projective planes*, Trans. Amer. Math. Soc., 54 (1943), pp. 229–277.

[40] Y. Hiramine, M. Matsumoto, and T. Oyama, *On some extension of* 1*-spread sets*, Osaka J. Math., 24 (1987), pp. 123–137.

[41] J. W. P. Hirschfeld and J. A. Thas, *General Galois geometries*, Oxford Mathematical Monographs, The Clarendon Press Oxford University Press, New York, 1991. Oxford Science Publications.

[42] H. Huang and N. L. Johnson, 8 *semifield planes of order* 8^2, Discrete Math., 80 (1990), pp. 69–79.

[43] D. R. Hughes and E. Kleinfeld, *Seminuclear extensions of Galois fields*, Amer. J. Math., 82 (1960), pp. 389–392.

[44] D. R. Hughes and F. C. Piper, *Projective planes*, Springer-Verlag, New York, 1973. Graduate Texts in Mathematics, Vol. 6.

[45] V. Jha and N. L. Johnson, *An analog of the Albert-Knuth theorem on the orders of finite semifields, and a complete solution to Cofman's subplane problem*, Algebras Groups Geom., 6 (1989), pp. 1–35.

[46] N. L. Johnson, *Sequences of derivable translation planes*, Osaka J. Math., 25 (1988), pp. 519–530.

[47] N. L. Johnson, V. Jha, and M. Biliotti, *Handbook of finite translation planes*, vol. 289 of Pure and Applied Mathematics (Boca Raton), Chapman & Hall/CRC, Boca Raton, FL, 2007.

[48] N. L. Johnson, G. Marino, O. Polverino, and R. Trombetti, *Semifields of order q^6 with left nucleus $\mathbb{F}_{q^3}$ and center $\mathbb{F}_q$*, Finite Fields Appl., 14 (2008), pp. 456–469.

[49] ——, *On a generalization of cyclic semifields*, J. Algebraic Combin., 29 (2009), pp. 1–34.

[50] W. M. Kantor, *Ovoids and translation planes*, Canad. J. Math., 34 (1982), pp. 1195–1207.

[51] ——, *Some generalized quadrangles with parameters q^2, q*, Math. Z., 192 (1986), pp. 45–50.

[52] ——, *Commutative semifields and symplectic spreads*, J. Algebra, 270 (2003), pp. 96–114.

[53] ——, *Isomorphisms of symplectic planes*, Adv. Geom., 7 (2007), pp. 553–557.

[54] ——, *HMO-planes*, Adv. Geom., 9 (2009), pp. 31–43.

[55] W. M. Kantor and R. A. Liebler, *Semifields arising from irreducible semilinear transformations*, J. Aust. Math. Soc., 85 (2008), pp. 333–339.

[56] W. M. Kantor and M. E. Williams, *Symplectic semifield planes and $\mathbb{Z}_4$-linear codes*, Trans. Amer. Math. Soc., 356 (2004), pp. 895–938 (electronic).

[57] E. Kleinfeld, *Techniques for enumerating Veblen-Wedderburn systems*, J. Assoc. Comput. Mach., 7 (1960), pp. 330–337.

[58] D. E. Knuth, *Finite semifields and projective planes*, J. Algebra, 2 (1965), pp. 182–217.

[59] M. Lavrauw, *Scattered spaces with respect to spreads, and eggs in finite projective spaces*, Eindhoven University of Technology, Eindhoven, 2001. Dissertation, Technische Universiteit Eindhoven, Eindhoven, 2001.

[60] ——, *Semifield flocks, eggs, and ovoids of $Q(4,q)$*, Adv. Geom., 5 (2005), pp. 333–345.

[61] ——, *The two sets of three semifields associated with a semifield flock*, Innov. Incidence Geom., 2 (2005), pp. 101–107.

[62] ——, *Sublines of prime order contained in the set of internal points of a conic*, Des. Codes Cryptogr., 38 (2006), pp. 113–123.

[63] ——, *On the isotopism classes of finite semifields*, Finite Fields Appl., 14 (2008), pp. 897–910.

[64] ——, *Finite semifields with a large nucleus and higher secant varieties to segre varieties*, Adv. Geom., to appear, (2010).

[65] M. Lavrauw and V. de Voorde G., *On linear sets on a projective line*, Des. Codes Cryptogr., 56 (2010), pp. 89–104.

[66] M. Lavrauw, G. Marino, O. Polverino, and R. Trombetti, *$\mathbb{F}_q$-pseudoreguli of $PG(3,q^3)$ and scattered semifields of order q^6.* Finite Fields Appl. (2010), doi:10.1016/j.ffa.2010.12.001

[67] R. Lidl and H. Niederreiter, *Finite fields*, vol. 20 of Encyclopedia of Mathematics and its Applications, Addison-Wesley Publishing Company Advanced Book Program, Reading, MA, 1983. (Now distributed by Cambridge University Press).

[68] G. Lunardon, *Flocks, ovoids of* $\mathrm{Q}(4,q)$ *and designs*, Geom. Dedicata, 66 (1997), pp. 163–173.

[69] ——, *Translation ovoids*, J. Geom., 76 (2003), pp. 200–215. Combinatorics, 2002 (Maratea).

[70] ——, *Symplectic spreads and finite semifields*, Des. Codes Cryptogr., 44 (2007), pp. 39–48.

[71] G. Lunardon, G. Marino, O. Polverino, and R. Trombetti, *Translation dual of a semifield*, J. Combin. Theory Ser. A, 115 (2008), pp. 1321–1332.

[72] ——, *Symplectic spreads and quadric veronesean.* submitted, 2010.

[73] H. Lüneburg, *Translation planes*, Springer-Verlag, Berlin, 1980.

[74] D. M. Maduram, *Transposed translation planes*, Proc. Amer. Math. Soc., 53 (1975), pp. 265–270.

[75] G. Marino, O. Polverino, and R. Trombetti, *On $\mathbb{F}_q$-linear sets of* $\mathrm{PG}(3,q^3)$ *and semifields*, J. Combin. Theory Ser. A, 114 (2007), pp. 769–788.

[76] ——, *On semifields of type (q^{2n},q^n,q^2,q^2,q), n odd*, Innov. Incidence Geom., 6/7 (2007/08), pp. 271–289.

[77] G. Marino and R. Trombetti, *A new semifield of order* 2^{10}, Discrete Math., 310 (2010), pp. 3108–3113.

[78] G. Menichetti, *On a Kaplansky conjecture concerning three-dimensional division algebras over a finite field*, J. Algebra, 47 (1977), pp. 400–410.

[79] ——, *n-dimensional algebras over a field with a cyclic extension of degree n*, Geom. Dedicata, 63 (1996), pp. 69–94.

[80] K. Nyberg, *Differentially uniform mappings for cryptography.* Helleseth, Tor (ed.), Advances in cryptology - EUROCRYPT '93. Workshop on the theory and application of cryptographic techniques, Lofthus, Norway, May 23-27, 1993. Proceedings. Berlin: Springer. Lect. Notes Comput. Sci. 765, 55-64 (1994).

[81] S. E. Payne, *Generalized quadrangles as group coset geometries*, in Proceedings of the Eleventh Southeastern Conference on Combinatorics, Graph Theory and Computing (Florida Atlantic Univ., Boca Raton, Fla., 1980), Vol. II, vol. 29, 1980, pp. 717–734.

[82] ——, *A new infinite family of generalized quadrangles*, in Proceedings of the sixteenth Southeastern international conference on combinatorics, graph theory and computing (Boca Raton, Fla., 1985), vol. 49, 1985, pp. 115–128.

[83] S. E. Payne and J. A. Thas, *Finite generalized quadrangles*, EMS Series of Lectures in Mathematics, European Mathematical Society (EMS), Zürich, second ed., 2009.

[84] T. Penttila and B. Williams, *Ovoids of parabolic spaces*, Geom. Dedicata, 82 (2000), pp. 1–19.

[85] O. Polverino, *Linear sets in finite projective spaces*, Discrete Math., 310 (2010), pp. 3096–3107.

[86] I. F. Rúa, E. F. Combarro, and J. Ranilla, *Classification of 64-element finite semifields*, J. Algebra, 322 (2009), pp. 4011–4029.

[87] R. Sandler, *Autotopism groups of some finite non-associative algebras*, Amer. J. Math., 84 (1962), pp. 239–264.

[88] B. Segre, *Teoria di Galois, fibrazioni proiettive e geometrie non desarguesiane*, Ann. Mat. Pura Appl. (4), 64 (1964), pp. 1–76.

[89] J. A. Thas, *The m-dimensional projective space* $S_m(M_n(\mathrm{GF}(q)))$ *over the total matrix algebra* $M_n(\mathrm{GF}(q))$ *of the* $n \times n$*-matrices with elements in the Galois field* $\mathrm{GF}(q)$, Rend. Mat. (6), 4 (1971), pp. 459–532.

[90] ——, *Generalized quadrangles and flocks of cones*, European J. Combin., 8 (1987), pp. 441–452.

[91] ——, *Generalized quadrangles of order* (s,s^2). *II*, J. Combin. Theory Ser. A, 79 (1997), pp. 223–254.

[92] J. A. Thas and S. E. Payne, *Spreads and ovoids in finite generalized quadrangles*, Geom. Dedicata, 52 (1994), pp. 227–253.

[93] J. A. Thas, K. Thas, and H. Van Maldeghem, *Translation generalized quadrangles*, vol. 26 of Series in Pure Mathematics, World Scientific Publishing Co. Pte. Ltd., Hackensack, NJ, 2006.

[94] R. J. Walker, *Determination of division algebras with* 32 *elements*, in Proc. Sympos. Appl. Math., Vol. XV, Amer. Math. Soc., Providence, R.I., 1963, pp. 83–85.

[95] Z. Zha, G. M. Kyureghyan, and X. Wang, *Perfect nonlinear binomials and their semifields*, Finite Fields Appl., 15 (2009), pp. 125–133.

In: Current Research Topics in Galois Geometry
Editors: J. De Beule and L. Storme
ISBN: 978-1-61209-523-3

Chapter 7

CODES OVER RINGS AND RING GEOMETRIES

Thomas Honold[1,*] ***and Ivan Landjev***[2,†]

[1] Zhejiang Provincial Key Laboratory of Information Network Technology *and* Department of Information and Electronic Engineering, Zhejiang University, 38 Zheda Road, 310027 Hangzhou, China

[2] New Bulgarian University, 21 Montevideo str., 1618 Sofia, Bulgaria, *and* Institute of Mathematics and Informatics, Bulgarian Academy of Sciences, 8 Acad. G. Bonchev str., 1113, Sofia, Bulgaria

Abstract

In this article, we bring together some recent results on special sets of points in coordinate projective geometries over finite chain rings. There is a clear coding theoretic relevance of these results due to the strong connection between multisets of points in the chain ring geometries and so-called fat linear codes over finite chain rings. In Section 1, we introduce axiomatically projective and affine Hjelmslev spaces. An important class of such spaces, obtained as coordinate geometries over finite chain rings, is given in Section 2. In Section 3, we define multisets of points in projective Hjelmslev geometries and fat linear codes over finite chain rings. Furthermore, we state a result saying that these are essentially one and the same object. In Sections 4 and 5, we survey the known results on arcs and blocking sets in projective Hjelmslev planes. We include tables of the sizes of the largest known arcs in projective Hjelmslev planes over some small chain rings.

Key Words: projective Hjelmslev geometry, projective Hjelmslev plane, finite chain ring, arcs, blocking sets, fat linear codes, Rédei type blocking sets, Witt vectors

AMS Subject Classification: 51C05, 51E26, 51E21, 51E22, 94B05, 94B27

1. Projective and Affine Hjelmslev Spaces

We start by introducing projective Hjelmslev spaces. The following axiomatic approach is due to Kreuzer [36–38,40]. Let $\Pi = (\mathcal{P}, \mathcal{L}, I)$, $I \subseteq \mathcal{P} \times \mathcal{L}$, be an incidence structure. The

*E-mail address: honold@zju.edu.cn
†E-mail address: ivan@math.bas.bg; i.landjev@nbu.bg

sets $\mathcal{P}$ and $\mathcal{L}$ are referred to as sets of points and lines, respectively. A *neighbour relation* $\frown$ is defined on $\mathcal{P}$ and $\mathcal{L}$ by the following conditions:

(N1) $\forall x, y \in \mathcal{P} : x \frown y \iff \exists S, T \in \mathcal{L}, S \neq T : \{(x,S),(x,T),(y,S),(y,T)\} \subseteq I$;

(N2) $\forall S, T \in \mathcal{L} : S \frown T \iff$ for every point x with $(x,S) \in I$ there is a point y with $(y,T) \in I$ and $x \frown y$, and, conversely, for every y with $(y,T) \in I$ there is a point x with $(x,S) \in I$ and $y \frown x$.

Given two points x, y with $x \not\frown y$ we denote by $\overline{x,y}$ the unique line incident with both of them if such a line does exist. For a point x and a line S, we write $x \frown S$ if there exists a point y with $(y,S) \in I, x \frown y$.

Definition 1.1. An incidence structure $\Pi = (\mathcal{P}, \mathcal{L}, I)$ with neighbour relation $\frown$ is said to be a *projective Hjelmslev space* if it satisfies the following axioms:

(H1) For any two points $x, y \in \mathcal{P}$ there exists a line S with $(x,S) \in I$, $(y,S) \in I$.
(H2) Every line $S \in \mathcal{L}$ contains at least three points which are pairwise non-neighbours.
(H3) Two lines S and T with $S \cap T \neq \emptyset$ are neighbours iff $|S \cap T| \geq 2$.
(H4) For any $x, y, z \in \mathcal{P}$, $x \frown y$ and $y \frown z$ imply $x \frown z$.
(H5) For any two lines S, T and any three points x, y, z with $(x,S) \in I$, $(y,S) \in I$, $(x,T) \in I$, $(z,T) \in I$, $x \not\frown y$, $x \not\frown z$, $y \frown z$, we have $S \frown T$.

(H6) For a point x not incident with $S \in \mathcal{L}$ with $x \frown S$, there always exist $y, z \in \mathcal{P}$ with $y \not\frown S$, $(z,S) \in I$ and $(x, \overline{y,z}) \in I$.

(H7) Let $x \in \mathcal{P}$, $S \in \mathcal{L}$ with $x \not\frown S$ and let $y, z \in S$. For every $(y', \overline{x,y}) \in I$ and every $(z', \overline{x,z}) \in I$ there exists a line T with $(y',T) \in I$, $(z',T) \in I$ and $S \cap T \neq \emptyset$.

The point set $\mathcal{T} \subseteq \mathcal{P}$ is called a *Hjelmslev subspace* of Π if for every two distinct points $x, y \in \mathcal{P}$, there exists a line $L \in \mathcal{L}(\mathcal{T}) = \{L \in \mathcal{L} \mid L \subseteq \mathcal{T}\}$ with $(x,L) \in I$, $(y,L) \in I$. We write $x \frown \mathcal{T}$ if there exists a point $y \in \mathcal{T}$ with $x \frown y$. Every Hjelmslev subspace $\mathcal{T}$ forms a projective Hjelmslev space $(\mathcal{T}, \mathcal{L}(\mathcal{T}), I_{\mathcal{T}})$ of its own, where $I_{\mathcal{T}} = I \cap (\mathcal{T} \times \mathcal{L}(\mathcal{T}))$. For every $\mathcal{X} \subseteq \mathcal{P}$ we define the *hull* $\langle \mathcal{X} \rangle$ as the intersection of all Hjelmslev subspaces containing $\mathcal{X}$. The set $\mathcal{X} \subseteq \mathcal{P}$ is said to be *independent* if for any $x \in \mathcal{X}$ we have $x \not\frown \langle \mathcal{X} \setminus \{x\} \rangle$.

Definition 1.2. The point set $\mathcal{B}$ is a *basis* of Π if $\langle \mathcal{B} \rangle = \mathcal{P}$ and $\mathcal{B}$ is independent.

The dimension of a projective Hjelmslev space Π is defined as $\dim \Pi = |\mathcal{B}| - 1$. In what follows, we consider only finite-dimensional Hjelmslev spaces.

2. Coordinate Hjelmslev Geometries

An important class of projective Hjelmslev spaces can be obtained as coordinate geometries from modules over so-called finite chain rings. We review only the most basic properties of this class of finite rings, and refer the reader for a detailed treatment to [7, 45, 46, 48].

An associative ring R with identity ($1 \neq 0$) is called a left (right) chain ring if the lattice of left (resp., right) ideals of R forms a chain. In the finite case, $|R| < \infty$, this condition is left-right symmetric and equivalent to R being a local principal ideal ring. In what follows the Jacobson radical rad(R) of R (which we assume to be a finite chain ring from now on) will be denoted by N, so that $R/N \cong \mathbb{F}_q$ is a finite field and $N = R\theta = \theta R$ for any $\theta \in N \setminus N^2$. Furthermore, there exists an integer $m \geq 1$ (called the length or nilpotency index of R) such that $N^{m-1} \neq \{0\}$, $N^m = \{0\}$, and every left or right ideal of R belongs to the chain $R > N > N^2 > \cdots > N^{m-1} > \{0\}$ of two-sided ideals $N^i = R\theta^i = \theta^i R$. The finite chain rings of length $m = 1$ are just the fields $\mathbb{F}_q$ and thus trivial from our point of view. For the smallest non-trivial case $m = 2$, a detailed description and classification of the corresponding rings will be given in Section 4..

Let M_R be a finite free (right) module over R of rank $\operatorname{rk} M \geq 3$. Denote by $\mathcal{P}$ and $\mathcal{L}$ the set of all free rank 1, respectively free rank 2, submodules of M_R and by $I \subseteq \mathcal{P} \times \mathcal{L}$ set-theoretical inclusion. The incidence structure $(\mathcal{P}, \mathcal{L}, I)$ satisfies (H1)–(H7) and, therefore, is a projective Hjelmslev space. If $\operatorname{rk} M = k$, this incidence structure is referred to as the (right) $(k-1)$-dimensional projective Hjelmslev geometry over the chain ring R and is denoted by $\mathrm{PHG}(R_R^k)$. (Since $M_R \cong R_R^k$, this is no essential restriction.)

Let R be a chain ring with $|R| = q^m$, $R/N \cong \mathbb{F}_q$. We consider the projective Hjelmslev space $\Pi = (\mathcal{P}, \mathcal{L}, I) = \mathrm{PHG}(R_R^k)$. Two points $x = \mathbf{x}R$ and $y = \mathbf{y}R$ are called *i-neighbours*, $i = 0, 1, \ldots, m$, if $|x \cap y| \geq q^i$. This fact is denoted by $x \frown_i y$. Two lines S and T are i-neighbours if for every point x on S there exists a point y on T with $x \frown_i y$, and conversely, for every y on T there exists x on S with $y \frown_i x$. Every two points (lines) are 0-neighbours; 1-neighbourhood is the same as the neighbour relation defined by (N1) and (N2).

For every $i \in \{0, 1, \ldots, m\}$, the relation $\frown_i$ is an equivalence relation on $\mathcal{P}$, as well as on $\mathcal{L}$. The equivalence classes of this relation are denoted by $[x]^{(i)}$, $x \in \mathcal{P}$, respectively $[S]^{(i)}$, $S \in \mathcal{L}$. The set of all equivalence classes of $\frown_i$ on points, resp. lines, is denoted by $\mathcal{P}^{(i)}$, resp. $\mathcal{L}^{(i)}$. We denote by $\pi^{(i)}$ the natural homomorphism $\pi^{(i)} : R \to R/R\theta^i$, where $R\theta = \operatorname{rad} R$. By $\overline{\pi}^{(i)}$, we denote the mapping induced by $\pi^{(i)}$ on the Hjelmslev subspaces of Π.

Below we state some facts on the combinatorics and the structure of the projective Hjelmslev geometries $\mathrm{PHG}(R_R^k)$ (cf. [2, 10, 12, 22, 34–36, 38, 50]).

Fact 2.1. *Let $\Pi = (\mathcal{P}, \mathcal{L}, I) = \mathrm{PHG}(R_R^k)$, where R is a chain ring with $|R| = q^m$, and $R/N \cong \mathbb{F}_q$. For every two integers s, t with $0 \leq t \leq s \leq k$, the number of all $(s-1)$-dimensional Hjelmslev subspaces through a fixed $(t-1)$-dimensional subspace is equal to*

$$q^{(s-t)(k-s)(m-1)} \begin{bmatrix} k-t \\ s-t \end{bmatrix}_q,$$

where

$$\begin{bmatrix} k \\ s \end{bmatrix}_q = \frac{(q^k-1)(q^{k-1}-1)\cdots(q^{k-s+1}-1)}{(q^s-1)(q^{s-1}-1)\cdots(q-1)}.$$

Moreover, the number of points that are i-th neighbours to a fixed point is $q^{(k-1)(m-i)}$ for all $i = 1, \ldots, m$.

The next few results explain the structure of the geometries $\mathrm{PHG}(R_R^k)$ in some more detail.

Define a new incidence relation $J^{(i)} \subseteq \mathcal{P}^{(i)} \times \mathcal{L}^{(i)}$ by

$$([x]^{(i)}, [S]^{(i)}) \in J^{(i)} \Leftrightarrow \exists x' \in [x]^{(i)}, \exists S' \in [S]^{(i)} : (x', S') \in I.$$

Fact 2.2. *The incidence structure* $(\mathcal{P}^{(i)}, \mathcal{L}^{(i)}, J^{(i)})$ *is isomorphic to the projective Hjelmslev geometry* $\mathrm{PHG}((R/R\theta^i)^k_{R/R\theta^i})$. *In particular,* $(\mathcal{P}^{(1)}, \mathcal{L}^{(1)}, J^{(1)})$ *is isomorphic to* $\mathrm{PG}(k-1,q)$.

Let Δ_1, Δ_2 be two Hjelmslev subspaces with $\dim\Delta_1 \leq \dim\Delta_2$. We write $\Delta_1 \frown_i \Delta_2$ if $\overline{\pi}^{(i)}(\Delta_1) \subseteq \overline{\pi}^{(i)}(\Delta_2)$. Note that under this definition $\frown_i$ is not symmetric. We consider again $\Pi = (\mathcal{P}, \mathcal{L}, I) = \mathrm{PHG}(R^k_R)$, where R is a chain ring with $|R| = q^m$, $R/N \cong \mathbb{F}_q$. Let us fix a Hjelmslev subspace Σ with $\dim\Sigma = u-1$ and an integer i, $1 \leq i \leq m-1$. Denote by $\mathcal{P}_i(\Sigma)$ the set of all points x with $x \frown_i \Sigma$. Now set

$$\mathfrak{P} = \left\{\Delta \cap [x]_{m-i} \mid x \in \mathcal{P}_i(\Sigma), \dim\Delta = u-1,\ \Delta \frown_i \Sigma,\ \Delta \cap [x]_{m-i} \neq \emptyset\right\}. \tag{1}$$

It can be proved that the sets $\Delta \cap [x]_{m-i}$ are either disjoint or coincide for the various choices of Δ.

Let $S \in \mathcal{L}$. We say that the "point" $\widetilde{x} = \Delta \cap [x]_{m-i} \in \mathfrak{P}$ is incident with the line S if

$$\Delta \cap [x]_{m-i} \cap S \neq \emptyset.$$

This defines an incidence relation $I' \subseteq \mathfrak{P} \times \mathcal{L}$. For two lines S and T we write $S \sim T$ if S and T are incident with the same points of $\mathfrak{P}$. Clearly $\sim$ is an equivalence relation on $\mathcal{L}$. Denote by $\mathfrak{L}$ a set of representatives from the different equivalence classes of lines under $\sim$, which have nonempty intersection with at least one of the sets $\Delta \cap [x]_{m-i}$. Let $\mathfrak{J}$ be the incidence relation induced by I' on $\mathfrak{P} \times \mathfrak{L}$. With the above notation, we have the following result.

Fact 2.3. *The incidence structure* $(\mathfrak{P}, \mathfrak{L}, \mathfrak{J})$ *can be embedded isomorphically into* $\mathrm{PHG}((R/R\theta^{m-i})^k_{R/R\theta^{m-i}})$. *The missing part consists of the points of a* $(k-u-1)$*-dimensional Hjelmslev subspace.*

The $(k-1)$*-dimensional affine Hjelmslev geometry* $\mathrm{AHG}(R^{k-1}_R)$ is defined as the incidence structure obtained from $\mathrm{PHG}(R^k_R)$ by deleting a neighbour class of hyperplanes. Equivalently, it can be defined as the incidence structure having as points all $(k-1)$ tuples $(\alpha_1, \ldots, \alpha_{k-1})$, $\alpha_i \in R$, and as lines all cosets of free rank 1 submodules of R^{k-1}_R. If in the discussion preceding Fact 2.3, we take Σ to be a point, say x, then $\mathfrak{P} = [x]_i$ and we get the following result.

Fact 2.4.

$$([x]_i, \mathfrak{L}, \mathfrak{J}) \cong \mathrm{AHG}\left((R/R\theta^{m-i})^{k-1}_{R/R\theta^{m-i}}\right).$$

In particular,

$$([x]_{m-1}, \mathfrak{L}) \cong \mathrm{AG}(k-1,q).$$

3. Multisets of Points in Projective Hjelmslev Geometries and Linear Codes over Finite Chain Rings

3.1. Multisets of Points in $\mathrm{PHG}(R^k_R)$

Let $\Pi = \mathrm{PHG}(R^k_R) = (\mathcal{P}, \mathcal{L}, I)$ be a finite-dimensional projective Hjelmslev geometry over the chain ring R.

Definition 3.1. A *multiset in* Π is a mapping $\mathfrak{K}\colon \mathcal{P} \to \mathbb{N}_0$.

The mapping $\mathfrak{K}$ is extended to the subsets of $\mathcal{P}$ by

$$\mathfrak{K}(Q) = \sum_{x \in Q} \mathfrak{K}(x) \quad \text{for } Q \subseteq \mathcal{P}. \tag{2}$$

The integer $\mathfrak{K}(x)$ is called the *multiplicity* of the point x. The integer $\mathfrak{K}(\mathcal{P}) = \sum_{x\in\mathcal{P}} \mathfrak{K}(x)$ is called the *cardinality* or *length* of the multiset $\mathfrak{K}$ and is denoted by $|\mathfrak{K}|$. The *support* $\operatorname{supp}\mathfrak{K}$ of $\mathfrak{K}$ is defined by $\operatorname{supp}\mathfrak{K} = \{x \in \mathcal{P} \mid \mathfrak{K}(x) > 0\}$. For a multiset $\mathfrak{K}$ in Π we define its *hull* $\langle\mathfrak{K}\rangle \leq R_R^k$ by

$$\langle\mathfrak{K}\rangle = \sum_{\mathbf{x}R \in \operatorname{supp}\mathfrak{K}} \mathbf{x}R. \tag{3}$$

Clearly, $\langle\mathfrak{K}\rangle$ can be considered as the set of all points $x = \mathbf{x}R$ with $x \leq \langle\mathfrak{K}\rangle$.

Given a set of points $Q \subseteq \mathcal{P}$, we define the *characteristic multiset* χ_Q by

$$\chi_Q(x) = \begin{cases} 1 & \text{if } x \in Q \\ 0 & \text{otherwise.} \end{cases}$$

All multisets $\mathfrak{K}$ satisfying $\mathfrak{K}(x) \in \{0,1\}$ for every $x \in \mathcal{P}$ arise in this manner from their supports. Such multisets are said to be *projective* and may be tacitly identified with their supports.

The multiset $\mathfrak{K}$ induces in a natural way multisets $\mathfrak{K}^{(i)}$ in $\overline{\pi}^{(i)}(\Pi)$ by

$$\mathfrak{K}^{(i)}\colon \mathcal{P}^{(i)} \to \mathbb{N}_0\colon [x]_i \mapsto \mathfrak{K}([x]_i)$$

for $i = 0, 1, \ldots, m$. Note that $\pi^{(i)}(\langle\mathfrak{K}\rangle) = \langle\mathfrak{K}^{(i)}\rangle$.

Definition 3.2. Denote by κ_i the rank of the R-module $\langle\mathfrak{K}^{(i)}\rangle$.

In geometric language, $\kappa_i - 1$ is the dimension of the smallest Hjelmslev subspace of $\overline{\pi}^{(i)}(\Pi)$ containing all points of $\operatorname{supp}\mathfrak{K}^{(i)}$.

Definition 3.3. Two multisets $\mathfrak{K}$ in Π and $\mathfrak{K}'$ in Π' are said to be *equivalent* if there exists a bijective R-semilinear mapping $\psi\colon \langle\mathfrak{K}\rangle_R \to \langle\mathfrak{K}'\rangle_R$ such that $\mathfrak{K}(x) = \mathfrak{K}'(\psi(x))$ for every point $x \in \langle\mathfrak{K}\rangle$.

3.2. Linear Codes over Finite Chain Rings

Let R be a chain ring with $|R| = q^m$, $R/N \cong \mathbb{F}_q$, let θ be a generator of N, and consider the set R^n of all n-tuples over R. The set R^n has the structure of an (R-R)-bimodule with respect to component-wise addition and left/right multiplication by elements from R. We say that θ^i is the *period* of the vector $\mathbf{x} \in R^n$ if i is the smallest non-negative integer with $\theta^i\mathbf{x} = \mathbf{0}$ (equivalently, with $\mathbf{x} \in R^n\theta^{m-i}$). We denote this by $\theta^{m-i} \parallel \mathbf{x}$. The set of vectors in R^n of period θ^m is denoted by $(R^n)^*$. Since $R\theta^i = \theta^i R$ for all $i \geq 0$, the concept of period is left-right symmetric even for non-commutative chain rings.

Definition 3.4. A *code* C *of length* n *over* R is a non-empty subset of R^n. The vectors of C are called *codewords*. The code C is *left* (resp., *right*) *linear* if it is an R-submodule of ${}_RR^n$ (resp., of R^n_R). A *linear* code is one which is either left or right linear.

Definitions and results in the sequel will be stated for left linear codes, most of them having obvious right counterparts.

A partition $\lambda \vdash n$ of an integer n is a sequence of non-negative integers $\lambda_0 \geq \lambda_1 \geq \lambda_2 \geq \ldots$ with $\sum_{i\geq 0} \lambda_i = n$. The trailing zeros of this sequence will be suppressed. The following theorem generalizes the structure theorem for finite abelian p-groups (see e.g. [44, Ch. 15,§ 2]):

Theorem 3.5 ([24]). *Every linear code C over a chain ring R is a direct sum of cyclic R-modules. The partition $\lambda = (\lambda_1, \ldots, \lambda_k) \vdash \log_q |C|$ satisfying*

$$_RC \cong R/N^{\lambda_1} \oplus \cdots \oplus R/N^{\lambda_k} \tag{4}$$

is uniquely determined by $_RC$. Moreover, the partition $\mu = \lambda' \vdash \log_q |C|$ conjugate to λ has components $\mu_i = \dim_{R/N}(\theta^{i-1}C/\theta^i C)$.

Definition 3.6. The *shape* of a linear code C over R is the partition

$$\lambda = (\lambda_1, \ldots, \lambda_k) \vdash \log_q |C|,$$

which satisfies $_RC \cong R/N^{\lambda_1} \oplus \cdots \oplus R/N^{\lambda_k}$. The partition λ' conjugate to λ is called the *conjugate shape* of C. The integer $k = \lambda'_1 = \dim_{R/N}(C/\theta C)$ is called the *rank* of C and is denoted by $\operatorname{rk} C$. A subset $\{\mathbf{x}_1, \ldots, \mathbf{x}_k\} \subseteq C \setminus \{\mathbf{0}\}$ is called a *basis* of C if $_RC = R\mathbf{x}_1 \oplus \cdots \oplus R\mathbf{x}_k$.

Definition 3.7. Let $C \leq {_RR^n}$ be a linear code of rank $\operatorname{rk} C = k$. A *generator matrix* of C is a $k \times n$-matrix having as its rows a basis of C, so that, in particular, $C = \{\mathbf{xG}; \mathbf{x} \in R^k\}$.

For two vectors $\mathbf{u} = (u_1, \ldots, u_n) \in R^n$ and $\mathbf{v} = (v_1, \ldots, v_n) \in R^n$ we define their inner product $\mathbf{u} \cdot \mathbf{v}$ by

$$\mathbf{u} \cdot \mathbf{v} := u_1 v_1 + \cdots + u_n v_n. \tag{5}$$

Given a code $C \subseteq R^n$, we define

$$C^\perp = \{\mathbf{y} \in R^n \mid \mathbf{x} \cdot \mathbf{y} = 0 \text{ for every } \mathbf{x} \in C\},$$

$$^\perp C = \{\mathbf{y} \in R^n \mid \mathbf{y} \cdot \mathbf{x} = 0 \text{ for every } \mathbf{x} \in C\}.$$

The linear code $C^\perp \leq R^n_R$ (resp., $^\perp C \leq {_RR^n}$) is called the *right* (resp., *left*) *orthogonal code* of C.

Theorem 3.8 ([24]). *Let $C \leq {_RR^n}$ be a linear code of shape $\lambda = (\lambda_1, \ldots, \lambda_n)$ and rank $\lambda'_1 = k$.*

1. *The orthogonal code $C^\perp$ has shape $(m - \lambda_n, m - \lambda_{n-1}, \ldots, m - \lambda_1)$ and conjugate shape $(n - \lambda'_m, n - \lambda'_{m-1}, \ldots, n - \lambda'_1)$. In particular, C is free as an R-module if and only if $C^\perp$ is free if and only if $\operatorname{rk}(C^\perp) = n - k$;*

2. $^\perp(C^\perp) = C$;

3. *if in addition $C' \leq {_RR^n}$ then $(C \cap C')^\perp = C^\perp + C'^\perp$, and $(C + C')^\perp = C^\perp \cap C'^\perp$.*

Corollary 3.9. *Let $\mathbf{G} \in \mathrm{M}_{m,n}(R)$ be any matrix. The linear codes $\mathcal{C} \leq {}_R R^n$ and $\mathcal{D} \leq R^m_R$ generated by the rows and columns of* **G**, *respectively, have the same shape.*

Definition 3.10. A *parity check matrix* of a linear code $\mathcal{C} \leq {}_R R^n$ is an $(n-\lambda'_m) \times n$-matrix whose rows form a basis of the orthogonal code $\mathcal{C}^{\perp}$

Note that if **H** is a parity-check matrix of $\mathcal{C}$, then by Part 2 of Theorem 3.8 we have $\mathbf{x} \in \mathcal{C}$ if and only if $\mathbf{x} \cdot \mathbf{H}^T = \mathbf{0}$. The number of (and periods of the) rows of **H** are determined by Part 1 of Theorem 3.8.

For $\mathbf{x} = (x_1, \ldots, x_n) \in R^n$ we set

$$a_i(\mathbf{x}) = |\{j \mid 1 \leq j \leq n \text{ and } \theta^i \parallel x_j\}|.$$

Definition 3.11. The sequence $(a_0(\mathbf{x}), \ldots, a_m(\mathbf{x}))$ is called the *type* of the word $\mathbf{x} \in R^n$.

Definition 3.12. An *automorphism of the code* R^n is a bijective mapping $\phi \colon R^n \to R^n$ which satisfies $a_i(\mathbf{x}-\mathbf{y}) = a_i(\phi(\mathbf{x}) - \phi(\mathbf{y}))$ for all $\mathbf{x}, \mathbf{y} \in R^n$ and all $i \in \{0, 1, \ldots, m\}$.

Definition 3.13. Two codes $\mathcal{C}_1, \mathcal{C}_2 \subseteq R^n$ are said to be *isomorphic* (resp., *semilinearly isomorphic*) if there exists a code automorphism (resp., semilinear code automorphism) ϕ of R^n with $\phi(\mathcal{C}_1) = \mathcal{C}_2$.

3.3. Equivalence of Multisets of Points and Linear Codes

Definition 3.14. A linear code $\mathcal{C} \leq {}_R R^n$ is said to be *fat* if for every $i \in \{1, \ldots, n\}$ there exists a codeword $\mathbf{c} = (c_1, c_2, \ldots, c_n) \in \mathcal{C}$ with $c_i \in R^{\times}$ (i.e. c_i is a unit in R).

Let $\mathcal{C} \leq {}_R R^n$ be a fat linear code. Let $S = (\mathbf{c}_1, \ldots, \mathbf{c}_k)$ be a sequence of (not necessarily independent) generators for ${}_R\mathcal{C}$ and let $\mathbf{G} \in \mathrm{M}_{k,n}(R)$ be the $k \times n$-matrix with rows $\mathbf{c}_1, \ldots, \mathbf{c}_k$. Denote the columns of **G** by $\mathbf{g}_1, \ldots, \mathbf{g}_n$. Since $\mathcal{C}$ is fat and $\mathbf{c}_1, \ldots, \mathbf{c}_k$ generate $\mathcal{C}$, the vectors $\mathbf{g}_j$ have period θ^m and thus define points $\mathbf{g}_j R$ in the projective (right) Hjelmslev geometry $(\mathcal{P}, \mathcal{L}, I) = \mathrm{PHG}(R^k_R)$. We define the *multiset $\mathfrak{K}_S$ induced by the generating sequence S of $\mathcal{C}$* as

$$\mathfrak{K}_S \colon \mathcal{P} \to \mathbb{N}_0 \colon x \mapsto |\{j \mid x = \mathbf{g}_j R\}|. \tag{6}$$

We say that the *multiset $\mathfrak{K}_S$ and the code $\mathcal{C} = R\mathbf{c}_1 + \cdots + R\mathbf{c}_k$ are associated.* By the definition of $\mathfrak{K}_S$, we have $|\mathfrak{K}_S| = n$. Furthermore, the modules $\langle \mathfrak{K}_S \rangle$ and ${}_R\mathcal{C}$ have the same shape and, in particular, the same cardinality; see [24].

The following theorem is a generalization of a similar result by Dodunekov and Simonis [11] about linear codes over finite fields.

Theorem 3.15. *For every multiset $\mathfrak{K}$ of length n in $PHG(R^k_R)$ there exists a fat linear code $\mathcal{C} \leq {}_R R^n$ and a generating sequence $S = (\mathbf{c}_1, \cdots, \mathbf{c}_k)$ of $\mathcal{C}$ which induces $\mathfrak{K}$. Two multisets $\mathfrak{K}_1$ in $\mathrm{PHG}(R^{k_1}_R)$ and $\mathfrak{K}_2$ in $\mathrm{PHG}(R^{k_2}_R)$ associated with fat (left) linear codes $\mathcal{C}_1$ and $\mathcal{C}_2$ over R, respectively, are equivalent if and only if the codes $\mathcal{C}_1$ and $\mathcal{C}_2$ are semilinearly isomorphic.*

Definition 3.16. Let $\mathfrak{K} \colon \mathcal{P} \to \mathbb{N}_0$ be a multiset in $\Pi = \mathrm{PHG}(R^k_R)$. A hyperplane Δ in Π is said to have the $\mathfrak{K}$-*type* $(a_0, a_1, \ldots, a_m)$, where

$$a_i = \sum_{x \bigcirc_i \Delta, x \not\bigcirc_{i+1} \Delta} \mathfrak{K}(x) \quad \text{for } 0 \leq i \leq m.$$

By duality (cf. Theorem 3.8), every hyperplane Δ in $\mathrm{PHG}(R_R^k)$ can be considered as a set of points, whose homogeneous coordinates $(x_1,\ldots,x_k)$ satisfy a linear equation

$$r_1x_1 + r_2x_2 + \ldots + r_kx_k = 0,$$

where at least one of the r_i's is a unit in R. Let C be a fat linear code associated with $\mathfrak{K}$, and let $\mathbf{G}_S$ be a $k \times n$-matrix whose sequence S of rows generates C and satisfies $\mathfrak{K}_S = \mathfrak{K}$. All codewords of C which belong to the cyclic submodule $R(r_1,\ldots,r_k)\mathbf{G}_S \leq {}_R C$ are called *codewords associated with the hyperplane* Δ (relative to the choice of the generating sequence S). There is a connection between the $\mathfrak{K}$-type of a hyperplane in Π and the number of codewords of a given type in C associated with that hyperplane.

Theorem 3.17. *Let $\mathfrak{K}$ be a multiset in $\mathrm{PHG}(R_R^k)$ and let C be a fat linear code over R associated with $\mathfrak{K}$. For each hyperplane Δ of $\mathfrak{K}$-type $(0,\ldots,0,a_j,a_{j+1},\ldots,a_m)$, with $a_j \neq 0$, $0 \leq j \leq m$, there exist exactly $q^{m-s} - q^{m-s-1}$ codewords in C of type*

$$\underbrace{(0,\ldots,0}_{s},a_j,\ldots,a_{m+j-s-1},\sum_{i=m+j-s}^{m} a_i) \qquad (j \leq s \leq m-1) \tag{7}$$

which are associated with Δ.

For a multiset $\mathfrak{K}$ in $\mathrm{PHG}(R_R^k)$, the numbers $\kappa_i = \mathrm{rk}\langle \mathfrak{K}^{(i)} \rangle$ (Definition 3.2) determine the shape of every fat linear code $C \leq {}_R R^n$ associated with $\mathfrak{K}$.

Theorem 3.18. *Let $\mathfrak{K}$ be a multiset in $\mathrm{PHG}(R_R^k)$ associated to the fat linear code C. Then C has conjugate shape $\lambda' = (\kappa_m,\kappa_{m-1},\ldots,\kappa_1)$, and, in particular,*

$$|C| = q^{\kappa_1+\kappa_2+\cdots+\kappa_m}.$$

3.4. Some Classes of Codes Defined Geometrically

Consider the Hjelmslev geometry $\Pi = (\mathcal{P},\mathcal{L},I) = \mathrm{PHG}(R_R^k)$. The linear code C associated with the multiset $\mathfrak{K}$ defined by $\mathfrak{K}(x) = 1$ for all $x \in \mathcal{P}$, is called the *k-dimensional simplex code over R* and is denoted by $\mathrm{Sim}(k,R)$. The code $\mathrm{Sim}(k,R)$ has length $q^{(k-1)(m-1)}\begin{bmatrix}k\\1\end{bmatrix}_q$, and by Theorem 3.18 it has shape m^k (i.e. its shape consists of k parts equal to m), in particular $|\mathrm{Sim}(k,R)| = q^{km}$. All hyperplanes Δ in Π have the same $\mathfrak{K}$-type $(a_0,a_1,\ldots,a_m)$, where

$$\begin{aligned}
a_0 &= q^{(k-1)(m-1)}\left(\begin{bmatrix}k\\1\end{bmatrix}_q - \begin{bmatrix}k-1\\1\end{bmatrix}_q\right) = q^{(k-1)m},\\
a_j &= q^{(k-2)(m-1)}\begin{bmatrix}k-1\\1\end{bmatrix}_q\left(q^{m-j} - q^{m-j-1}\right), \quad j = 1,\ldots,m-1,\\
a_m &= q^{(k-2)(m-1)}\begin{bmatrix}k-1\\1\end{bmatrix}_q.
\end{aligned}$$

The dual code $\mathrm{Sim}(k,R)^\perp$ is called the *k-th order Hamming code over R* and is denoted by $\mathrm{Ham}(k,R)$. It is free of rank $q^{(k-1)(m-1)}\begin{bmatrix}k\\1\end{bmatrix}_q - k$, in particular $|\mathrm{Ham}(k,R)| = q^{mq^{(k-1)(m-1)}\begin{bmatrix}k\\1\end{bmatrix}_q - mk}$. A parity check matrix and a generator matrix for $\mathrm{Ham}(k,R)$ may be obtained similarly to the special case of R being a field.

4. Arcs in Projective Hjelmslev Planes

4.1. The Maximal Arc Problem

Definition 4.1. A multiset $\mathfrak{K}$ in $(\mathcal{P}, \mathcal{L}, I)$ is called a *(k,n)-arc* if

(i) $\mathfrak{K}(\mathcal{P}) = k$.

(ii) $\mathfrak{K}(L) \leq n$ for every line $L \in \mathcal{L}$.

According to this definition, a (k,n)-*arc* is also a (k,n')-*arc* for every integer $n' \geq n$. For this reason we shall usually assume that n is chosen to be minimal, i.e. there exists at least one line $L_0 \in \mathcal{L}$ with $\mathfrak{K}(L_0) = n$ (but there are exceptions). Moreover, sometimes we say "n-arc" in place of "(k,n)-arc" without referring to the cardinality of $\mathfrak{K}$.

Of course, Definition 4.1 also makes sense for other incidence structures. In the classical cases of $\mathrm{PG}(2,q)$ or $\mathrm{AG}(2,q)$ (which can be considered as special cases of projective Hjelmslev planes) a lot of research has been done on arcs and many results are known. For an overview, see [15] and also [9]. Some of these results will be used in the sequel.

The arcs considered in this section will be projective and can be identified with sets of points, as described earlier.

Furthermore, for the rest of this survey, we will confine ourselves to the case of finite chain rings R of length 2, i.e. the case $|R| = q^2$, $R/N \cong \mathbb{F}_q$. The classification of all those rings is known and summarized in the following result.

Fact 4.2 (cf. [8,46,48,49]). *Suppose R is a finite chain ring with $|R| = q^2$, $R/N \cong \mathbb{F}_q$, where $q = p^r$. Then*

(i) either R has characteristic p^2 and is isomorphic to the Galois ring $\mathrm{GR}(q^2, p^2)$ of order q^2 and characteristic p^2, defined as $\mathrm{GR}(q^2, p^2) = \mathbb{Z}_{p^2}[X]/(h)$ where $h \in \mathbb{Z}_{p^2}[X]$ is a (monic) polynomial of degree r which is irreducible modulo p, or

(ii) R has characteristic p and for some $\sigma \in \mathrm{Aut}(\mathbb{F}_q)$ is isomorphic to the ring $\mathbb{F}_q[X;\sigma]/(X^2)$ of σ-dual numbers over $\mathbb{F}_q$, defined as the set of all $a_0 + a_1X \in \mathbb{F}_q[X]$ with operations $(a_0 + a_1X) + (b_0 + b_1X) := a_0 + b_0 + (a_1 + b_1)X$, $(a_0 + a_1X)(b_0 + b_1X) := a_0b_0 + \big(a_0b_1 + a_1\sigma(b_0)\big)X$.

Moreover, the $r+1$ different rings listed in (i), (ii) are pairwise non-isomorphic.

In the sequel, we will also refer to these rings as $\mathbb{G}_q := \mathrm{GR}(q^2, p^2)$, $\mathbb{S}_q := \mathbb{F}_q[X]/(X^2)$, and $\mathbb{T}_q^{(i)} := \mathbb{F}_q[X;\sigma^i]/(X^2)$ for $1 \leq i \leq r-1$, where σ denotes the Frobenius automorphism of $\mathbb{F}_q$. Furthermore we will use the abbreviations $\mathbb{T}_q = \mathbb{T}_q^{(1)}$ and $\mathbb{T}_q^{\circ} = \mathbb{T}_q^{(r-1)}$.[1]

Denote by $m_n(R_R^3)$ the maximal value of k for which there exists a projective (k,n)-arc in $\mathrm{PHG}(R_R^3)$. The problem of determining the exact values of $m_n(R_R^3)$ for various values of n and for various rings R is central and has a clear coding theoretic relevance.

---

[1] The latter reflects the fact that $\mathbb{T}_q^{(r-1)}$ is isomorphic to the opposite ring of $\mathbb{T}_q^{(1)}$. Note that the smallest case where a symbol $\mathbb{T}_q^{(i)}$ cannot be avoided is $q^2 = 256$.

4.2. A General Upper Bound on the Size of an Arc

The following theorems provide upper bounds on the size of a (k,n)-arc in $\mathrm{PHG}(R_R^3)$ [43].

Theorem 4.3. *Let $\mathfrak{K}$ be a (k,n)-arc in $\mathrm{PHG}(R_R^3)$ where $|R| = q^2$, $R/N \cong \mathbb{F}_q$. Suppose there exists a neighbour class of points $[x]$ with $\mathfrak{K}([x]) = u$ and let u_i, $i = 0,1,\dots,q$, be the maximum number of points on a line from the i-th parallel class in the affine plane defined on $[x]$. Then*

$$k \le q(q+1)n - q\sum_{i=0}^{q} u_i + u.$$

Proof. Let $\{L_i \mid i = 0,1,\dots,q\}$ be a set of $q+1$ lines no two of which are neighbours and such that $\mathfrak{K}([x]\cap L_i) = u_i$. For every $i \in \{0,\dots,q\}$, denote by $L_i^{(j)}$, $j = 1,\dots,q$, the q lines in $\mathrm{PHG}(R_R^3)$ that coincide with L_i on $[x]$. The sum of the multiplicities of the points from $L_i^{(j)}$ not in $[x]$ does not exceed $n - u_i$, which gives the estimate

$$\begin{aligned}
k &= \mathfrak{K}([x]) + \sum_{i=0}^{q}\sum_{j=1}^{q} \mathfrak{K}(L_i^{(j)} \setminus ([x]\cap L_i)) \\
&\le u + \sum_{i=0}^{q}\sum_{j=1}^{q}(n-u_i) \\
&= u + \sum_{i=0}^{q} q(n-u_i) \\
&= u + q(q+1)n - q\sum_{i=0}^{q} u_i.
\end{aligned}$$

□

Typically, the numbers u_i are unknown. We can use some simple estimates to get a more convenient form for the above upper bound. From the obvious inequality $u_i \ge \lceil u/q \rceil$, we get

$$k \le q(q+1)(n - \lceil u/q \rceil) + u. \tag{8}$$

Fix a point $y \in [x]$ and let $S_0,\dots,S_q$ be lines through y, no two of which are neighbours. Without loss of generality, we assume that $L_i \frown S_i$ for $i = 0,\dots,q$. Set $s_i = \mathfrak{K}([x]\cap S_i) - \mathfrak{K}(y)$. Clearly, $\mathfrak{K}(y) + s_i \le u_i$. Then

$$\begin{aligned}
k &\le q(q+1)n - q\sum_{i=0}^{q} u_i + u \\
&\le q(q+1)n - q\sum_{i=0}^{q}(\mathfrak{K}(y) + s_i) + u \\
&= q(q+1)n - q(q+1)\mathfrak{K}(y) - q(u - \mathfrak{K}(y)) + u \\
&= q^2(n - \mathfrak{K}(y)) + q(n-u) + u.
\end{aligned}$$

Since we may certainly assume that $\mathfrak{K}(y) \ge 1$, the last inequality simplifies to

$$k \le q^2(n-1) + q(n-u) + u.$$

Theorem 4.4.

$$m_n(R_R^3) \le \max_{1\le u\le \min\{\mu_n(q),q^2\}} \min\{u(q^2+q+1), \\ q^2(n-1)+q(n-u)+u, q(q+1)(n-\lceil u/q\rceil)+u\},$$

where $\mu_n(q)$ denotes the maximal size of a (k,n)-arc in $\mathrm{AG}(2,q)$.

For small values of n, we can derive somewhat better bounds.

Theorem 4.5.

$$m_2(R_R^3) \le \begin{cases} q^2+q+1 & \text{for } q \text{ even,} \\ q^2 & \text{for } q \text{ odd.} \end{cases} \tag{9}$$

In case of equality, we have

(i) for q even, $\mathfrak{K}([x]) = 1$ for every $[x] \in \mathcal{P}^{(1)}$;

(ii) for q odd, $\mathfrak{K}([x]) \le 1$ for every $[x] \in \mathcal{P}^{(1)}$. Moreover, the neighbour classes with $\mathfrak{K}([x]) = 0$ form a line in the factor plane $(\mathcal{P}^{(1)}, \mathcal{L}^{(1)}, \mathcal{J}^{(1)}) \cong \mathrm{PG}(2,q)$.

Theorem 4.6. $m_3(R_R^3) \le 2q^2 - q + 3$, *for every $q \ge 5$.*

Note that in the cases $q = 2,3$, the exact value of $m_3(R_R^3)$ is known. It is 10 for the rings of cardinality 4, 19 for $R = \mathbb{Z}_9$, and 18 for $R = \mathbb{F}_3[X]/(X^2)$. For $q = 4$, we have the bounds $29 \le m_3(R_R^3) \le 30$ for all three rings $\mathbb{G}_4$, $\mathbb{S}_4$, $\mathbb{T}_4$.

4.3. Constructions for Arcs

In this section, we present general constructions for arcs in projective Hjelmslev planes. Throughout this section, R will be a chain ring with $|R| = q^2$, $R/N \cong \mathbb{F}_q$, and $\Pi = (\mathcal{P}, \mathcal{L}, I)$ will be the projective Hjelmslev plane $\mathrm{PHG}(R_R^3)$.

Example 4.7. For values of n close to q^2+q, the exact value of $m_n(R_R^3)$ can be easily computed. For every chain ring R, with $|R| = q^2$, $R/N \cong \mathbb{F}_q$, and every integer $s = 0,1,\dots,q$

$$m_{q^2+s}(R_R^3) = q^4 + q^2 s + qs.$$

Denote by $\mathcal{F}$ the point set obtained in the following way. Fix a line L. Take in $\mathcal{F}$ the points of the line L (if $s < q$) plus $q-s-1$ additional line segments parallel to $L \cap [x_i]$ in each of the neighbour classes $[x_i]$ incident with $[L]$ in the factor geometry. The multiset $\chi_{\mathcal{P}} - \chi_{\mathcal{F}}$ is easily checked to be the desired arc. The upper bound is obtained from Theorem 4.4.

Example 4.8. Now we describe a general construction for (q^4-q^2-2q+1, q^2-1)-arcs in $\mathrm{PHG}(R_R^3)$ that does not depend on the underlying ring. Remarkably, this construction is better than the "triangle construction" which yields a (q^4-2q^2+1, q^2-1)-arc $\chi_{\mathcal{P}} - \chi_{\mathcal{F}}$ as the complement of a "triangle" $\mathcal{F}$ ($\mathcal{F}$ consists of a neighbour class of lines and two additional lines that are not neighbours).

Fix a point class $[x_0]$ and a line class $[L_0]$ incident with $[x_0]$ in the factor plane. Set

$$[L_0] = \{[x_i] \mid i = 0,\dots,q\}.$$

Furthermore, denote by $[L_i]$, $1 \le i \le q$, the other line classes through $[x_0]$ in the factor plane. Consider the set $\mathfrak{K}$ containing the following points:

1) The complement of a $(2q-1)$-blocking set in the affine plane induced on $[x_0]$ (which is isomorphic to $\mathrm{AG}(2,q)$). Thus $\mathfrak{K}$ contains $(q-1)^2$ points from $[x_0]$.

2) The line segments from the point classes $[x_1],\dots,[x_q]$ together with q additional lines (containing the segments in $[x_i]$, $i=1,\dots,q$) form a structure isomorphic to $\mathrm{AG}(2,q)$. In every class $[x_i]$, choose $q-2$ line segments (having the direction of $[L_0]$) such that the resulting $q(q-2)$ line segments form the complement of a blocking set in $\mathrm{AG}(2,q)$.

3) From each of the remaining point classes $[y]$, select the following points. If $[y]\in[L_i]$, take the q^2-q points from $q-1$ parallel line segments having the direction of the line $[\overline{yx_i}]$.

The total number of points is $(q-1)^2+q\cdot q(q-2)+q^2(q^2-q)=q^4-q^2-2q+1$. A line in $[L_0]$ meets $[x_0]\cap\mathfrak{K}$ in at most $q-1$ points and at most $q-1$ of the sets $[x_i]\cap\mathfrak{K}$, $i=1,\dots,q$, in q points, i.e., it contains at most $q-1+(q-1)q=q^2-1$ points from $\mathfrak{K}$.

A line in the class $[\overline{yx_0}]$, $y\not\triangleright L_0$, meets $[x_0]\cap\mathfrak{K}$ in at most $q-1$ points and each of the other q sets $[z]\cap\mathfrak{K}$ in exactly $q-1$ points. Hence, such a line contains at most $q-1+q(q-1)=q^2-1$ points from $\mathfrak{K}$.

Finally, a line in the class $[\overline{yx_i}]$, $y\not\triangleright L_0$, $i\neq 0$, meets one set $[z]\cap\mathfrak{K}$ in at most q points, $q-1$ such sets in $q-1$ points and one set (the set $[x_i]\cap\mathfrak{K}$) in $q-2$ points. Therefore, such a line contains at most $q+(q-1)^2+q-2=q^2-1$ points.

Thus the arc defined above has the desired parameters.

This construction can be further improved if we take the blocking set on $[L_0]$ to consist of two lines that meet on $[x_0]$. Furthermore, we replace the $q-1$ points from $[x_0]$ that form a line segment in a direction different from that of L_0 by $q-2$ collinear points in $[x_1]$ that again have a direction different from that of L_0 and are not already part of the blocking set on $[L_0]$. It is an easy check that the size of the arc is increased by 1 and we get a $((q^3+q^2-2)(q-1),q+1)$-arc.

For $q^2=9,16,25$, this construction gives: $m_8(R_R^3)\geq 68$, for $q^2=9, m_8(R_R^3)\geq 234$, for $q^2=16$, and $m_8(R_R^3)\geq 592$, for $q^2=25$.

The exact formula for $m_n(R_R^3)$ in the range $q^2\leq n\leq q^2+q$ presented in Example 4.7 may also be written as $m_n(R_R^3)=q^4+q^3+q^2-(q^2+q-n)(q^2+q)$. From this point of view it says that the complementary $((q^2+q-n)(q^2+q),q^2+q-n)$-blocking set has the same cardinality as the (generally non-projective) sum of q^2+q-n lines. It seems reasonable to conjecture that the lower bound $m_n(R_R^3)\geq q^4+q^3+q^2-(q^2+q-n)(q^2+q)$ holds for all n. (For small values of n, this lower bound is even rather weak.) The following theorem extends the range of integers n, for which the lower bound is known to hold, to $q^2-\lfloor q/2\rfloor\leq n\leq q^2+q$.

Theorem 4.9. *For every chain ring R with $|R|=q^2$, $R/\operatorname{rad}R\cong\mathbb{F}_q$, and every integer $s=1,2,\dots,\lfloor q/2\rfloor$, the following inequality holds:*

$$m_{q^2-s}(R_R^3)\geq q^4-q^2s-qs. \tag{10}$$

Proof. We will prove the existence of a $(t(q^2+q),t)$-blocking set in $\mathrm{PHG}(R_R^3)$ for $q+1\leq t\leq\lfloor 3q/2\rfloor$ except in the case $(q,t)=(3,4)$, which is covered by the subsequent Example 4.10.

Choose point classes $[x_0]$, $[x_1]$, $[x_2]$ and line classes $[L_0]$, $[L_1]$, $[L_2]$ which form a triangle in the factor plane $\mathrm{PG}(2,q)$, indexed in such a way that $[x_i]$ is incident with $[L_{i-1}]$ and $[L_i]$ (indices taken modulo 3). There exist (unique) integers t_1, t_2, t_3 satisfying $t = t_1 + t_2 + t_3$ and $1 \le t_1 \le t_2 \le t_3 \le t_1 + 1$. In each point class $[x]$ incident with $[L_i]$ but different from the vertices $[x_i]$ and $[x_{i+1}]$, choose t_i parallel line segments in the direction of $[L_i]$. In each class $[x_i]$ choose $t_{i-1} + t_i$ parallel line segments in the direction of $[L_i]$. This is possible, since $t_{i-1} + t_i \le t - t_1 = t - \lfloor t/3 \rfloor = \lceil 2t/3 \rceil \le q$. It is clear that the resulting point set in $\mathrm{PHG}(R_R^3)$ blocks every line outside $[L_1] \cup [L_2] \cup [L_3]$ exactly t times. [2] Every line $L \in [L_i]$ is blocked $t_i + t_{i+1}$ times by the line segments in $[x_{i+1}]$. Since $t_3 = \lceil t/3 \rceil \le \lceil q/2 \rceil < q$, we have $q + t_i + t_{i+1} > t$. In order to have $\mathfrak{K}(L) \ge t$, it is therefore enough to ensure that L is blocked at least once by the line segments chosen in $[L_i] \setminus [x_{i+1}]$. The q^2 line segments in $[L_i] \setminus [x_{i+1}]$ (as points) together with the q^2 lines in $[L_i]$ and the q point classes $[y_1] = [x_i], [y_2], \ldots, [y_q]$ (as lines) form an incidence structure isomorphic to $\mathrm{AG}(2,q)$. Our task is to arrange the $t_{i-1} + t_i$ line segments in $[y_1]$ and the t_i line segments in each class $[y_j]$, $2 \le j \le q$, in such a way that they form a blocking set in $\mathrm{AG}(2,q)$.[3] Since $t_i \ge 1$, we may assume that q of these line segments, one from each class $[y_j]$, are collinear. The remaining $t_{i-1} + t_i - 1 + (q-1)(t_i - 1)$ line segments can be used to block the $q-1$ lines parallel to this line (and thus construct the required blocking set), provided there are at least $q-1$ of them. If $t_i > 1$, we are done. If $t_i = 1$, then either $(q,t) = (3,4)$ or $(q,t) = (4,5)$. The first case has been already excluded. In the second case, we have $t_1 = 1$, $t_2 = t_3 = 2$. We change the direction of the 3 line segments in $[x_1]$ from $[L_1]$ to $[L_0]$. Then each line in $[L_1]$ is blocked 6 times, while for $[L_2]$, $[L_3]$ we have $t_2 > 1$, $t_3 > 1$ and thus are done. □

Example 4.10. The following construction produces a $(48,4)$-blocking set in the projective Hjelmslev planes $\mathrm{PHG}(\mathbb{Z}_9^3)$ and $\mathrm{PHG}(\mathbb{S}_3^3)$, with $\mathbb{S}_3^3 = \mathbb{F}_3[X]/(X^2)$. The factor plane $\mathrm{PG}(2,3)$ contains an oval (quadrangle) which has 4 tangents and 6 external points (intersection points of the tangents). Each external point is on exactly two tangents. In each point class $[x]$ external to the oval, place a double line segment in one of the two tangent directions. Choose the directions in such a way that no tangent is chosen more than twice. In each point class $[x]$ on the oval, place a single line segment in the tangent direction. For those tangent directions $[T]$ which were chosen twice in the above process, arrange the 5 line segments in $[T]$ with direction $[T]$ in such a way that they block every line in $[T]$. As is easily verified, the resulting point set forms a $(48,4)$ blocking set. The complementary $(69,8)$-arc was originally discovered during a computer search [32]. The computational data suggested the preceding construction.

Example 4.11. *The general cascade construction.*

The following general cascade construction has been proposed in [25]. Let $\mathfrak{K}_0$ be a (k_0, n_0)-arc in $\mathrm{PG}(2,q)$. Let $\operatorname{supp}\mathfrak{K} = \{x_1, \ldots, x_{k_0}\}$ and let $\{X_1, \ldots, X_{k_0}\}$ be a set of k_0 lines in $\mathrm{PG}(2,q)$ such that $x_i \in X_i$. Then for each pair of integers $\alpha, s \in \{1, \ldots, q\}$, there exists an

[2] The construction so far can also be seen as taking the sum of $t = t_1 + t_2 + t_3$ lines in $\mathrm{PHG}(R_R^3)$, where t_i lines are chosen from $[L_i]$ in such a way that they have a line segment in $[x_{i+1}]$ in common. To make the resulting multiset projective, the t_i-fold line segment in $[x_{i+1}]$ is replaced by t_i line segments in $[x_{i+1}]$ having direction $[L_{i+1}]$ and not already chosen during the first step.

[3] Note that the special lines $[y_1], [y_2], \ldots, [y_q]$ are blocked by construction, since $t_{i-1} + t_i > t_i \ge 1$.

arc in $\mathrm{PHG}(R_R^3)$ with parameters $(\alpha s k_0, \max_{0\le i\le k_0} v_i)$, where $v_0 = \alpha n_0$ and $v_i = s + \alpha |X_i \cap \operatorname{supp} \mathfrak{K}_0| - \alpha$, for $i = 1, \ldots, k_0$.

Below a special instance of this construction is described. Take $q^2 = 25$, $s = 5$, and $\mathfrak{K}_0$ to be an $(11,3)$-arc in $\mathrm{PG}(2,5)$. There exist two such arcs and for both of them $a_1 + a_2 = 11$. Select the lines X_i to be the 1- and 2-lines of $\mathfrak{K}_0$. It is easily checked that $\max_{0\le i\le k_0} v_i = \max\{5+\alpha, 3\alpha\}$. For $\alpha = 2$, we get a $(110,7)$-arc, while for $\alpha = 3$, we get a $(165,9)$-arc.

Example 4.12. Take $\mathfrak{K}_0$ to be the trivial $(q^2+q+1, q+1)$-arc in $\mathrm{PG}(2,q)$ consisting of all the points of the plane. Index the point and line classes in $\mathrm{PHG}(R_R^3)$ in such a way that $[x_i]$ is incident with $[L_i]$ in the factor geometry, $i = 1, \ldots, q^2+q+1$. Select a line segment in each neighbour class consisting of q points that are collinear with a line from $[L_i]$. Denote this set of points by $\mathcal{F}$. The arc $\chi_{\mathcal{F}}$ has parameters $(q(q^2+q+1), 2q)$. More importantly, it gives rise to a strongly regular graph in the following way (as described in [6]). Let $\mathcal{C} \le {}_R R^n$, $n = q(q^2+q+1)$, be a linear code associated with $\mathcal{F}$. Since every line of $\mathrm{PHG}(R_R^3)$ is incident with either q or $2q$ points of $\mathcal{F}$, there are only two $\mathcal{F}$-types of lines, $(a_0, a_1, a_2) = (q^3, q^2, q)$ and $(q^3, q^2-q, 2q)$, which in turn yield three types of non-zero codewords in $\mathcal{C}$, namely $\big(a_0(\mathbf{x}), a_1(\mathbf{x}), a_2(\mathbf{x})\big) = (q^3, q^2, q)$, $(q^3, q^2-q, 2q)$ and $(0, q^3, q^2+q)$ with corresponding frequencies $q^5 - q^2$, $(q-1)(q^5-q^2)$ and $q^3 - 1$. Now take $G = (V, E)$ as the Cayley graph of $(\mathcal{C}, +)$ with respect to the set $\mathcal{C}_1 \subset \mathcal{C}$ of codewords of type $(q^3, q^2-q, 2q)$, i.e. G has vertex set $V = \mathcal{C}$ and edge set $E = \big\{(\mathbf{x},\mathbf{y}) \mid \mathbf{x}-\mathbf{y} \in \mathcal{C}_1\big\}$.[4] As shown in [6], the graph G is strongly regular with parameters

$$v = q^6, \quad k = q^5 - q^2, \quad \lambda = q^4 + q^3 - 3q^2, \quad \mu = q^4 - q^2.$$

Moreover, $\mathcal{C}$ can be mapped (cf. [23]) onto a (possibly non-linear) two-weight code over $\mathbb{F}_q$.

In the next section, we give an algebraic construction for $(k,2)$-arcs.

4.4. $(k,2)$-Arcs

For $(k,2)$-arcs we have the bound (9). In some cases this bound is achieved. There exists a $(7,2)$-arc in the plane over $\mathbb{G}_2 = \mathbb{Z}_4$, but there is no such arc in the plane over $\mathbb{S}_2 = \mathbb{F}_2[x]/(x^2)$. There exist $(9,2)$-arcs in the projective Hjelmslev planes over both chain rings with 9 elements. For larger chain rings, it is possible to get large $(k,2)$-arcs with more than one point in some of the neighbour classes.

Remarkably, $(q^2+q+1, 2)$-arcs exist in the planes over the Galois rings $\mathbb{G}_{2^r} = \mathrm{GR}(4^r, 4)$ for all r. Below, we explain the construction in a more general setting [14,26,27].

Let $q = p^r > 1$ be a prime power and $\mathbb{G}_q = \mathrm{GR}(q^2, p^2)$ be the Galois ring of cardinality q^2 and characteristic p^2. For any $k \in \mathbb{N}$, the ring $\mathbb{G}_{q^k}$ is the unique Galois extension of $\mathbb{G}_q$ of degree k and conversely, $\mathbb{G}_{q^k}$ contains a unique subring isomorphic to $\mathbb{G}_q$. It is known that $\mathbb{G}_{q^k}$ is free of rank k as a module over $\mathbb{G}_q$. Hence, $\mathbb{G}_{q^k}$ can be viewed as the underlying module of the $(k-1)$-dimensional projective Hjelmslev geometry over $\mathbb{G}_q$. We denote this geometry by $\mathrm{PHG}(\mathbb{G}_{q^k}/\mathbb{G}_q)$.

The group $\mathbb{G}_q^\times$ of units of $\mathbb{G}_q$ contains a unique cyclic subgroup T_q of order $q-1$, called the group of Teichmüller units. This applies to both $\mathbb{G}_q$ and its extension ring $\mathbb{G}_{q^k}$, and we have $\mathrm{T}_{q^k} = \langle \eta \rangle$, $\mathrm{T}_q = \langle \eta^{(q^k-1)/(q-1)} \rangle$ for any element $\eta \in \mathbb{G}_{q^k}^\times$ of order $q^k - 1$.

[4] Since $\mathcal{C}_1 = -\mathcal{C}_1$, this actually defines an undirected graph.

Definition 4.13. The set $\{\mathbb{G}\eta^j \mid 0 \le j < (q^k-1)/(q-1)\}$ is called the *Teichmüller set of* $\mathrm{PHG}(\mathbb{G}_{q^k}/\mathbb{G}_q)$ and is denoted by $\mathfrak{T}_{q,k}$.

Since $\{\eta^j \mid 0 \le j < (q^k-1)/(q-1)\}$ is a set of coset representatives for T_q in T_{q^k}, the Teichmüller set $\mathfrak{T}_{q,k}$ contains exactly one point from each neighbour class. In case of $\mathbb{G}_2 = \mathbb{Z}_4$, k odd, the linear code over $\mathbb{Z}_4$ associated with $\mathfrak{T}_{2,k}$ (via the columns of a generator matrix) is isomorphic to the shortened quaternary Kerdock code; cf. [13,47].

Recall that a set of points is called a cap if no three points of this set are collinear.

Theorem 4.14. *Let* $\mathbb{G}_q = \mathrm{GR}(q^2, p^2)$ *be a Galois ring of characteristic* p^2 *and let* $k \ge 3$ *be an integer.*

- *If every prime divisor of* k *is larger than* p, *then the Teichmüller set* $\mathfrak{T}_{q,k}$ *is a cap in* $\mathrm{PHG}(\mathbb{G}_{q^k}/\mathbb{G}_q)$.
- *If* k *is even,* $\mathfrak{T}_{q,k}$ *is never a cap.*

In particular, the Teichmüller set $\mathfrak{T}_{2^r,3}$ *forms a* $(2^{2r}+2^r+1, 2)$*-arc in the projective Hjelmslev plane* $\mathrm{PHG}(\mathbb{G}_{2^{3r}}/\mathbb{G}_{2^r}) \cong \mathrm{PHG}(\mathbb{G}_{2^r}^3)$ *over the Galois ring* $\mathbb{G}_{2^r}$.

For projective Hjelmslev planes over chain rings R containing a subring isomorphic to the residue field of R, the following result holds [26].

Theorem 4.15. *Let* R *be a chain ring with* $|R| = 2^{2r}$, $R/N \cong \mathbb{F}_{2^r}$, $\mathrm{char}\, R = 2$. *Then there exists no* $(2^{2r}+2^r+1, 2)$*-arc in the projective Hjelmslev plane* $\mathrm{PHG}(R_R^3)$.

At present, it is not known whether $(2^{2r}+2^r, 2)$-arcs do exist over chain rings of nilpotency index 2 and characteristic 2, except for the two smallest cases. The answer is positive for $q = 2$, but negative for $q = 4$; see [30].

Recently it was proved that for odd prime characteristics the bound (9) is tight as well [17].

Theorem 4.16. *Let* $R = \mathbb{F}_q[X;\sigma]/(X^2)$ *be a chain ring of length* 2 *and prime characteristic. There exists a* $(q^2, 2)$*-arc in the projective Hjelmslev plane* $\mathrm{PHG}(R_R^3)$.

Further it is known that the maximum size of a 2-arc in the plane over $\mathbb{Z}_{25}$ is 21; see [31,33]. Below, we give a $(21,2)$-arc in $\mathrm{PHG}(\mathbb{Z}_{25}^3)$ taken from the online tables [1]. The points are represented by the columns of a 3×21-matrix over $\mathbb{Z}_{25}$.

$$\begin{pmatrix} 0 & 1 & 5 & 1 & 1 & 15 & 1 & 1 & 10 & 1 & 1 & 1 & 1 & 1 & 1 & 0 & 1 & 1 & 1 & 1 & 1 \\ 0 & 5 & 1 & 7 & 15 & 1 & 0 & 3 & 1 & 11 & 18 & 24 & 2 & 13 & 22 & 1 & 1 & 20 & 4 & 12 & 14 \\ 1 & 0 & 6 & 17 & 24 & 4 & 1 & 3 & 18 & 7 & 15 & 7 & 8 & 22 & 11 & 15 & 11 & 23 & 1 & 24 & 3 \end{pmatrix}$$

4.5. Dual Constructions

Let $\Pi = (\mathcal{P}, \mathcal{L}, I) = \mathrm{PHG}(R_R^3)$ be a coordinate projective Hjelmslev plane over a finite chain ring R. Using duality properties of the inner product $R^3 \times R^3 \to R$: $(\mathbf{x}, \mathbf{y}) \mapsto \mathbf{x} \cdot \mathbf{y} = x_1y_1 + x_2y_2 + x_3y_3$, one can show that the dual plane $\Pi^* = (\mathcal{L}, \mathcal{P}, I^*)$ is isomorphic to the left coordinate plane $\mathrm{PHG}({}_RR^3)$ or, what is the same, to the projective Hjelmslev plane $\mathrm{PHG}(S_S^3)$ over the opposite ring $S = R^\circ$. This duality can be exploited in some cases for new constructions of arcs with good parameters [28].

Example 4.17. There exist maximal $((q^4-q)/2, q^2/2)$-arcs in the projective Hjelmslev planes over the Galois rings $\mathbb{G}_q$, $q=2^r$. These arcs are obtained by taking $\mathfrak{K}$ as the set of passants (0-lines) of a $(q^2+q+1,2)$-arc in the corresponding dual plane. The new arcs have intersection numbers 0 and $q^2/2$ with the lines of the dual plane and so are maximal. Since $\mathbb{G}_q = \mathbb{G}_q^\circ$, the result follows.

In the smallest case $q=2$, the $(7,2)$-arc in $\mathrm{PHG}(\mathbb{Z}_4^3)$ is self-dual. In all other cases, Example 4.17 gives new arcs not covered by previous constructions, for example a $(126,8)$-arc in the plane over $\mathbb{G}_4$.

Theorem 4.18. *Let R be a chain ring with $|R|=2^{2r}$, $R/N \cong \mathbb{F}_{2^r}$, $\operatorname{char} R = 2$. Then*

$$m_{q^2/2}(R_R^3) \le q^4/2 - q/2 - 1. \tag{11}$$

Since $(q^2+q+1,2)$-arcs and $((q^4-q)/2, q^2/2)$-arcs are dual to each other, this theorem is a corollary of Theorem 4.15.

4.6. Constructions Using Automorphisms

By the Fundamental Theorem of Projective Hjelmslev Geometry [39], every collineation of a coordinate projective Hjelmslev plane $\Pi = \mathrm{PHG}(R_R^3)$ over a finite chain ring is induced by a semilinear automorphism of the underlying module R_R^3, and the collineation group of the plane $\mathrm{PHG}(R_R^3)$ is isomorphic to the projective semilinear group $\mathrm{P\Gamma L}(3,R) = \Gamma\mathrm{L}(3,R)/\mathrm{Z}(R)$, where $\mathrm{Z}(R)$ denotes the center of the ring R.

Automorphisms of Π can be used to considerably shorten searches for arcs with good parameters and make computer constructions of such arcs feasible which would otherwise be out of reach. As a simple example, we mention the fact that one can always assume the standard quadrangle $(1,0,0)R$, $(0,1,0)R$, $(0,0,1)R$, $(1,1,1)R$ to be part of $\mathfrak{K}$, since $\mathrm{PGL}(3,R)$ acts transitively on ordered quadrangles in Π.

The construction of discrete objects using incidence preserving group actions pioneered by Kerber et al. [3,29] can also be applied to the construction of arcs in projective Hjelmslev planes. To make the resulting computational tasks feasible for larger planes, one restricts attention to arcs which are invariant under certain automorphisms of Π, for example (lifted) Singer cycles of the factor plane $\mathrm{PG}(2,q)$. This method has been used successfully in [21, 32] for the construction of new arcs with good parameters, accounting for many entries (lower bounds) in the tables of Section 4.7.. The authors of [32] also maintain online tables of optimal arcs in projective Hjelmslev planes of small sizes [1].

Suppose now that Π is a projective Hjelmslev plane over a Galois ring $\mathbb{G}_q$, represented as $\mathrm{PHG}(\mathbb{G}_{q^3}/\mathbb{G}_q)$ (cf. Section 4.4.). A generator η of the Teichmüller subgroup T_{q^3} of $\mathbb{G}_{q^3}^\times$ induces a collineation $\sigma \in \mathrm{Aut}(\Pi)$ of order q^2+q+1, which acts as a Singer cycle on the factor plane $\mathrm{PG}(2,q)$. There is obviously a one-to-one correspondence between σ-invariant multisets in Π and multisets in a fixed point neighbour class of Π, for example $[\mathbb{G}_q 1]$. For a σ-invariant multiset $\mathfrak{K}$ in Π, it is possible to compute the $\mathfrak{K}$-types of all lines in Π from certain combinatorial data of the corresponding multiset $\mathfrak{k}$ in $[\mathbb{G}_q 1] \cong \mathrm{AG}(2,q)$. As shown in [19], suitable choices of $\mathfrak{k}$ yield σ-invariant arcs with good parameters. As an example of this construction, we mention a family of arcs in the planes over $\mathbb{G}_p$, where p is an odd prime, which includes an optimal $(39,5)$-arc in the plane over $\mathbb{Z}_9$. A multiset $\mathfrak{k}$ in $\mathrm{AG}(2,p)$ is called a *triangle set* if it is affinely equivalent to the set $\{(x,y) \in \mathbb{F}_p^2 \mid x+y < p-1\}$. Here $\mathbb{F}_p = \{0,1,\ldots,p-1\}$ is considered as a subset of $\mathbb{Z}$.

Theorem 4.19 ([19]). *For every odd prime p, there exists a σ-invariant $\big((p^4-p)/2,(p^2+p)/2-1\big)$-arc in the projective Hjelmslev plane over the Galois ring $\mathbb{G}_p$. The arc is induced from an appropriately chosen triangle set in $[\mathbb{G}_p 1] \cong \mathrm{AG}(2,p)$.*

Finally we want to note that arcs in projective Hjelmslev planes with extremal parameters may be of interest also from a group theoretic point-of-view (just like their classical counterparts). This is exemplified by the following result.

Proposition 4.20 ([16]). *The set $\mathfrak{H}$ of hyperovals (maximal $(7,2)$-arcs) of $\mathrm{PHG}(2,\mathbb{Z}_4)$ has cardinality 256. The automorphism group G of $\mathrm{PHG}(2,\mathbb{Z}_4)$ acts transitively on $\mathfrak{H}$ and the stabilizer $G_{\mathfrak{h}}$ of a hyperoval $\mathfrak{h} \in \mathfrak{H}$ has order 168. Furthermore, G has a normal subgroup H which acts regularly on $\mathfrak{H}$.*

4.7. Tables for Arcs in Geometries over Small Chain Rings

In the tables below, we summarize our knowledge about the values of $m_n(R_R^3)$ for the chain rings R with $|R| = q^2 \leq 25$, $R/\operatorname{rad} R \cong \mathbb{F}_q$ (cf. also [1,18,20,21]). We give information about all values of n with $2 \leq n \leq q^2-1$. The cases $n = q^2,\ldots,q^2+q$ are covered by Example 4.7. We want to remark the fact that we have lots of examples with $m_n(R_R^3) \neq m_n(S_S^3)$ for non-isomorphic chain rings R, S with $|R| = |S|$, $R/\operatorname{rad} R \cong S/\operatorname{rad} S$ (cf. Theorems 4.14, 4.15 and 4.16 and the results in Section 4.5.). However, in all these examples $\operatorname{char} R \neq \operatorname{char} S$, and we do not have a single example of chain rings R and S of the same order, length and characteristic, in which the values of $m_n(R_R^3)$ and $m_n(S_S^3)$ are different.

Table 1. Values of $m_n(R_R^3)$ for Hjelmslev planes of order $q^2 = 4$ and $q^2 = 9$

n/R	$\mathbb{Z}_4$	$\mathbb{S}_2$	$\mathbb{Z}_9$	$\mathbb{S}_3$
2	7	6	9	9
3	10	10	19	18
4			30	30
5			39	38
6			49	50
7			60	60
8			69	69

5. Blocking Sets in Projective Hjelmslev Planes

5.1. General Results

Definition 5.1. A multiset $\mathfrak{K}$ in $(\mathcal{P},\mathcal{L},I)$ is called a *(k,n)-blocking multiset* if

(i) $\mathfrak{K}(\mathcal{P}) = k$;

Table 2. Values of $m_n(R_R^3)$ for Hjelmslev planes of order $q^2 = 16$

n/R	$\mathbb{G}_4$	$\mathbb{S}_4$	$\mathbb{T}_4$
1	1	1	1
2	21	18	18
3	29 – 30	29 – 30	29 – 30
4	52	52	52
5	68	68	68
6	84	81 – 83	81 – 83
7	97 – 101	99 – 101	96 – 101
8	126	120 – 125	120 – 125
9	140	140	140
10	152 – 160	152 – 160	152 – 160
11	166 – 169	166 – 169	166 – 169
12	186 – 189	186 – 189	186 – 189
13	203 – 208	202 – 208	202 – 208
14	224 – 228	216 – 228	219 – 228
15	236 – 248	236 – 248	236 – 248

Table 3. Values of $m_n(R_R^3)$ for Hjelmslev planes of order $q^2 = 25$

n/R	$\mathbb{Z}_{25}$	$\mathbb{S}_5$	n/R	$\mathbb{Z}_{25}$	$\mathbb{S}_5$
1	1	1	13	310 – 311	297 – 311
2	21	25	14	319 – 341	318 – 341
3	40 – 43	42 – 43	15	355 – 367	355 – 367
4	66 – 70	64 – 70	16	375 – 395	375 – 395
5	85 – 102	90 – 102	17	400 – 425	405 – 425
6	114 – 130	130	18	425 – 455	433 – 455
7	142 – 156	152 – 156	19	465 – 466	455 – 466
8	162 – 186	162 – 186	20	490 – 496	490 – 496
9	186 – 208	190 – 208	21	515 – 525	515 – 525
10	210 – 238	225 – 238	22	540 – 555	540 – 555
11	235 – 265	250 – 265	23	565 – 585	565 – 585
12	264 – 295	280 – 295	24	595 – 615	595 – 615

(ii) $\mathfrak{K}(L) \geq n$ for every line $L \in \mathcal{L}$.

Similarly to Definition 4.1, we assume in addition that there exists at least one line L_0 with $\mathfrak{K}(L_0) = n$. A (k,n)-blocking multiset $\mathfrak{K}$ is called *irreducible* if it does not contain a $(k-1,n)$-blocking multiset, i.e. decreasing the multiplicity of any point $p \in \operatorname{supp}\mathfrak{K}$ by one yields a multiset $\mathfrak{K}'$ with $\mathfrak{K}'(L) = n-1$ for some line $L \in \mathcal{L}$. Blocking sets (i.e. projective blocking multisets) and projective arcs are complementary concepts in the sense that the complement of a projective (k,n)-arc in $\mathcal{P}$ is a $(q^4+q^3+q^2-k, q^2+q-n)$-blocking set and vice versa.

First, let us consider blocking sets in planes over general chain rings R with $|R| = q^m$, $R/N \cong \mathbb{F}_q$. For (k,n)-blocking sets in such planes, we have the following theorem [41].

Theorem 5.2. *Let R be a chain ring with $|R| = q^m$, $R/N \cong \mathbb{F}_q$, and let $\mathfrak{K}$ be a (k,n)-blocking multiset with $1 \le n \le q$, in $\Pi = \mathrm{PHG}(R_R^3)$. Then $k \ge nq^{m-1}(q+1)$. If $\mathfrak{K}$ is a (k,n)-blocking multiset with $k = nq^{m-1}(q+1)$, $n < q/p$, where $p = \operatorname{char} \mathbb{F}_q$, then there exist lines, $L_1, L_2, \ldots, L_n$ say, such that*

$$\mathfrak{K}^{(1)}([x]) = q^{m-1} |\{j \mid j \in \{1,\ldots,n\}, ([x],[L_j]) \in J^{(1)}\}|.$$

The second part of the theorem says that the induced multiset $\mathfrak{K}^{(1)}/q^{m-1}$ is a sum of lines. It is impossible to generalize this to the stronger condition: "$\mathfrak{K}^{(i)}/q^{m-i}$ is a sum of lines for some $i > 1$".

For the most interesting case of $(k,1)$-blocking sets, we have $k \ge q^{m-1}(q+1)$ and in case of equality the support of such a blocking set is necessarily a line. By taking a line L and from each class $[x]_{m-1}$ incident with $[L]_{m-1}$ in $(\mathcal{P}^{(m-1)}, \mathcal{L}^{(m-1)}, J^{(m-1)})$ exactly $n-1$ further line segments in the direction of L, one obtains for each $n \in \{1,2,\ldots,q\}$ an $(n, nq^{m-1}(q+1))$-blocking set, showing that the extremal cases $k = nq^{m-1}(q+1)$ of Theorem 5.2 can be realized by projective multisets.

Under certain conditions, some subplanes of $\mathrm{PHG}(R_R^3)$ form a blocking set.

Theorem 5.3. *Let R be a chain ring with $|R| = q^m$, $R/N \cong \mathbb{F}_q$, where q^m is a perfect square. Let there exist a subring S of R that is a chain ring with $|S| = q^{m/2}$ and such that R is free over S. Then the multiset $\mathfrak{K}$ defined by*

$$\mathfrak{K}(x) = \begin{cases} 1 & \text{if } x \text{ is a point from } \mathrm{PHG}(S_S^3), \\ 0 & \text{otherwise,} \end{cases}$$

is a blocking set in $\mathrm{PHG}(R_R^3)$.

In the special case when R is a chain ring with $|R| = q^2$, $R/N \cong \mathbb{F}_q$, that contains a subring S isomorphic to the residue field $\mathbb{F}_q$, $\mathrm{PHG}(R_R^3)$ contains a subplane Π' isomorphic to $\mathrm{PG}(2,q)$ and the projective multiset $\mathfrak{K}$ defined by $\operatorname{supp} \mathfrak{K} = \Pi'$ is an irreducible $(q^2+q+1,1)$-blocking set. These blocking sets are introduced in [5] in a slightly different context. They are defined as the orbit of a fixed point with coordinates from the field $\mathbb{F}_q$ under a Singer cycle of $\mathrm{PG}(2,q)$. As shown in [6], linear codes associated with these multisets can be mapped (cf. [23]) to two-weight *linear* codes over $\mathbb{F}_q$. These in turn give rise to a family of strongly regular graphs with parameters

$$v = q^6, \quad k = q^4 - q, \quad \lambda = q^3 + q^2 - 3q, \quad \mu = q^2 - q.$$

Let us now consider planes over chain rings with $|R| = q^2$, $R/N \cong \mathbb{F}_q$. It is of interest to find the smallest size of an irreducible blocking set which is not a line. Unlike the situation in the classical projective planes where there is a gap between the size of a line and the size of the smallest non-trivial blocking sets (see e.g. [4]), there exist irreducible blocking sets of size q^2+q+1 in all planes $\mathrm{PHG}(R_R^3)$.

Theorem 5.4. *Let $\mathfrak{K}$ be an irreducible $(q^2+q+1,1)$-blocking set in $\mathrm{PHG}(R_R^3)$, $|R| = q^2$, $R/\operatorname{rad} R \cong \mathbb{F}_q$. Then $\mathfrak{K}$ is of one of the following types:*

(1) *a projective plane of order q;*

(2) *for lines L_0 and L_1 with $L_0 \frown L_1$, and a point $z \in L_0 \setminus L_1$*

$$\mathfrak{K}(x) = \begin{cases} 1 & \text{if } x \in (L_0 \setminus [z]) \cup \{z\} \text{ or } x \in L_1 \cap [z] \\ 0 & \text{otherwise.} \end{cases} \tag{12}$$

If $R = \mathrm{GR}(q^2, p^2)$, then there is no $(q^2+q+1, 1)$-blocking set of type (1).

Let us note that the blocking set described in (12) is in some sense trivial since $\mathfrak{K}^{(1)} = q \cdot \chi_{[L]} + \chi_{[z]}$ consists of a q-fold line and a further point on this line. We would like to construct non-trivial blocking sets also for the planes over the Galois rings $\mathbb{G}_q$. This can be done by generalizing the familiar technique of Rédei type blocking sets to projective Hjelmslev planes.

5.2. Rédei Type Blocking Sets

As before let $\Pi = \mathrm{PHG}(R_R^3)$, where R is a chain ring of nilpotency index 2. Fix a generator θ of $\operatorname{rad} R$ and a set $\Gamma \subset R$ of representatives for the residue classes in $R/\operatorname{rad} R \cong \mathbb{F}_q$. Suppose that $\Gamma = \{\gamma_0, \gamma_1, \ldots, \gamma_{q-1}\}$ with $\gamma_0 = 0$, $\gamma_1 = 1$, and hence $\operatorname{rad} R = \{\gamma_i\theta \mid 0 \le i \le q-1\} = \{\theta\gamma_j \mid 0 \le j \le q-1\}$. Thus each $c \in \operatorname{rad} R$ has unique representations $c = \gamma_i\theta = \theta\gamma_j$, where in the non-commutative cases i, j may be different.

As already noted, the affine plane $\mathrm{AHG}(R_R^2)$ is obtained by deleting a neighbour class of lines (the "class at infinity") together with all points incident with a line in this class. With no loss of generality we can take the class $[z=0]$ as the class at infinity. This class consists of all lines with equations of the form $aX + bY + Z = 0$, where $a, b \in \operatorname{rad} R$. All points incident with lines in this class have homogeneous coordinates (x, y, z) with $z \in \operatorname{rad} R$. The points outside this class have coordinates $(x, y, 1)$, $x, y \in R$. Now the points of the affine plane $\mathrm{AHG}(R_R^2)$ are identified with the pairs (x, y), where $x, y \in R$. The lines of $\mathrm{AHG}(R_R^2)$ have equations $Y = aX + b$ or $X = cY + d$, $a, b, d \in R$, $c \in \operatorname{rad} R$. We say that a line of the first type has slope a. A line with equation $X = cY + d$ is said to have slope ∞_j, if $c = \theta\gamma_j$, $j = 0, 1, \ldots, q-1$.

The infinite points on a fixed line L from the neighbour class of infinite lines can be identified with the slopes. So, (a) (resp. (∞_j)) will denote the infinite point from L of the lines with slope a (resp. ∞_j). The q^2 lines with a fixed slope form a parallel class of lines in $\mathrm{AHG}(R_R^2)$, and the line set of $\mathrm{AHG}(R_R^2)$ is partitioned into q^2+q such parallel classes.

Definition 5.5. Let U be a set of q^2 points in $\mathrm{AHG}(R_R^2)$. We say that the infinite point (a) is determined by U if there exist different points $u, v \in U$ such that u, v and (a) are collinear in $\mathrm{PHG}(R_R^3)$.

Note that in view of the assumption $|U| = q^2$, the point (a) is determined by U iff there exists a line in $\mathrm{AHG}(R_R^2)$ with slope a which is disjoint from U.

Theorem 5.6 ([42]). *Let U be a set of q^2 points in $\mathrm{AHG}(R_R^2)$. Denote by D the set of infinite points determined by U and by $D^{(1)}$ the set of neighbour classes on the infinite line containing points from D. If $|D| < q^2+q$, then there exists an irreducible blocking set in*

$\mathrm{PHG}(R_R^3)$ of size $q^2+q+1+|D|-|D^{(1)}|$ that contains U. In particular, if D contains representatives from all neighbour classes on the infinite line, then $B=U\cup D$ is an irreducible blocking set of size $q^2+|D|$ in $\mathrm{PHG}(R_R^3)$.

The above construction gives blocking sets of size at most $2q^2+q-1$. We are interested in sets U that are of the form

$$U=\{(x,f(x))\mid x\in R\}$$

for some suitably chosen function $f\colon R\to R$. Let x and y be two different elements from R. We have the following possibilities:

1) if $x-y\notin \operatorname{rad}R$, then $(x,f(x))$ and $(y,f(y))$ determine the point (a), where
$$a=(f(x)-f(y))(x-y)^{-1}.$$

2) if $x-y\in\operatorname{rad}R\setminus\{0\}$, and $f(x)-f(y)\notin\operatorname{rad}R$, the points $(x,f(x))$ and $(y,f(y))$ determine the point (∞_j) if
$$(x-y)(f(x)-f(y))^{-1}=\theta\gamma_j,\ \gamma_j\in\Gamma.$$

3) if $x-y\in\operatorname{rad}R\setminus\{0\}$, and $f(x)-f(y)\in\operatorname{rad}R$, say $x-y=\theta a$, $f(x)-f(y)=\theta b$, $a,b\in\Gamma$ and

   a) if $b\neq 0$, then $(x,f(x))$ and $(y,f(y))$ determine all points (c) with $c\in ab^{-1}+\operatorname{rad}R$;
   b) if $b=0$, then $(x,f(x))$ and $(y,f(y))$ determine the infinite points $(\infty_0),\ldots,(\infty_{q-1})$.

Example 5.7. Let R be a chain ring with $|R|=q^2$, $R/\operatorname{rad}R\cong\mathbb{F}_q$ that contains a proper subring isomorphic to its residue field $\mathbb{F}_q$ (i.e. one of the rings $\mathbb{S}_q$ or $\mathbb{T}_q^{(i)}$).

Define
$$f\colon R\to R\colon\ a+\theta b\mapsto b+\theta a. \tag{13}$$
It can be checked that the set of points $U=\{(x,f(x))\mid x\in R\}$ determines $q+1$ infinite points.

We can compute the parameters of the Rédei-type blocking sets given by (13) also for the plane over the Galois ring $\mathbb{G}_q=\mathrm{GR}(q^2,p^2)$. In this case, U determines exactly q^2-q+2 directions, and the size of the corresponding Rédei-type blocking set is $2q^2-q+2$.

Below we will give two further of examples Rédei-type blocking sets in the plane over $\mathbb{G}_q$. For these examples, we need to collect a few additional facts about $\mathbb{G}_q$.

In the case of $\mathbb{G}_q$ (and Galois rings in general), there are canonical choices for θ and Γ, which we will adopt for the rest of this paper. Since $\operatorname{rad}\mathbb{G}_q=p\mathbb{G}_q$, we can take $\theta=p$ as a generator of $\operatorname{rad}\mathbb{G}_q$. Furthermore, since the augmented Teichmüller subgroup $\Gamma_q:=\mathrm{T}_q\cup\{0\}$ (for the definition of T_q, see Section 4.4.) forms a system of coset representatives modulo $\operatorname{rad}\mathbb{G}_q$, we can take $\Gamma=\Gamma_q$.

Every $a\in\mathbb{G}_q$ can be written in exactly one way as $a=a_0+a_1p$ with $a_0,a_1\in\Gamma_q$.[5]

[5]This is true regardless of the particular choice of Γ. However, for the following Fact 5.8 the choice $\Gamma=\mathrm{T}_q\cup\{0\}$ is essential.

Fact 5.8. *The ring $\mathbb{G}_q$ is isomorphic to the ring $W_2(\mathbb{F}_q)$ of so-called* Witt vectors of length 2 over $\mathbb{F}_q$, *which is defined as the set of all pairs $(a,b) \in \mathbb{F}_q \times \mathbb{F}_q$ with the following addition and multiplication:*

$$\begin{aligned}(a_0,a_1)+(b_0,b_1) &= (a_0+b_0, a_1+b_1-\sum_{j=1}^{p-1}\frac{1}{p}\binom{p}{j}a_0^j b_0^{p-j}),\\ (a_0,a_1)\cdot(b_0,b_1) &= (a_0b_0, a_0^p b_1 + b_0^p a_1).\end{aligned}$$

The map $\phi\colon \mathbb{G}_q \to W_2(\mathbb{F}_q)\colon a_0+a_1p \mapsto (\overline{a_0}, \overline{a_1}^p)$, where $\overline{a} = a + \operatorname{rad}\mathbb{G}_q$, provides a ring isomorphism.

For the definition of Witt vectors see [51], and for a proof of 5.8 see [49]. Working with Witt vectors instead of the original representation of $\mathbb{G}_q = \mathbb{Z}_{p^2}[X]/(h)$ has the advantage that all computations are now done in $\mathbb{F}_q$.

Example 5.9. Let $q = p^r$, where p is odd. We are going to define f as a function on $W_2(\mathbb{F}_q)$. For $x = (a_0, a_1)$, set

$$f(x) = \begin{cases} (a_0,a_1) & \text{if } a_0 \text{ is a square in } \mathbb{F}_q,\\ (-a_0,-a_1) & \text{if } a_0 \text{ is a non-square in } \mathbb{F}_q.\end{cases} \tag{14}$$

Theorem 5.10 ([42]). *Let $R = \mathrm{GR}(q^2,p^2)$, $q = p^m$, p odd. The set $U = \{(x,f(x)) \mid x \in R\}$, where the function is defined in (14), determines*

$$\frac{q^2}{2}+\frac{3}{2}q$$

directions in $\mathrm{AHG}(R_R^2)$. Furthermore, there exists a Rédei type blocking set in $\mathrm{PHG}(R_R^3)$ of size

$$\frac{3}{2}q^2+2q-\frac{1}{2}.$$

In our last example, we will construct a Rédei type blocking set over the Galois ring $S = \mathbb{G}_{q^m}$, where $m \geq 1$ is arbitrary, using the fact that S is a Galois extension of $R = \mathbb{G}_q$. Recall that the trace function $\mathrm{Tr}_{S/R}\colon S \to R$ is defined as

$$\mathrm{Tr}_{S/R}(x) := \sum_{\sigma\in\mathrm{Aut}(S/R)} \sigma(x) = \sum_{i=0}^{m-1}(x_0^{q^i}+x_1^{q^i}p) \quad \text{for} \quad x \in S, \tag{15}$$

where $x = x_0 + x_1 p$ with $x_0, x_1 \in \Gamma_{q^m}$.

Example 5.11. As above let $R = \mathbb{G}_q$ and $S = \mathbb{G}_{q^m}$. We define a Rédei type blocking set in $\mathrm{PHG}(S_S^3)$ by setting $f(x) = \mathrm{Tr}_{S/R}(x)$.

Theorem 5.12 ([42]). *Let $R = \mathrm{GR}(q^2,p^2)$ and let S be an extension of R of degree m. The set $U = \{(x,f(x)) \mid x \in S\}$ defined by the function $f(x) = Tr_{S/R}(x)$ determines*

$$\frac{q^m-1}{q-1}q^m$$

directions in $\mathrm{AHG}(S_S^2)$. There exists a Rédei type blocking set in $\mathrm{PHG}(S_S^3)$ of size

$$q^{2m}+q^m+1+\frac{q^m-1}{q-1}q^m-q^{m-1}.$$

Acknowledg nents

The authors wish to thank Jan de Beule and Leo Storme for encouragement and support during various stages of the present survey and Michael Kiermaier for help with the tables in Section 4.7. and with Examples 4.10 and 4.17.

The first author was supported by the Open Project of Zhejiang Provincial Key Laboratory of Information Network Technology and by the National Natural Science Foundation of China under Grant No. 60872063. The second author was supported by the Project *Combined algorithmic and theoretical study of combinatorial structures* between the Research Foundation – Flanders (Belgium) (FWO) and the Bulgarian Academy of Sciences, as well as by the Strategic Development Fund of the New Bulgarian University under Contract 357/14.05.2009.

References

[1] Onlinetabl esofarcsinprojecti veHjel mslevplanes. http: //www.algori thmuni-bayreuth. de/en/research/ Coding_Theory/PHG_arc_table/index. html.

[2] B. Artmann, *Hjelmslev-Ebenen mit verfeinerten Nachbarschaftsrelationen*, Math. Z., 112 (1969), pp. 163–180.

[3] A. Betten, M. Braun, H. Fripertinger, A. Kerber, A. Kohnert, and A. Wassermann, *Error-correcting linear codes. Classification by isometry and applications*, vol. 18 of Algorithms and Computation in Mathematics, Springer-Verlag, Berlin, 2006.

[4] A. Blokhuis, P. Sziklai, and T. Szőnyi, *Blocking sets in projective spaces*, in Current research topics in Galois geometry, Nova Sci. Publ., New York, 2011, ch. 3, pp. 61–84.

[5] E. Byrne, M. Greferath, and T. Honold, *Two-weight codes over finite Frobenius rings and strongly regular graphs*, in Proceedings of the Fourth International Workshop on Optimal codes and related topics, Pamporovo, Bulgaria, 2005, pp. 64–73.

[6] ——, *Ring geometries, two-weight codes, and strongly regular graphs*, Des. Codes Cryptogr., 48 (2008), pp. 1–16.

[7] W. E. Clark and D. A. Drake, *Finite chain rings*, Abh. Math. Sem. Univ. Hamburg, 39 (1973), pp. 147–153.

[8] A. Cronheim, *Dual numbers, Witt vectors, and Hjelmslev planes*, Geom. Dedicata, 7 (1978), pp. 287–302.

[9] F. De Clerck and N. Durante, *Constructions and characterizations of classical sets in* PG(n,q), in Current research topics in Galois geometry, Nova Sci. Publ., New York, 2011, ch. 1, pp. 1–32.

[10] P. Dembowski, *Finite geometries*, Ergebnisse der Mathematik und ihrer Grenzgebiete, Band 44, Springer-Verlag, Berlin, 1968.

[11] S. Dodunekov and J. Simonis, *Codes and projective multisets*, Electron. J. Combin., 5 (1998), Research Paper 37, 23 pp. (electronic).

[12] D. A. Drake, *On n-uniform Hjelmslev planes*, J. Combinatorial Theory, 9 (1970), pp. 267–288.

[13] A. R. Hammons, Jr., P. V. Kumar, R. Calderbank, N. J. A. Sloane, and P. Solé, *The $\mathbf{Z}_4$-linearity of Kerdock, Preparata, Goethals, and related codes*, IEEE Trans. Inform. Theory, 40 (1994), pp. 301–319.

[14] L. Hemme, T. Honold, and I. Landjev, *Arcs in projective Hjelmslev spaces obtained from Teichmüller sets.* Seventh international workshop on algebraic and combinatorial coding theory, ACCT-7, Bansko, Bulgaria, June 18–24, 2000. Proceedings. Sofia: Bulgarian Academy of Sciences, Institute of Mathematics and Informatics. 177-182, 2000.

[15] J. W. P. Hirschfeld, *Projective geometries over finite fields*, Oxford Mathematical Monographs, The Clarendon Press Oxford University Press, New York, second ed., 1998.

[16] T. Honold and M. Kiermaier, *Classification of maximal arcs in small projective Hjelmslev geometries*, in Proceedings of the Tenth International Workshop on Algebraic and Combinatorial Coding Theory (ACCT-10), Zvenigorod, Russia, 2006, pp. 112–117.

[17] ——, *The existence of maximal $(q^2, 2)$-arcs in projective Hjelmslev planes over chain rings of odd prime characteristic.* Submitted for publication.

[18] ——, *The maximal size of* 6- *and* 7-*arcs in projective Hjelmslev planes over chain rings of order* 9. Submitted for publication.

[19] ——, *Singer arcs in uniform projective Hjelmslev planes over Galois rings.* In preparation.

[20] T. Honold, M. Kiermaier, and I. Landjev, *New arcs of maximal size in projective Hjelmslev planes of order nine*, C. R. Acad. Bulgare Sci., 63 (2010), pp. 171–180.

[21] T. Honold, M. Kiermaier, A. Kohnert, I. Landjev, and J. Zwanzger, *New Results on Arcs in Projective Hjelmslev planes over Small Chain Rings.* In preparation.

[22] T. Honold and I. Landjev, *Projective Hjelmslev Geometries*, in Proceedings of the International Workshop on Optimal Codes, Sozopol, 1998, pp. 97–115.

[23] ——, *Linearly representable codes over chain rings*, Abh. Math. Sem. Univ. Hamburg, 69 (1999), pp. 187–203.

[24] ——, *Linear codes over finite chain rings*, Electron. J. Combin., 7 (2000), Research Paper 11, 22 pp. (electronic).

[25] ——, *On arcs in projective Hjelmslev planes*, Discrete Math., 231 (2001), pp. 265–278. 17th British Combinatorial Conference (Canterbury, 1999).

[26] ——, *On maximal arcs in projective Hjelmslev planes over chain rings of even characteristic*, Finite Fields Appl., 11 (2005), pp. 292–304.

[27] ——, *Caps in projective Hjelmslev spaces over finite chain rings of nilpotency index 2*, Innov. Incidence Geom., 4 (2006), pp. 13–25.

[28] ——, *The dual construction for arcs in projective Hjelmslev planes*, Adv. Math. Commun., 5 (2011), pp. 11–21.

[29] A. Kerber, *Applied finite group actions*, vol. 19 of Algorithms and Combinatorics, Springer-Verlag, Berlin, second ed., 1999.

[30] M. Kiermaier, *Arcs und Codes über endlichen Kettenringen*, Master's thesis, Technische Universität München, 2006.

[31] M. Kiermaier and M. Koch, *New complete arcs in projective Hjelmslev planes over chain rings of order 25*, in Proceedings of the Sixth International Workshop on Optimal codes and related topics, Varna, Bulgaria, 2009, pp. 106–113.

[32] M. Kiermaier and A. Kohnert, *New arcs in projective Hjelmslev planes over Galois rings*, in Proceedings of the Fifth International Workshop on Optimal codes and related topics, White Lagoon, Bulgaria, 2007, pp. 112–117.

[33] M. Kiermaier, M. Koch, and S. Kurz, *2-arcs of maximal size in the affine and the projective Hjelmslev plane over* $\mathbf{Z}_{25}$, Adv. Math. Commun., to appear.

[34] E. Kleinfeld, *Finite Hjelmslev planes*, Illinois J. Math., 3 (1959), pp. 403–407.

[35] W. Klingenberg, *Projektive und affine Ebenen mit Nachbarelementen*, Math. Z., 60 (1954), pp. 384–406.

[36] A. Kreuzer, *Hjelmslev-Räume*, Results Math., 12 (1987), pp. 148–156.

[37] ——, *Projektive Hjelmslev-Räume*, doctoral dissertation, Technische Universität München, 1988.

[38] ——, *Hjelmslevsche Inzidenzgeometrie – Ein Bericht*, in Beiträge zur Geometrie und Algebra Nr. 17., Technische Universität München, 1990, pp. 31 – 45.

[39] ——, *Fundamental theorem of projective Hjelmslev spaces*, Mitt. Math. Ges. Hamburg, 12 (1991), pp. 809–817. Mathematische Wissenschaften gestern und heute. 300 Jahre Mathematische Gesellschaft in Hamburg, Teil 3 (Hamburg, 1990).

[40] ——, *A system of axioms for projective Hjelmslev spaces*, J. Geom., 40 (1991), pp. 125–147.

[41] I. Landjev, *On blocking sets in projective Hjelmslev planes*, Adv. Math. Commun., 1 (2007), pp. 65–81.

[42] I. Landjev and S. Boev, *Blocking sets of Rédei type in projective Hjelmslev planes*, Discrete Math., 310 (2010), pp. 2061–2068.

[43] I. Landjev and T. Honold, *Arcs in projective Hjelmslev planes*, Discrete Math. Appl., 11 (2001), pp. 53–70.

[44] S. Lang, *Algebra*, Addison-Wesley Publishing Company Advanced Book Program, Reading, MA, second ed., 1984.

[45] B. R. McDonald, *Finite rings with identity*, Marcel Dekker Inc., New York, 1974. Pure and Applied Mathematics, Vol. 28.

[46] A. A. Nechaev, *Finite principal ideal rings*, Russian Academy of Sciences. Sbornik. Mathematics, 20 (1973), pp. 364–382.

[47] ——, *Kerdock's code in cyclic form*, Discrete Math. Appl., 1 (1991), pp. 365–384.

[48] ——, *Finite rings with applications*, in Handbook of algebra. Vol. 5, M. Hazewinkel, ed., Elsevier/North-Holland, Amsterdam, 2008, pp. 213–320.

[49] R. Raghavendran, *Finite associative rings*, Compositio Math., 21 (1969), pp. 195–229.

[50] F. D. Veldkamp, *Geometry over rings*, in Handbook of incidence geometry, North-Holland, Amsterdam, 1995, pp. 1033–1084.

[51] E. Witt, *Zyklische Körper und Algebren der Charakteristik p vom Grad p^n. Struktur diskret bewerteter perfekter Körper mit vollkommenem Restklassenkörper der Charakteristik p*, J. reine angew. Math., 176 (1936), pp. 126–140.

In: Current Research Topics in Galois Geometry
Editors: J. De Beule and L. Storme
ISBN: 978-1-61209-523-3

Chapter 8

Galois Geometries and Coding Theory

***Ivan Landjev*[1,*] *and Leo Storme*[2,†]**
[1] New Bulgarian University, 21 Montevideo str., 1618 Sofia, Bulgaria,
and
Institute of Mathematics and Informatics, Bulgarian Academy of Sciences,
8 Acad. G. Bonchev str., 1113, Sofia, Bulgaria
[2] Ghent University, Department of Mathematics,
Krijgslaan 281-S22, 9000 Ghent, Belgium

Abstract

Many problems on linear codes can be retranslated into equivalent problems on specific substructures in Galois geometries. This implies that geometrical methods can be used to investigate problems on linear codes, and vice versa that coding-theoretical methods can be used to investigate problems in Galois geometries. We present in this article a number of the most interesting links between linear codes and substructures in Galois geometries. We start with some basic facts from coding theory to make the article self-contained. Then we present the important links between n-arcs in Galois geometries and linear MDS codes, minihypers and linear codes meeting the Griesmer bound, links between the extendability of linear codes and blocking sets, saturating sets and the covering radius of linear codes, and conclude with the linear codes arising from the incidence matrices of Galois geometries, illustrating their relevance for Galois geometries by giving an upper bound on the sizes of sets of points in $\mathrm{PG}(N,q)$ having in each of their points a tangent hyperplane.

Key Words: Linear codes, Arcs, MDS codes, Minihypers, Griesmer bound, Saturating sets, Covering radius, Extendability, Incidence Matrices.

AMS Subject Classification: 05B25, 51E15, 51E20, 51E21, 51E22, 94B05.

---

*E-mail address: ivan@math.bas.bg; i.landjev@nbu.bg
†E-mail address: ls@cage.ugent.be

1.0 Linear Codes over Finite Fields

1.01. General Definitions

Let $\mathbb{F}_q^n$ denote the vector space of all n-tuples over the q-element field $\mathbb{F}_q$. Every k-dimensional subspace C of $\mathbb{F}_q^n$ is called a *q-ary linear code C of length n and dimension k*, or an $[n,k]_q$ code [54]. The inner product of the vectors $u=(u_1,\dots,u_n)$ and $v=(v_1,\dots,v_n)$ from $\mathbb{F}_q^n$ is defined by

$$u \cdot v = u_1 v_1 + \cdots + u_n v_n.$$

Two vectors are said to be *orthogonal* if their inner product is 0. The set of all vectors of $\mathbb{F}_q^n$ orthogonal to all codewords from C is called the *dual code* $C^{\perp}$ of C:

$$C^{\perp} = \{x \in \mathbb{F}_q^n | x \cdot y = 0 \text{ for all } y \in C\}.$$

Clearly, the code $C^{\perp}$ is a linear $[n,n-k]_q$ code.

A k-by-n matrix G having as rows the vectors of a basis of C is called a *generator matrix* of C. A generator matrix H of the code $C^{\perp}$, dual to C, is a *parity check matrix* for C.

The number of non-zero positions in a vector $x \in \mathbb{F}_q^n$ is called the *Hamming weight* $w(x)$ of x. The *Hamming distance* $d(x,y)$ between two vectors $x,y \in \mathbb{F}_q^n$ is defined by

$$d(x,y) = w(x-y).$$

The *minimum distance* of a linear code C is

$$d(C) = \min\{d(x,y) | x,y \in C, x \neq y\} = \min\{w(c) | c \in C, c \neq 0\}.$$

A q-ary linear code of length n, dimension k, and minimum distance d, is referred to as an $[n,k,d]_q$ or $[n,k,d]$ code.

Theorem 1.1.1. *Let C be a linear code over $\mathbb{F}_q$ with parity check matrix H. If any $\delta - 1$ columns of H are linearly independent over $\mathbb{F}_q$, then $d(C) \geq \delta$. The minimum distance of C is d if and only if any $d-1$ columns of H are linearly independent over $\mathbb{F}_q$ and there exist d linearly dependent columns in H.*

A central problem in coding theory is to optimize one of the parameters n,k, or d of a linear code, given the other two. This leads to the following three optimization problems:

(A) Find $n_q(k,d)$, the smallest value of n for which there exists an $[n,k,d]_q$ code.

(B) Find $K_q(n,d)$, the largest value of k for which there exists an $[n,k,d]_q$ code.

(C) Find $D_q(n,k)$, the largest value of d for which there exists an $[n,k,d]_q$ code.

If we know the exact values of one of the functions defined in (A)–(C) for all pairs of arguments, we can find the exact values of the remaining two functions.

A code of length $n_q(k,d)$, dimension k, and minimum distance d, is said to be *optimal* with respect to n. Similarly, codes with parameters $[n,K_q(n,d),d]_q$ and $[n,k,D_q(n,k)]_q$ are called *optimal* with respect to k and d. It may turn out that a code, which is optimal with respect to one of the parameters n, k, d, is not optimal with respect to (one of) the other two parameters. However, a code which is optimal with respect to the length is also optimal with respect to the dimension and the minimum distance. This follows by the easy observation that the function $n_q(k,d)$ is strictly increasing in both of its arguments, i.e. $n_q(k+1,d) > n_q(k,d)$ and $n_q(k,d+1) > n_q(k,d)$. Hence, a code which minimizes n for given k and d, maximizes k for given n and d, and at the same time maximizes d for given n and k. Thus the function defined in (A) is the most sensitive of all three. In Section 4.0, we present a natural lower bound on $n_q(k,d)$, the so-called *Griesmer bound.*

1.12. Automorphisms of Linear Codes

Let C_1 and C_2 be two linear $[n,k,d]_q$ codes. They are said to be *semi-linearly equivalent* if the codewords of C_2 can be obtained from the codewords of C_1 via a sequence of transformations of the following types:

(i) permutation on the set of coordinate positions;

(ii) multiplication of the elements in a given position by a non-zero element of $\mathbb{F}_q$;

(iii) application of a field automorphism to the elements in all coordinate positions.

1.13. The Spectrum of a Linear Code

Given an $[n,k,d]_q$ code C, we denote by A_i the number of codewords of weight i in C. The sequence of integers $(A_0,\ldots,A_n,\ldots)$ is called the *spectrum* of C. Sometimes, it is convenient to work with the so-called *Hamming weight enumerator* of C defined by

$$W_C(X,Y) = \sum_{i=0}^{n} A_i X^{n-i} Y^i.$$

1.14. Generalized Hamming Weights

Let C be a linear $[n,k]_q$ code. The set supp(C) of those coordinate positions, where not all the codewords of C are zero, is called the *support* of C. The *support of a codeword* is the support of the one-dimensional subcode generated by this codeword. The *r-th generalized Hamming weight* $d_r(C)$ is defined to be the cardinality of the minimal support of an $[n,r]_q$ subcode of C, $1 \le r \le k$, i.e.,

$$d_r(C) = \min\left\{ |\mathrm{supp}(D)| \;\middle|\; D \text{ is an } [n,r]_q \text{ subcode of } C \right\}.$$

Obviously, $d_1(C)$ is the minimum distance of C. The following theorems give some fundamental properties of the generalized Hamming weights.

Theorem 1.2.2 (Wei [71]). *For every linear $[n,k]_q$ code C,*

$$0 < d_1(C) < d_2(C) < \cdots < d_k(C) \le n.$$

Theorem 1.3.3 (Wei [71]). *Let H be a parity check matrix of the linear code C, then $d_r(C) = \delta$ if and only if*

(a) *any* $\delta - 1$ *columns of* H *have rank larger than or equal to* $\delta - r$;

(b) *there exist* δ *columns in* H *of rank* $\delta - r$.

Theorem 1.4.4 (Wei [71]). *Let* C *be a linear* $[n,k]_q$ *code and let* $C^\perp$ *be its dual code, then*

$$\{d_r(C) \,|\, r = 1,\ldots,k\} \cup \{n+1-d_r(C^\perp) \,|\, r = 1,\ldots,n-k\} \;=\; \{1,2,\ldots,n\}.$$

Theorem 1.5.5 (The generalized Singleton bound. Wei [71]).

$$d_r(C) \;\leq\; n-k+r, \;\; r = 1,\ldots,k.$$

2.0 Arcs in Galois Geometries

2.01. Multiarcs and Minihypers

Let $\mathcal{P}$ be the set of points of the projective geometry $\mathrm{PG}(N,q)$. Every mapping $\mathcal{K} : \mathcal{P} \to \mathbb{N}$ from the points of $\mathrm{PG}(N,q)$ to the non-negative integers is called a *multiset* in $\mathrm{PG}(N,q)$. This mapping is extended in a natural way to the subsets $\mathcal{Q}$ of $\mathcal{P}$ by $\mathcal{K}(\mathcal{Q}) = \sum_{P\in\mathcal{Q}} \mathcal{K}(P)$. The integer $\mathcal{K}(P)$ is called the *multiplicity* of the point P and $n = \sum_{P\in\mathcal{P}} \mathcal{K}(P)$ is called the *cardinality* of $\mathcal{K}$. The *support* $\mathrm{supp}(\mathcal{K})$ of a multiset $\mathcal{K}$ is the set of all points of positive multiplicity. A multiset is said to be *projective* if $\mathcal{K}(P) \in \{0,1\}$ for all points P. Projective multisets can be considered as sets of points by identifying them with their support.

Given a finite set $\mathcal{Q}$ of points in $\mathrm{PG}(N,q)$, we define the *characteristic multiset* $\chi_{\mathcal{Q}}$ by:

$$\chi_{\mathcal{Q}}(P) = \begin{cases} 1 & \text{if } P \in \mathcal{Q}, \\ 0 & \text{if } P \notin \mathcal{Q}. \end{cases}$$

A multiset in $\mathrm{PG}(N,q)$ is called an $(n,w;N,q)$*-multiarc* or $(n,w;N,q)$*-arc* if

(a) $\mathcal{K}(\mathcal{P}) = n$;

(b) $\mathcal{K}(H) \leq w$ for any hyperplane H, and there exists a hyperplane H_0 with $\mathcal{K}(H_0) = w$.

A multiset in $\mathrm{PG}(N,q)$ is called an $(n,w;N,q)$*-blocking multiset* or $(n,w;N,q)$*-minihyper* if

(a) $\mathcal{K}(\mathcal{P}) = n$;

(b) $\mathcal{K}(H) \geq w$ for any hyperplane H, and there exists a hyperplane H_0 with $\mathcal{K}(H_0) = w$.

We will speak of (n,w)*-multiarcs* or (n,w)*-minihypers* if the geometry $\mathrm{PG}(N,q)$ we consider is clear from the context.

The *characteristic function* of a subspace of dimension u in $\mathrm{PG}(N,q)$, $u \leq N$, is a projective minihyper with parameters (v_{u+1}, v_u), where $v_N = \frac{q^N-1}{q-1}$.

2.02. Equivalence of Multisets

Two multisets $\mathcal{K}$ in $\mathrm{PG}(N,q)$ and $\mathcal{K}'$ in $\mathrm{PG}(N',q')$ are said to be *equivalent* if there exists a collineation $\psi : \langle \mathrm{supp}(\mathcal{K}) \rangle \to \langle \mathrm{supp}(\mathcal{K}') \rangle$, such that $\mathcal{K}(P) = \mathcal{K}'(\psi(P))$ for every point $P \in \langle \mathrm{supp}(\mathcal{K}) \rangle$. Here $\langle \mathcal{Q} \rangle$, where $\mathcal{Q} \subseteq \mathcal{P}$ is the subspace of $\mathrm{PG}(N,q)$ generated by the points of $\mathcal{Q}$.

2.03. Arcs and Codes

There exists a familiar correspondence between the linear codes of *full length* (i.e. codes in which no coordinate position is identically zero) and the multiarcs in the projective geometries $\mathrm{PG}(N,q)$. Let C be an $[n,k]_q$ linear code of full length and let

$$G = [c_1,\ldots,c_k]^t = [g_1,\ldots,g_n], \quad c_i \in \mathbb{F}_q^n,\ g_i \in \mathbb{F}_q^k,$$

be a generator matrix of C. We define the multiarc $\mathcal{K}_S$ induced by the sequence of codewords $S = [c_1,\ldots,c_k]$ of C by

$$\mathcal{K}_S : \mathcal{P} = \mathrm{PG}(k-1,q) \to \mathbb{N} : P \mapsto \left|\{j \mid P = \lambda_j g_j, \text{ for some } \lambda_j \in \mathbb{F}_q \setminus \{0\}\}\right|.$$

The code C and the multiarc $\mathcal{K}_S$ are said to be *associated to each other*. A multiarc associated with an $[n,k,d]_q$ code has parameters $(n,n-d;k-1,q)$. Clearly, a linear code can be associated to different arcs, but we have the following theorem.

Theorem 2.1.1. *For every multiset $\mathcal{K}$ of cardinality n in $\mathrm{PG}(k-1,q)$, there exists a linear code C of full length in $\mathbb{F}_q^n$ and a generating sequence S of C that induce $\mathcal{K}$. Two multiarcs $\mathcal{K}_1$ and $\mathcal{K}_2$ in $\mathrm{PG}(k-1,q)$ associated with the linear codes C_1 and C_2, respectively, are equivalent if and only if the codes C_1 and C_2 are semi-linearly equivalent.*

Theorem 2.1.1 can be further generalized for linear codes over finite chain rings and arcs in projective Hjelmslev geometries (see e.g. [48]).

Let C be a linear code and let $\mathcal{K}$ be a multiarc associated with C. Denote by s the maximal multiplicity of a point from $\mathcal{P}$. The minihyper $\mathcal{F} = s\chi_{\mathcal{P}} - \mathcal{K}$ is called a *minihyper associated with C* (respectively, a *minihyper associated with $\mathcal{K}$*). Note that different multiarcs can be associated with the same minihyper. Since

$$\mathcal{F} = s\chi_{\mathcal{P}} - \mathcal{K} = (s+a)\chi_{\mathcal{P}} - (\mathcal{K} + a\chi_{\mathcal{P}}), \quad a \in \mathbb{N},$$

the multiarcs $\mathcal{K}$ and $\mathcal{K}' = \mathcal{K} + a\chi_{\mathcal{P}}$ give rise to the same minihyper. Conversely, if $\mathcal{K}$ and $\mathcal{K}'$ are two multiarcs that give rise to the same minihyper, then $\mathcal{K}' - \mathcal{K} = a\chi_{\mathcal{P}}$, $a \in \mathbb{Z}$. Minihypers will be studied in more detail in Section 4.0, in relation with the problem of linear codes meeting the Griesmer bound.

Given an $(n,w;k-1,q)$-arc, denote by a_i the number of hyperplanes H with $\mathcal{K}(H) = i$. The sequence $(a_i)_{i\geq 0}$ is called the *spectrum* of $\mathcal{K}$. If C is a linear code associated with $\mathcal{K}$ with spectrum $(A_i)_{i\geq 0}$, then $a_i = A_{n-i}/(q-1)$ for $i = 0,\ldots,n$.

2.14. Weight Hierarchy and Generalized Spectra for Arcs

Given an (n,w)-arc $\mathcal{K}$ in $\mathrm{PG}(N,q)$, we define w_r as the maximal multiplicity of an r-dimensional subspace of $\mathrm{PG}(N,q)$:

$$w_r = w_r(\mathcal{K}) = \max_{\Delta} \mathcal{K}(\Delta),$$

where Δ runs over all r-dimensional subspaces of $\mathrm{PG}(N,q)$. By definition, $w_{N-1} = w$ and $w_N = n$.

For the numbers w_r, we have the following straightforward result.

Theorem 2.2.2. *Let $\mathcal{K}$ be a non-degenerate (n,w)-arc in $\mathrm{PG}(N,q)$ (i.e. an arc with $\langle supp(\mathcal{K})\rangle = \mathrm{PG}(N,q)$), then*

$$0 < w_0 < w_1 < \cdots < w_{N-1} < w_N = n.$$

The ordered $(N+1)$-tuple $(w_0, w_1, \ldots, w_{N-1}, w_N)$ is called the *weight hierarchy* of $\mathcal{K}$.

2.25. Constructions for Arcs

Sum of Multisets

Let $\mathcal{K}_1$ and $\mathcal{K}_2$ be multiarcs in $\mathrm{PG}(N,q)$ with parameters (n_1,w_1) and (n_2,w_2), and let $a,b \in \mathbb{Q}$ be rational numbers, not both zero, such that $a\mathcal{K}_1(P) + b\mathcal{K}_2(P)$ is a non-negative integer for every point P. Then $\mathcal{K} = a\mathcal{K}_1 + b\mathcal{K}_2$ is a multiarc with parameters (n,w), where $n = an_1 + bn_2$ and $w \leq aw_1 + bw_2$. The following special cases are of particular interest:

- $\mathcal{K} = a\mathcal{K}_1$ - a replicated arc.
- $\mathcal{K} = -\mathcal{K}_1 + b\chi_{\mathcal{P}}$, where $b = \max_P \mathcal{K}_1(P)$ (usually, $\mathcal{K}_1$ is considered as a minihyper – the minihyper associated with $\mathcal{K}$).

A very important instance of the sum of multisets construction is the following. Let S_i, $i = 1,\ldots,h$, be subspaces of $\mathrm{PG}(N,q)$ with $\dim S_i = \lambda_i$, then the multiset $\mathcal{F} = \sum_{i=1}^{h} \chi_{S_i}$ is a minihyper with parameters

$$\left(\sum_{i=1}^{h} v_{\lambda_i+1}, \sum_{i=1}^{h} v_{\lambda_i}; N, q\right).$$

If $s = \max_{P\in\mathcal{P}}(\sum_{i=1}^{h}\chi_{S_i}(P))$, then the multiset $\mathcal{K} = s\chi_{\mathcal{P}} - \mathcal{F}$ is a multiarc with parameters

$$\left(sv_{N+1} - \sum_{i=1}^{h} v_{\lambda_i+1}, sv_N - \sum_{i=1}^{h} v_{\lambda_i}; N, q\right).$$

We will discuss this construction in more detail in Section 4.0 when we describe the Belov-Logachev-Sandimirov construction for linear codes meeting the Griesmer bound.

Restriction to a Subspace

Let $\mathcal{K} : \mathcal{P} \to \mathbb{N}$ be an (n,w)-multiarc in $\mathrm{PG}(N,q)$ and let U be an u-dimensional subspace of $\mathrm{PG}(N,q)$. The restriction of $\mathcal{K}$ to U is defined by

$$\mathcal{K}|_U : \mathcal{P}(U) \to \mathbb{N} : P \mapsto \mathcal{K}(P).$$

Then $\mathcal{K}|_U$ is an (n',w')-multiarc in $\mathrm{PG}(u,q)$, with $n' = \mathcal{K}(U)$. For the value of w', we can give only a bound.

Projections of Arcs

Let $\mathcal{K}$ be an $(n,w;N,q)$-multiarc. Fix an u-dimensional subspace U in $\mathrm{PG}(N,q)$. Let furthermore V be a v-dimensional subspace in $\mathrm{PG}(N,q)$, with $u+v=N-1$ and $U\cap V=\emptyset$. Define the projection $\varphi=\varphi_{U,V}$ from U onto V by

$$\varphi_{U,V}:\mathcal{P}\setminus U\to V:P\mapsto V\cap\langle U,P\rangle,$$

where $\mathcal{P}$ is the point set of $\mathrm{PG}(N,q)$. Note that $\varphi_{U,V}$ maps $(u+s)$-dimensional subspaces containing U into $(s-1)$-dimensional subspaces contained in V. The induced multiarc $\mathcal{K}^{\varphi}$ is defined on the points of V by

$$\mathcal{K}^{\varphi}:\mathcal{P}(V)\to\mathbb{N}:P\mapsto\sum_{Q\in\mathcal{P}\setminus U:\ \varphi_{U,V}(Q)=P}\mathcal{K}(Q).$$

If S is a t'-dimensional subspace in V, then $\mathcal{K}^{\varphi}(S)=\mathcal{K}(\langle S,U\rangle)-\mathcal{K}(U)$. Here, $\langle S,U\rangle$ denotes the projective subspace of $\mathrm{PG}(N,q)$ generated by S and U. Clearly, $\mathcal{K}^{\varphi}$ is an $(n-\mathcal{K}(U),w'-\mathcal{K}(U))$-multiarc in $V\cong\mathrm{PG}(v,q)$, with $w'\leq w$. Similarly, if $\mathcal{K}$ is an (n,w)-minihyper, then $\mathcal{K}^{\varphi}$ is an $(n-\mathcal{K}(U),w'-\mathcal{K}(U))$-minihyper in V, with $w'\geq w$.

The Dual Construction for Arcs

This construction is a generalization of familiar geometrical constructions to the case where multiple points are allowed. It has been introduced by Brouwer and van Eupen [10] for linear codes and formulated for multiarcs by Dodunekov and Simonis [26]. Let $\mathcal{K}$ be an $(n,w;N,q)$-multiarc and set $W=\{\mathcal{K}(H)\mid H\in\mathcal{H}\}$, where $\mathcal{H}$ is the set of all hyperplanes in $\mathrm{PG}(N,q)$. Let $\sigma:W\to\mathbb{N}$ be a fixed mapping. The multiarc

$$\mathcal{K}^{\sigma}:\mathcal{H}\to\mathbb{N}:H\mapsto\sigma(\mathcal{K}(H))$$

is called the σ-*dual* multiarc to $\mathcal{K}$. Let $(a_i)_{i\geq 0}$ be the spectrum of $\mathcal{K}$. Then the parameters of $\mathcal{K}^{\sigma}$ are (n',w'), where

$$n'=\sum_{i\in W}\sigma(i)a_i,\quad w'=\max_P\mathcal{K}^{\sigma}(P)=\max_P\sum_{H:P\in H,H\in\mathcal{H}}\mathcal{K}^{\sigma}(H).$$

Let $\sigma(x)=\alpha x+\beta$, $\alpha,\beta\in\mathbb{Q}$, be a linear function, which takes on non-negative integer values for each $x\in W$. Then

$$\mathcal{K}^{\sigma}=\mathcal{K}^{(\alpha,\beta)}=\alpha\mathcal{K}+\beta\chi|_{\mathcal{H}}.$$

Theorem 2.3.3. *Let $\mathcal{K}$ be an (n,w)-multiarc in* $\mathrm{PG}(N,q)$. *Then $\mathcal{K}^{(\alpha,\beta)}$ has parameters*

$$\begin{aligned} n' &= \alpha n\frac{q^N-1}{q-1}+\beta\frac{q^{N+1}-1}{q-1},\\ w' &= \max_P\left\{\alpha\left(n\frac{q^{N-1}-1}{q-1}+q^{N-1}\mathcal{K}(P)\right)+\beta\frac{q^N-1}{q-1}\right\},\end{aligned}$$

where the maximum is taken over all points P in $\mathrm{PG}(N,q)$.

Theorem 2.3.3 has been used repeatedly in the construction of various optimal arcs and codes [42, 52].

3.0 Arcs and Linear MDS Codes

3.01. Introduction to Arcs and Linear MDS Codes

We now present the most famous example of the links between coding theory and Galois geometries, i.e., the link between linear MDS codes and arcs in Galois geometries. The chapter on linear MDS codes is described in [54] as *one of the most fascinating in all of coding theory*, and this is motivated by the many nice results on linear MDS codes obtained via the geometrical links with the arcs in Galois geometries. We first present the linear MDS codes, and then the arcs in Galois geometries.

Theorem 3.1.1 (The Singleton bound). *For a linear $[n,k,d]_q$ code C, $d \leq n-k+1$.*

Definition 3.2.2. A linear $[n,k,d=n-k+1]_q$ code is called a linear *Maximum Distance Separable (MDS)* code.

The following theorem gives the fundamental properties of linear MDS codes, which will enable us to make the links to the geometrically equivalent arcs in Galois geometries.

Theorem 3.3.3. *Let C be a linear $[n,k,d]_q$ code, then the following properties are equivalent:*

1. *C is a linear $[n,k,n-k+1]_q$ MDS code,*
2. *every k columns of a generator matrix G of C are linearly independent,*
3. *every $n-k$ columns of a parity check matrix H of C are linearly independent,*
4. *$C^\perp$ is a linear $[n,n-k,k+1]_q$ MDS code.*

Independently, the following concept of arcs was defined in Galois geometries [43].

Definition 3.4.4. An *n-arc* in $\mathrm{PG}(k-1,q)$ is a set of n points, every k of which are linearly independent. An n-arc in $\mathrm{PG}(k-1,q)$ is called *complete* if and only if it is not contained in an $(n+1)$-arc of $\mathrm{PG}(k-1,q)$.

Definition 3.4.4 immediately makes the link with Theorem 3.3.3 (2), which gives the following equivalence.

Theorem 3.5.5. *The set $K=\{g_1,\ldots,g_n\}$ is an n-arc in $\mathrm{PG}(k-1,q)$ if and only if the $(k\times n)$ matrix $G=(g_1\cdots g_n)$ defines a linear $[n,k,n-k+1]_q$ MDS code C.*

The equivalence of Theorem 3.3.3 (1) and Theorem 3.3.3 (4) now leads to the following geometrical result.

Theorem 3.6.6. *Let $K=\{g_1,\ldots,g_n\}$ be an n-arc in $\mathrm{PG}(k-1,q)$ defining the linear $[n,k,n-k+1]_q$ MDS code with generator matrix $G=(g_1\cdots g_n)$, then there exists an n-arc $\tilde{K}=\{h_1,\ldots,h_n\}$ in $\mathrm{PG}(n-k-1,q)$ such that $\tilde{K}$ defines the dual $[n,n-k,k+1]_q$ MDS code $C^\perp$ via the parity check matrix $H=(h_1\cdots h_n)$ of C.*

So the existence of an n-arc K in $PG(k-1,q)$ implies the existence of an n-arc $\tilde{K}$ in $PG(n-k-1,q)$. We say that an n-arc K in $PG(k-1,q)$ and an n-arc $\tilde{K}$ in $PG(n-k-1,q)$ are *C-dual* if and only if they define dual linear MDS codes.

The standard example of an n-arc in $PG(k-1,q)$ is the normal rational curve.

Definition 3.7.7. A *normal rational curve* K in $PG(k-1,q)$, $2 \leq k \leq q-1$, is a $(q+1)$-arc projectively equivalent to the set of points $\{(1,t,\ldots,t^{k-1})|t \in \mathbb{F}_q^+\}$; $\mathbb{F}_q^+ = \mathbb{F}_q \cup \{\infty\}, \infty \notin \mathbb{F}_q, t = \infty$ defines the point $(0,\ldots,0,1)$.

The normal rational curves define the classical examples of linear MDS codes, i.e., the *Generalized Doubly-Extended Reed-Solomon (GDRS)* codes. These GDRS codes are used to encode music on compact disc.

We also wish to mention that a particular non-GDRS $[8,4,5]_{256}$ code is used in the *Advanced Encryption Standard* (see e.g. [55]).

3.72. The Largest Arcs in Galois Geometries

The maximum number of points in an n-arc of $PG(k-1,q)$ is denoted by $m(k-1,q)$. The problem of determining the exact value of $m(k-1,q)$ and of characterizing the $m(k-1,q)$-arcs in $PG(k-1,q)$ has been in the focus of research on arcs and linear MDS codes. We now state the main results on this central point of research.

3.73. Arcs in $PG(2,q)$

Table 1. $m(2,q)$

q	$m(2,q)$	
q even	$q+2$	[8]
q odd	$q+1$	[8]

An $m(2,q)$-arc in $PG(2,q)$, q odd, is called an *oval*, and an $m(2,q)$-arc in $PG(2,q)$, q even, is called a *hyperoval*. The following theorem of B. Segre inspired and motivated many researchers to investigate substructures in Galois geometries.

Theorem 3.8.8 ((Segre [61]). *For q odd, an oval is the set of rational points of a conic.*

The classical example of a hyperoval in $PG(2,q)$, q even, is a conic plus its *nucleus* (the intersection point of its tangents). A hyperoval of this type is called *regular*. As shown by Segre [62], for $q = 2,4,8$, every hyperoval is regular.

For $q = 2^h, h \geq 4$, there exist *irregular* hyperovals, that is, hyperovals which are not the union of a conic and its nucleus. Several infinite classes of irregular hyperovals are known. The problem of classifying the hyperovals in $PG(2,q)$, q even, is one of the hardest problems in Galois geometries. In general, the following result is valid.

Theorem 3.9.9 (Segre [62]). *Any hyperoval of $PG(2,q)$, $q = 2^h$ and $h > 1$, is projectively equivalent to a hyperoval*

$$\mathcal{D}(F) = \{(1,t,F(t))|t \in \mathbb{F}_q\} \cup \{(0,1,0),(0,0,1)\},$$

where F is a permutation polynomial over $\mathbb{F}_q$ of degree at most $q-2$, satisfying $F(0)=0$, $F(1)=1$, and such that $F_s(X)=(F(X+s)+F(s))/X$ is a permutation polynomial for each s in $\mathbb{F}_q$, satisfying $F_s(0)=0$.

We refer to [47, Table 2.2] for the known infinite classes of hyperovals. Particular examples include the *translation* hyperovals $\mathcal{D}(F)=\{(1,t,t^{2^i})|t\in\mathbb{F}_q\}\cup\{(0,1,0),(0,0,1)\}$, with $q=2^h$, $\gcd(i,h)=1$.

In the next tables, we state results on the $m(k-1,q)$-arcs in $\mathrm{PG}(k-1,q)$, for $k\geq 3$. In many cases, we have chosen to state only one bound for q odd and for q even, so that we can explain the results in more detail. For the best known bounds, we refer to the tables of [47]. A large number of these results rely on the problem of finding the size $m'(2,q)$ of the second largest complete arcs in $\mathrm{PG}(2,q)$. We first mention some of the known results, and then give a brief description on how these results were obtained.

In Table 2, for q subject to the conditions in the first column, the second column gives an upper bound on $m'(2,q)$, and the third column indicates when this upper bound is sharp. The fourth column gives the value of $m(2,q)$. So any n-arc, with $n>m'(2,q)$, is contained in an $m(2,q)$-arc. The last column describes the type of the $m(2,q)$-arc.

Table 2. Upper bounds on $m'(2,q)$

q	$m'(2,q)$	Sharp	$m(2,q)$	$m(2,q)$-arc	
$q=p^h$, $p\geq 5$	$\leq q-\sqrt{q}/2+5$		$q+1$	conic	[44]
$q=2^{2e}$, $e>1$	$=q-\sqrt{q}+1$	yes	$q+2$	hyperoval	[7,28,49,63]
$q=2^{2e+1}$, $e\geq 1$	$\leq q-\sqrt{2q}+2$	$q=8$	$q+2$	hyperoval	[69]

Techniques 3.10.10. *The results of Table 2 were found by using the following method. Consider an n-arc $K=\{\ell_1,\ldots,\ell_n\}$ of lines in* PG$(2,q)$, *i.e., a set of n lines, no three concurrent. Then every line ℓ_i contains $q+2-n$ points $P_{i1},\ldots,P_{i,q+2-n}$ not lying on a second line of K. It has been proven that all these $n(q+2-n)$ points P_{ij}, $i=1,\ldots,n$, $j=1,\ldots,q+2-n$, belong to an algebraic curve Γ of degree $q+2-n$ when q is even, and of degree $2(q+2-n)$ when q is odd* [43]. *By using bounds on the number of points and other fundamental properties of these algebraic curves, it was shown that Γ contains linear components over $\mathbb{F}_q$; linear components which extend the given n-arc K to a larger arc. So, here in this context, algebraic geometry plays a fundamental role.*

Based on the sharpness of the second bound in Table 2, the following conjecture has been stated.

Conjecture 3.11.11. $m'(2,q)=q-\sqrt{q}+1$ *for* $q=p^{2e}$, $q>9$.

For arcs in $\mathrm{PG}(2,q)$, $q=2^{2e}$, $q>4$, of size smaller than $q-\sqrt{q}+1$, there is the following result of Hirschfeld and Korchmáros [45].

Theorem 3.12.12. *A complete n-arc of* PG$(2,q)$, $q=2^{2e}$, $e>2$, *has size* $q+2$, $q-\sqrt{q}+1$, *or at most size* $q-2\sqrt{q}+6$.

3.124. Results in Higher Dimensions

In Tables 3 and 4, NRC stands for *normal rational curve*. In $PG(3,q)$, $q = 2^h$, $h > 2$, $L_e = \{(1,t,t^e,t^{e+1}) | t \in \mathbb{F}_q^+\}$, with $e = 2^v$, $\gcd(v,h) = 1$, and with $t = \infty$ defining the point $e_3 = (0,0,0,1)$. Let $e_0 = (1,0,\ldots,0), e_1 = (0,1,0,\ldots,0), \ldots, e_N = (0,\ldots,0,1)$, and let $e = (1,\ldots,1)$.

Table 3 shows the value of $m(N,q)$ for small dimensions N. The characterization of the $m(N,q)$-arcs L in $PG(N,q)$ is given in the fourth column.

Table 3. $m(N,q)$ and $m(N,q)$-arcs

q	N	$m(N,q)$	$m(N,q)$-arc L	
q odd , $q > 3$	3	$q+1$	NRC	[60]
$q = 2^h$, $q > 4$	3	$q+1$	L_e, $e = 2^v$, $\gcd(v,h) = 1$	[15, 16, 33]
q odd, $q > 5$	4	$q+1$	NRC for $q > 83$	[60]
$q = 2^h$, $q > 4$	4	$q+1$	NRC	[15, 17, 33]
q	$N \geq q-1$	$N+2$	$\{e_0,\ldots,e_N,e\}$	[14]

Table 4 summarizes the main results on the extendability of n-arcs in $PG(N,q)$ to larger arcs. An n-arc in $PG(N,q)$, satisfying the condition on n in the third column, can be extended uniquely to a $(q+1)$-arc L, whose description is given in column 4. The results are respectively due to Hirschfeld and Korchmáros for the first formula [44], and to Storme and Thas [66] for the last two formulas.

Table 4. Upper bounds on $m'(N,q)$

N	q	$n >$	L
≥ 2	$q = p^h$, $p \geq 5$	$q - \sqrt{q}/2 + N + 3$	NRC
3	$q = 2^h, h > 1$	$q - \sqrt{q}/2 + 9/4$	L_e
≥ 4	$q = 2^h$, $h > 2$	$q - \sqrt{q}/2 + N - 3/4$	NRC

Techniques 3.13.13. *The principal arguments for obtaining these extension results are as follows.*

For q odd, projection arguments can be used. Consider an n-arc $K = \{P_1,\ldots,P_n\}$ in $PG(N,q)$. *Project $K \setminus \{P_n\}$ from the point P_n onto a hyperplane Π not passing through P_n, then an $(n-1)$-arc K' in Π is obtained. By selecting the lower bound on n in the correct way, as was done in Table 4, by induction on N, it is known that this $(n-1)$-arc in Π is contained in a normal rational curve of Π; hence, K belongs to a cone with vertex P_n and base a normal rational curve in a hyperplane Π of* $PG(N,q)$. *But P_n is an arbitrary point of K, so in fact, K is contained in the intersection of n such cones. This implies that K itself is contained in a normal rational curve of* $PG(N,q)$.

For n-arcs in $PG(3,q)$, *q even, the arguments for q odd cannot be used, since there is no classification of hyperovals in* $PG(2,q)$, *q even. So a completely different technique had to be developed* [13]. *Here, consider an n-arc K of planes in* $PG(3,q)$, *then it is possible to associate an algebraic surface Φ of degree $q+3-n$ to K. For large n, it was shown* *that this surface Φ contains at least one plane Π, and this plane Π extends K to a larger $(n+1)$-arc.*

Similarly, to an n-arc K in $\mathrm{PG}(4,q)$, q even, an algebraic hypersurface Φ of degree $q+4-n$ can be associated. For large n, it is proven that Φ contains a hyperplane Π; where Π extends K to an $(n+1)$-arc.

Since it was proven that every $(q+1)$-arc in $\mathrm{PG}(4,q)$, q even, $q \geq 8$, is a normal rational curve (Table 3), for the characterization of large arcs in $\mathrm{PG}(N,q)$, q even, $N > 4$, the projection arguments used for q odd can now be applied here.

Techniques 3.14.14. *As indicated in the beginning of this section, an n-arc in $\mathrm{PG}(k-1,q)$ defines a C-dual n-arc in $\mathrm{PG}(n-k-1,q)$. This C-duality principle of arcs makes it possible to translate results on n-arcs to results on C-dual n-arcs. The results in the preceding tables on arcs in $\mathrm{PG}(N,q)$, with N small, immediately imply other results on arcs in $\mathrm{PG}(N,q)$, where N is close to q. We now present a number of these results.*

In Table 5, in the spaces $\mathrm{PG}(N,q)$, with N satisfying the bounds in the table, any n-arc, where n satisfies the bound in the second column, is contained in a normal rational curve. So, in these spaces, $m(N,q) = q+1$ and every $(q+1)$-arc is a normal rational curve.

Table 5. Arcs in $\mathrm{PG}(N,q)$, N close to q

q	$n \geq$	$m(N,q) = q+1$ and $(q+1)$-arc = NRC
$q = p^{2e}, p > 2, e \geq 1$	$N+4$	$q-3 \geq N > q-\sqrt{q}/4-39/16$
$q = p^h, p \geq 5$	$N+4$	$q-3 \geq N > q-\sqrt{q}/2+1$
$q = 2^h, h > 2$	$N+6$	$q-5 \geq N > q-\sqrt{q}/2-11/4$

Up to now, all presented results for q odd state that a $(q+1)$-arc in $\mathrm{PG}(k-1,q)$, q odd, $2 \leq k \leq q-1$, is a normal rational curve. So the conjecture arose that this is indeed the case. However, Glynn found a counterexample to this conjecture.

Theorem 3.15.15 (Glynn [30]). *In $\mathrm{PG}(4,9)$, a 10-arc is one of two types; it is either a normal rational curve or it is equivalent to the 10-arc $L = \{(1,t,t^2+\eta t^6,t^3,t^4) | t \in \mathbb{F}_9\} \cup \{(0,0,0,0,1)\}$, where $\eta^4 = -1$.*

We already have observed that there are fundamental differences between arcs in spaces of even characteristic and of odd characteristic; one of the differences consists of the $(q+2)$-arcs in $\mathrm{PG}(q-2,q)$, q even.

Theorem 3.16.16 (Thas [68]). *In $\mathrm{PG}(q-2,q)$, q even, $m(q-2,q) = q+2$.*

Theorem 3.17.17 (Storme and Thas [67]). *In $\mathrm{PG}(q-2,q)$, q even, a point $P = (a_0,\ldots,a_{q-2})$ extends the normal rational curve $K = \{(1,t,\ldots,t^{q-2}) | t \in \mathbb{F}_q^+\}$ to a $(q+2)$-arc if and only if $F(X) = \sum_{i=0}^{q-2} a_{q-2-i}X^{i+1}$ defines a $(q+2)$-arc $K' = \{(1,t,F(t)) | t \in \mathbb{F}_q\} \cup \{e_1,e_2\}$ in $\mathrm{PG}(2,q)$; in this case, K' is a C-dual $(q+2)$-arc of $K \cup \{P\}$.*

Table 6 presents the results of Storme and Thas [65] for the values of n for which there exist complete n-arcs in the respective spaces $\mathrm{PG}(N,q)$, q even, N large.

Table 6. Spectrum of complete arcs

q	N	$n \in$
$q=2^h, q \geq 32, q \neq 64$	$q-5 \geq N > q-\sqrt{q}/2-11/4$	$\{N+4, N+5, q+1\}$
$q=64$	$N=58$ or $N=59$	$\{N+4, q+1\}$
$q=2^h, q \geq 8$	$q-4$	$\{q, q+1\}$
$q=2^h, q \geq 8$	$q-3$	$\{q+1\}$
$q=2^h, q \geq 4$	$q-2$	$\{q+2\}$

3.175. Open Problems

1. The key tool in obtaining the results on the extendability of large n-arcs in $PG(2,q)$ to ovals and hyperovals is the link between n-arcs K of lines in $PG(2,q)$ and algebraic curves of degree $2(q+2-n)$ for q odd, and of degree $q+2-n$ for q even (Techniques 3.10.10). Continuing this study of algebraic curves is of great interest. See [46] for detailed information on algebraic curves over a finite field.

2. The problem of the classification of the hyperovals in $PG(2,q)$, q even, has been one of the earliest problems investigated in Galois geometries. The complete classification of hyperovals has not yet been obtained. So we propose to investigate the classification problem of hyperovals in $PG(2,q)$, q even.

3. The problem of constructing large n-arcs in $PG(2,q)$, different from ovals and hyperovals, still merits attention. For q a square, we know the existence of complete $(q-\sqrt{q}+1)$-arcs [7,28,49,63].

   Apart from this example, for general q, all the other known largest complete arcs have size at most approximately $(q+1+2\sqrt{q})/2$. The main constructions consist of half of the points of absolutely irreducible cubic curves, and of half of the points of a conic, to which some other points not lying on this conic are added. So we propose to investigate the construction of complete n-arcs in $PG(2,q)$, with $n > (q+1+2\sqrt{q})/2$.

4.0 Minihypers and the Griesmer Bound

4.01. A Geometrical Proof of the Griesmer Bound

Let $\mathcal{K}$ be an $(n,w;k-1,q)$-arc. We start with the observation that w points generate a subspace of projective dimension at most $w-1$, or, in other words, the maximal multiplicity of a subspace of dimension u is at least $u+1$. Hyperplanes have projective dimension $k-2$, therefore $w \geq k-1$. This is easily seen to be equivalent to the Singleton bound (Theorem 3.1.1). The Griesmer bound is a generalization of the Singleton bound.

Theorem 4.1.1 ([32,64]). *For every linear $[n,k,d]_q$ code,*

$$n \geq \sum_{i=0}^{k-1} \lceil \frac{d}{q^i} \rceil = g_q(k,d).$$

Proof. By induction on k. For the case $k = 2$, consider a multiarc $\mathcal{K}$ with parameters $(n, w; 1, q)$, $w = n - d$. On the projective line, fix a point P of maximal multiplicity w. There is a point Q on the projective line, $Q \neq P$, that has multiplicity $\mathcal{K}(Q) \geq \lceil (n-w)/q \rceil$. Since P is of maximal multiplicity, we have $w \geq \lceil \frac{n-w}{q} \rceil$ which implies $n \geq d + \lceil \frac{d}{q} \rceil$.

Assume the inequality has been proven for multiarcs in the projective geometries PG(N, q), $N \leq k-2$. Consider an $(n, w; k-1, q)$-multiarc $\mathcal{K}$. There exists a hyperplane H of multiplicity $n - d$. The sum of the multiplicities of the points outside of H is d. Since the number of points outside of H is q^{k-1}, there exists a point P ($P \notin H$) such that $\mathcal{K}(P) \geq \lceil d/q^{k-1} \rceil$. Consider a projection φ from P onto some hyperplane not incident with P. The induced multiarc $\mathcal{K}^{\varphi}$ has parameters $(n', w'; k-2, q)$, where $n' = n - \mathcal{K}(P)$ and $w' \leq w - \mathcal{K}(P)$. Note that $n' - w' \geq n - w = d$. Hence, by the induction hypothesis, $n - \mathcal{K}(P) = n' \geq \sum_{i=0}^{k-2} \lceil \frac{d}{q^i} \rceil$. □

Linear codes attaining the Griesmer bound, i.e. with parameters $[g_q(k,d), k, d]_q$, are called *Griesmer codes*.

4.12. Minihypers and the Belov-Logachev-Sandimirov Construction

The link between minihypers in PG$(k-1, q)$ and linear $[n, k, d]_q$ codes meeting the Griesmer bound is described in the following way.

For $(s-1)q^{k-1} < d \leq sq^{k-1}$, d can be written uniquely as $d = sq^{k-1} - \sum_{i=1}^{h} q^{\lambda_i}$ such that:

(a) $0 \leq \lambda_1 \leq \cdots \leq \lambda_h < k-1$,

(b) at most $q-1$ of the values λ_i are equal to a given value.

Using this expression for d, the Griesmer bound for a linear $[n, k, d]_q$ code can be expressed as:

$$n \geq sv_k - \sum_{i=1}^{h} v_{\lambda_i + 1}.$$

Hamada and Helleseth showed that in the case $d = sq^{k-1} - \sum_{i=1}^{h} q^{\lambda_i}$, there is a one-to-one correspondence between the set of all non-equivalent $[n, k, d]_q$ codes meeting the Griesmer bound and the set of all projectively distinct $(\sum_{i=1}^{h} v_{\lambda_i+1}, \sum_{i=1}^{h} v_{\lambda_i}; k-1, q)$-minihypers $\mathcal{F}$ [37].

Belov, Logachev, and Sandimirov [3] gave a construction method for Griesmer codes, which is easily described by using the corresponding minihypers.

Consider in PG$(k-1, q)$ a sum of ε_0 points $P_1, P_2, \ldots, P_{\varepsilon_0}$, ε_1 lines $\ell_1, \ell_2, \ldots, \ell_{\varepsilon_1}$, ..., ε_{k-2} $(k-2)$-dimensional subspaces $\pi_1^{(k-2)}, \ldots, \pi_{\varepsilon_{k-2}}^{(k-2)}$, with $0 \leq \varepsilon_i \leq q-1, i = 0, \ldots, k-2$, then such a sum defines a $(\sum_{i=0}^{k-2} \varepsilon_i v_{i+1}, \sum_{i=0}^{k-2} \varepsilon_i v_i; k-1, q)$-minihyper $\mathcal{F}$, where the multiplicity of a point R of PG$(k-1, q)$ equals the number of objects, in the description above, in which it is contained (See also the sum of multisets in Subsection 2.25..).

Now that the standard examples of minihypers are known, the characterization problem on minihypers, and equivalently on linear codes meeting the Griesmer bound, arises:

Characterize $(f,m;k-1,q)$*-minihypers* $\mathcal{F}$ *for given parameters* $f=\sum_{i=0}^{k-2}\varepsilon_i v_{i+1}$, $m=\sum_{i=0}^{k-2}\varepsilon_i v_i, k$, *and* q.

Fundamental research on this problem was performed by Hamada *et al.* who, in many articles, obtained a lot of results on minihypers and who developed a great amount of techniques useful in the study of minihypers. Their main results are in [36, 38].

Improvements to the results of [36, 38] were found by for instance De Beule, Metsch, and Storme.

Theorem 4.2.2 (De Beule, Metsch, and Storme [23]). *A projective* $(\sum_{i=0}^{k-2}\varepsilon_i v_{i+1}, \sum_{i=0}^{k-2}\varepsilon_i v_i; k-1,q)$*-minihyper, where* $\sum_{i=0}^{k-2}\varepsilon_i \leq \delta_0$ *with* δ_0 *equal to one of the values in Table* 4.1*, is a union of* ε_{k-2} *hyperplanes,* ε_{k-3} $(k-3)$*-dimensional spaces,* $\ldots, \varepsilon_1$ *lines, and* ε_0 *points, which all are pairwise disjoint, so is of Belov-Logachev-Sandimirov type.*

In the following table, $q=p^s$, p prime, $s\geq 1$.

Table 7. Upper bounds on δ_0

p	s	δ_0
p	1	$\leq (p+1)/2$
p	3	$\leq p^2$
p	even	$\leq \sqrt{q}$
2	$6m+1, m\geq 1$	$\leq 2^{4m+1}-2^{4m}-2^{2m+1}/2$
>2	$6m+1, m\geq 1$	$\leq p^{4m+1}-p^{4m}-p^{2m+1}/2+1/2$
2	$6m+3, m\geq 1$	$< 2^{4m+5/2}-2^{4m+1}-2^{2m+1}+1$
>2	$6m+3, m\geq 1$	$\leq p^{4m+2}-p^{2m+2}+2$
≥ 5	$6m+5, m\geq 0$	$< p^{4m+7/2}-p^{4m+3}-p^{2m+2}/2+1$

Regarding characterization results on weighted minihypers, we mention the following two results. The next theorem was first proven by Hamada for projective minihypers, while the second theorem is the weighted version of a result of Hamada, Helleseth, and Maekawa [36, 38].

Theorem 4.3.3 (Hamada [34,35] and Landjev and Storme [51]). *A* $(\sum_{i=1}^{h} v_{\lambda_i+1}, \sum_{i=1}^{h} v_{\lambda_i}; k-1,q)$*-minihyper, with* $k-1>\lambda_1>\lambda_2>\cdots>\lambda_h\geq 0$*, is the sum of a* λ_1*-dimensional space, a* λ_2*-dimensional space,* $\ldots$*, and a* λ_h*-dimensional space.*

Theorem 4.4.4 (De Beule, Metsch, and Storme [24]). *A* $(\sum_{i=0}^{k-2}\varepsilon_i v_{i+1}, \sum_{i=0}^{k-2}\varepsilon_i v_i; k-1,q)$*-minihyper, where* $\sum_{i=0}^{k-2}\varepsilon_i<\sqrt{q}+1$*, is a sum of* ε_{k-2} *hyperplanes,* ε_{k-3} $(k-3)$*-dimensional spaces,* $\ldots, \varepsilon_1$ *lines, and* ε_0 *points, so it is of Belov-Logachev-Sandimirov type.*

Techniques 4.5.5. *The results on the minihypers are obtained via a variety of techniques. First of all, minihypers are particular examples of blocking sets (see e.g.* [6]*). Hence, characterization results on minimal blocking sets play a crucial role in the characterization of minihypers. The characterization of a minihyper is in many cases obtained by building up the minihyper in inductive steps. As particular example, a Belov-Logachev-Sandimirov* $(\sum_{i=0}^{k-2}\varepsilon_i v_{i+1}, \sum_{i=0}^{k-2}\varepsilon_i v_i; k-1,q)$*-minihyper,* $\sum_{i=0}^{k-2}\varepsilon_i$ *small, which is* *the union of* ε_{k-2} *hyperplanes,* ε_{k-3} $(k-3)$*-dimensional spaces,* $\ldots,\varepsilon_1$ *lines, and* ε_0 *points, which are pairwise disjoint, can be characterized in the following way. First of all,* $(\varepsilon_1(q+1)+\varepsilon_0,\varepsilon_1;k-1,q)$*-minihypers,* $\varepsilon_1+\varepsilon_0$ *small, are characterized as the union of* ε_1 *lines and* ε_0 *points. Once this is done, this result is used to characterize* $(\varepsilon_2(q^2+q+1)+\varepsilon_1(q+1)+\varepsilon_0,\varepsilon_2(q+1)+\varepsilon_1;k-1,q)$*-minihypers,* $\varepsilon_2+\varepsilon_1+\varepsilon_0$ *small, as the union of* ε_2 *planes,* ε_1 *lines, and* ε_0 *points, which are pairwise disjoint. Namely, many hyperplane intersections of* $(\varepsilon_2(q^2+q+1)+\varepsilon_1(q+1)+\varepsilon_0,\varepsilon_2(q+1)+\varepsilon_1;k-1,q)$*-minihypers are* $(\varepsilon_1'(q+1)+\varepsilon_0',\varepsilon_1';k-2,q)$*-minihypers, which are already characterized. So, these hyperplane intersections are exactly known. This then is used to characterize the* $(\varepsilon_2(q^2+q+1)+\varepsilon_1(q+1)+\varepsilon_0,\varepsilon_2(q+1)+\varepsilon_1;k-1,q)$*-minihypers,* $\varepsilon_2+\varepsilon_1+\varepsilon_0$ *small, as the union of* ε_2 *planes,* ε_1 *lines, and* ε_0 *points. Once this is done, inductive arguments can be used to characterize* $(\sum_{i=0}^{k-2}\varepsilon_i v_{i+1}, \sum_{i=0}^{k-2}\varepsilon_i v_i;\ k-1,q)$*-minihypers,* $\sum_{i=0}^{k-2}\varepsilon_i$ *small, as the union of* ε_{k-2} *hyperplanes,* ε_{k-3} $(k-3)$*-dimensional spaces,* $\ldots,\varepsilon_1$ *lines, and* ε_0 *points, which are pairwise disjoint.*

Polynomial techniques play a central role in the study of blocking sets, see e.g. [2] *and* [6]*. So, it is worth considering the polynomial techniques for the study of minihypers. In particular, in obtaining the results of Theorem 4.2.2, also polynomial techniques were used.*

Recently, characterizations of minihypers involving Baer subgeometries in $\mathrm{PG}(N,q)$, q square, have been obtained. For particular results, we refer to [31].

5.0 Saturating Sets in Galois Geometries and Covering Radius

Definition 5.1.1. Let C be a linear $[n,k,d]_q$ code. The *covering radius* of the code C is the smallest integer R such that every n-tuple in $\mathbb{F}_q^n$ lies at Hamming distance at most R from a codeword in C.

The following theorem will be the basis for making the link with the geometrically equivalent objects of the *saturating sets in Galois geometries*.

Theorem 5.2.2. *Let* C *be a linear* $[n,k,d]_q$ *code with parity check matrix* $H=(h_1\cdots h_n)$.

Then the covering radius of C *is equal to* R *if and only if every* $(n-k)$*-tuple over* $\mathbb{F}_q$ *can be written as a linear combination of at most* R *columns of* H.

Definition 5.3.3. Let S be a subset of $\mathrm{PG}(N,q)$. The set S is called ρ-*saturating* when every point P from $\mathrm{PG}(N,q)$ can be written as a linear combination of at most $\rho+1$ points of S.

Taking into account Theorem 5.2.2, the preceding definition means that ρ-*saturating sets* S *in* $\mathrm{PG}(n-k-1,q)$ *determine the parity check matrices of linear* $[n,k,d]_q$ *codes with covering radius* $R=\rho+1$.

Example 5.4.4. The linear codes with covering radius $R=2$ and with minimum distance $d\geq 4$ are important examples. Such a code has a parity check matrix whose columns define an n-cap $K=\{h_1,\ldots,h_n\}$ in $\mathrm{PG}(n-k-1,q)$ (see e.g. [5]). The fact that the covering radius R is equal to two signifies that every point from $\mathrm{PG}(n-k-1,q)\setminus K$ can be written as a linear combination of two columns of H, i.e., that every point of $\mathrm{PG}(n-k-1,q)\setminus K$ is linearly dependent on two columns of H. This signifies also that no point of $PG(n-k-1,q)\setminus K$ extends the n-cap K in $PG(n-k-1,q)$ to an $(n+1)$-cap. Hence, this altogether proves that the columns of a parity check matrix H of a linear $[n,k,d]_q$ code, with $d\geq 4$ and $R=2$, define a complete n-cap of $PG(n-k-1,q)$. In this way, the complete caps K in a projective space $PG(N,q)$ are particular examples of 1-saturating sets.

In the study of ρ-saturating sets, one of the most important research problems is the problem of finding ρ-saturating sets of the smallest possible cardinality. The cardinality of a smallest possible set S from $PG(N,q)$ which is ρ-saturating is denoted by the parameter $k(N,q,\rho)$.

We now present a number of the known upper bounds on the parameter $k(N,q,\rho)$.

Regarding the parameters $k(N,q,1)$, good upper bounds on $k(N,q,1)$ have been found by the construction of small complete caps. In the following table, the first two results are of Davydov and Tombak [29], the next two results of Pambianco and Storme [59], and the last two results of Davydov and Östergård [22]. In the upper bounds of the following table, the parameter $n_2(N,q)$ denotes the smallest cardinality of a complete cap in $PG(N,q)$.

Table 8. Upper bounds on $k(N,q,1)$ and $n_2(N,q)$

N	q	$k(N,q,1),n_2(N,q)$
$2k$	2	$\leq 23\cdot 2^{k-3}-3$
$2k+1$	2	$\leq 30\cdot 2^{k-3}-3$
$2k$	$q=2^h\geq 4$	$\leq q^k+3(q^{k-1}+q^{k-2}+\cdots+q)+2$
$2k+1$	$q=2^h\geq 4$	$\leq 3(q^k+q^{k-1}+\cdots+q)+2$
$4k+2$	$q\geq 5$ odd	$\leq q^{2k+1}+n_2(2k,q)$
$4k+2$	$q\geq 9$ odd	$q^{2k+1}-(q+1)+n_2(2k,q)+n_2(2,q)$

Other upper bounds on $k(N,q,\rho)$ were given by Davydov and Östergård [18–21]. A number of these upper bounds are mentioned in the next theorem.

Theorem 5.5.5. (1) *For* $p\geq 2$ *and* $m\geq 2$, $k(2,p^m,1)\leq 2p^{m-1}+2$.
(2) *For* $q\geq 4$, $k(3,q,1)\leq 2q+1$.
(3) *For* $p\geq 2$ *and* $m\geq\rho+1$, $k(\rho+1,p^m,\rho)\leq(p-1)\binom{\rho+1}{2}+p^{m-\rho}(\rho+1)+1$.
(4) *For* $q\neq 3$, $k(5,q,2)\leq 3q+1$.

Techniques 5.6.6. (1) *Of particular interest to these results is the fact that these upper bounds were obtained by the construction of carefully selected subsets of points from* $PG(N,q)$. *The* 1*-saturating sets and* 2*-saturating sets of Theorem* 5.5.5 (2) *and Theorem* 5.5.5 (4) *are defined by the columns of the following two matrices* H_1 *and* H_2:

$$H_1=\left[\begin{array}{ccc|c|cccc}1&\cdots&1&0&0&0&\cdots&0\\a_1&\cdots&a_q&1&0&0&\cdots&0\\a_1^2&\cdots&a_q^2&0&0&1&\cdots&1\\0&\cdots&0&0&1&a_2&\cdots&a_q\end{array}\right]$$

and

$$H_2 = \left[\begin{array}{ccc|c|ccc|ccc|c} 1 & \cdots & 1 & 0 & 0 & \cdots & 0 & 0 & \cdots & 0 & 0 \\ a_1 & \cdots & a_q & 1 & 0 & \cdots & 0 & 0 & \cdots & 0 & 0 \\ a_1^2 & \cdots & a_q^2 & 0 & 1 & \cdots & 1 & 0 & \cdots & 0 & 0 \\ 0 & \cdots & 0 & 0 & a_2 & \cdots & a_q & a_1^2 & \cdots & a_q^2 & 0 \\ 0 & \cdots & 0 & 0 & 0 & \cdots & 0 & a_1 & \cdots & a_q & 1 \\ 0 & \cdots & 0 & 0 & 0 & \cdots & 0 & 1 & \cdots & 1 & 0 \end{array}\right],$$

with $\mathbb{F}_q = \{a_1 = 0, a_2, \dots, a_q\}$.

These two particular examples show that by taking the unions of particularly selected subsets of Galois geometries, such as lines *and* conics, *it is possible to obtain very good upper bounds on the parameter* $k(N,q,\rho)$. *In the matrix* H_1, *the first* q *columns are points of a conic and the last* q *columns are points of a line. In the matrix* H_2, *we recognize* q *points of two conics, and* $q-1$ *points of a line.*

(2) *In* [18, 19, 21], *Davydov and Östergård also show how, by means of inductive constructions, it is possible to construct infinite classes of* ρ*-saturating sets.*

5.61. Open Problems

1. In the articles of Davydov and Östergård, a lot of attention has been paid to 2-saturating and 3-saturating sets. It is of great interest to construct small ρ-saturating sets, with $\rho > 3$.

2. Which particular subsets of Galois geometries, or unions of carefully selected subsets of Galois geometries, define small ρ-saturating sets?

3. Which inductive construction methods lead to interesting infinite classes of small ρ-saturating sets?

6.0 Extension Results

6.01. The Extension Result of Hill and Lizak

We start this section by formulating two theorems on blocking sets that have become classical. Interestingly, these results are related to the extendability problem for linear codes. The first theorem is of Bose and Burton [9].

Theorem 6.1.1 (Bose and Burton [9]). *Let* $\mathcal{K}$ *be an* $(n,1)$*-blocking set in* PG(N,q) *with respect to the s-dimensional subspaces that has the smallest possible cardinality. Then* $n = v_{N-s+1}$ *and* $\mathcal{K} = \chi_F$, *where F is a fixed* $(N-s)$*-dimensional subspace of* PG(N,q).

An $(n,1)$-blocking set with respect to s-dimensional subspaces in PG(N,q) is called *non-trivial* if there exists no $(N-s)$-dimensional subspace δ with $\mathcal{K}(P) > 0$ for every point P on δ. The next result which was proven independently by Beutelspacher [4] and Heim [39] characterizes the smallest non-trivial blocking sets.

Theorem 6.2.2 (Beutelspacher and Heim [4,39]). *The smallest non-trivial $(n,1)$-blocking sets in* $\mathrm{PG}(N,q)$ *with respect to the s-dimensional subspaces are cones with an $(N-s-2)$-dimensional vertex and a non-trivial $(n',1;2,q)$-blocking set of minimum cardinality in a plane of* $\mathrm{PG}(N,q)$ *as base curve. Consequently,*

$$n = q^{N-s} + q^{N-s-1} + \cdots + 1 + q^{N-s-1} \cdot r(q),$$

where $q + r(q) + 1$ is the minimal size of a non-trivial blocking set in $\mathrm{PG}(2,q)$.

Now we turn to the extendability problem for linear codes and arcs. It has been long known that a binary $[n,k,d]$ code of odd minimum distance can be extended to an $[n+1,k,d+1]$ code by adding a parity check. This result has been generalized by Hill and Lizak in [40,41].

Theorem 6.3.3 (Hill and Lizak [40,41]). *Let C be an $[n,k,d]_q$ code with* $\gcd(d,q) = 1$ *and with all non-zero weights congruent to 0 or d* $\pmod q$. *Then C can be extended to an $[n+1,k,d+1]_q$ code.*

The geometrical version of this result is given below. We include a proof which relies on the Bose-Burton theorem (Theorem 6.1.1).

Theorem 6.4.4. *Let $\mathcal{K}$ be an $(n,w;k-1,q)$-arc with* $\gcd(n-w,q) = 1$. *Assume that the multiplicities of all hyperplanes are congruent to n or w* $\pmod q$. *Then $\mathcal{K}$ can be extended to an $(n+1,w;k-1,q)$-arc.*

Proof. Fix a hyperplane H_0 in $\mathrm{PG}(k-1,q)$ with $\mathcal{K}(H_0) = w$. For any subspace δ of codimension 2, $\delta \subset H_0$, consider the hyperplanes H_i, $i = 0,\ldots,q$, containing δ. Let α of them be of multiplicity congruent to $n \pmod q$. Then

$$n = \sum_{i=0}^{q} \mathcal{K}(H_i) - q\mathcal{K}(\delta) \equiv \alpha n + (q+1-\alpha)w \pmod q,$$

whence $(\alpha-1)(n-w) \equiv 0 \pmod q$ and $\alpha = 1$. Hence, the number of hyperplanes of multiplicity congruent to $n \pmod q$ equals the number of subspaces of codimension 1 in H_0 and forms a blocking set in the dual space. By Theorem 6.1.1, a blocking set in $\mathrm{PG}(k-1,q)$ with respect to the lines having cardinality $(q^{k-1}-1)/(q-1)$ consists of the points of a hyperplane. By duality, this implies that all hyperplanes of multiplicity congruent to $n \pmod q$ pass through a fixed point P. Moreover, these are all the hyperplanes through P. Hence, we can get an $(n+1,w;k-1,q)$-arc by increasing the multiplicity of P by 1. □

Using the result of Beutelspacher and Heim (Theorem 6.2.2), we can go a bit further.

Theorem 6.5.5 (Landjev and Rousseva [50]). *Let $\mathcal{K}$ be an $(n,w;k-1,q)$-arc, $q = p^s$, with spectrum $(a_i)_{i\geq 0}$. Let $w \not\equiv n \pmod q$ and*

$$\sum_{i \not\equiv w \pmod q} a_i < q^{k-2} + q^{k-3} + \cdots + 1 + q^{k-3} \cdot r(q), \tag{1}$$

where $q + r(q) + 1$ is the minimal size of a non-trivial blocking set of $\mathrm{PG}(2,q)$. *Then $\mathcal{K}$ is extendable to an $(n+1,w;k-1,q)$-arc.*

From this theorem, we can derive a useful corollary which roughly says that if for a given $(n,w;k-1,q)$-arc with $w \equiv n+1 \pmod q$, there are not too many hyperplanes of multiplicity $\not\equiv n, n+1 \pmod q$, then this arc is extendable.

Theorem 6.6.6 (Landjev and Rousseva [50]). *Let $\mathcal{K}$ be a non-extendable $(n,w;k-1,q)$-arc, $k \geq 3$, $q = p^s$, with $\gcd(n-w,q) = 1$ and with spectrum $(a_i)_{i\geq 0}$. Let H be a hyperplane with $\mathcal{K}(H) \equiv w \pmod q$ and denote by θ the maximal number of hyperplanes of multiplicity $\not\equiv w \pmod q$ that are incident with a subspace of codimension 2 contained in H. Then*

$$\sum_{i \not\equiv n,w \pmod q} a_i \geq q^{k-3} \cdot r(q)/(\theta - 1),$$

where $r(q)$ is the same as in Theorem 6.5.5.

This result is easily restated for linear codes.

Theorem 6.7.7. *Let C be a non-extendable $[n,k,d]_q$ code, $q = p^s$, with $\gcd(d,q) = 1$. If $(A_i)_{i\geq 0}$ is the spectrum of C, then $\sum_{i \not\equiv 0,d \pmod q} A_i \geq q^{k-3} \cdot r(q)$, where $r(q)$ is the same as in Theorem 6.5.5.*

6.72. Diversity and Extendability

In a series of papers, Maruta further generalized these results [56–58]. Let C be an $[n,k,d]_q$ code with $k \geq 3$ and with $\gcd(d,q) = 1$, and with spectrum $(A_i)_{i\geq 0}$. We define

$$\Phi_0 = \frac{1}{q-1} \sum_{q \mid i, i \neq 0} A_i, \quad \Phi_1 = \frac{1}{q-1} \sum_{i \not\equiv 0,d \pmod q} A_i.$$

The pair (Φ_0, Φ_1) is called the *diversity* of C. The theorem of Hill and Lizak (Theorem 6.3.3) states that every linear code with $\Phi_1 = 0$ is extendable.

Theorem 6.8.8. *Let $q \geq 5$ be an odd prime power and let $k \geq 3$ be an integer. For a linear $[n,k,d]_q$ code C with $d \equiv -2 \pmod q$ and with diversity (Φ_0, Φ_1) such that $A_i = 0$ for all $i \not\equiv 0, -1, -2 \pmod q$, the following results are equivalent:*

1. *C is extendable.*

2. *$(\Phi_0, \Phi_1) \in \{(v_{k-1}, 0), (v_{k-1}, 2q^{k-2}), (v_{k-1} + (\rho-2)q^{k-2}, 2q^{k-2})\} \cup \{(v_{k-1} + iq^{k-2}, (q-2i)2^{k-2}) \mid i = 1, \ldots, \rho - 1\}$, where $\rho = (q+1)/2$.*

Furthermore, if 1. and 2. are valid and if $(\Phi_0, \Phi_1) \neq (v_{k-1} + (\rho-2)q^{k-2}, 2q^{k-2})$, then C is doubly extendable.

6.83. Extension Results Depending on Divisibility and Quasi-divisibility

Let C be a linear $[n,k,d]_q$ code. Following Ward [70], we call the integer $\Delta > 1$ a *divisor* of C if it is a common factor of all weights of C. The code C is called *divisible* if it has a divisor $\Delta > 1$. This definition can be given in a geometrical setting. Let $\mathcal{K}$ be an $(n,w;k-1,q)$-arc. The integer $\Delta > 1$ is said to be a *divisor* of $\mathcal{K}$ if $\mathcal{K}(H) \equiv n \pmod \Delta$ for every hyperplane

H in PG$(k-1,q)$. The arc $\mathcal{K}$ is *divisible* if it has a divisor. Clearly, Δ is a divisor of the code C if and only if it is a divisor of the associated arc $\mathcal{K}$.

Ward proved in [70] a theorem on the divisibility of linear codes meeting the Griesmer bound. Below we reformulate Ward's result for *Griesmer arcs*, i.e. arcs associated with Griesmer codes.

Theorem 6.9.9 (Ward [70]). *Let $\mathcal{K}$ be a Griesmer (n,w)-arc in* PG$(k-1,p)$, *p prime, with $w \equiv n \pmod{p^e}$, $e \geq 1$. Then $\mathcal{K}(H) \equiv n \pmod{p^e}$ for every hyperplane H.*

For Griesmer arcs in projective geometries over non-prime fields (resp. Griesmer codes over such fields), we have the following weaker version of this result [70].

Theorem 6.10.10 (Ward [70]). *Let $\mathcal{K}$ be a Griesmer (n,w)-arc in* PG$(k-1,q)$, *where $q = p^m$, p prime, $m \geq 1$, and let $w \equiv n \pmod{q^e}$ for some integer $e \geq 1$. Then $\mathcal{K}(H) \equiv n \pmod{p^e}$ for every hyperplane H of* PG$(k-1,q)$.

Let w,n be integers with $w < n$, $\gcd(w,n) = 1$. The integer $\Delta > 1$ is called a *quasi-divisor* of the $(n,w;k-1,q)$-arc $\mathcal{K}$ if $\mathcal{K}(H) \equiv n$ or $w \pmod{\Delta}$ for all hyperplanes in PG$(k-1,q)$. An arc is called *quasi-divisible* if it has a quasi-divisor. The theorem of Hill-Lizak in its geometrical form (Theorem 6.4.4) says that if the $(n,w;k-1,q)$-arc is quasi-divisible with $\Delta = q$ and with $w \equiv n+1 \pmod{q}$, then it is extendable to a divisible $(n+1,w;k-1,q)$-arc. The following theorem is useful if we know all the induced arcs of a given arc.

Theorem 6.11.11 (Landjev and Rousseva [50]). *Let $\mathcal{K}$ be an (n,w)-arc in* PG$(k-1,q)$, *$k \geq 4$, with $n > w(q-1)$. Denote by φ_P the projection from the point P. If the induced arc $\mathcal{K}^{\varphi_P}$ has quasi-divisor q for every point P with $\mathcal{K}(P) > 0$, then $\mathcal{K}$ is extendable.*

7.0 Codes Arising from Incidence Matrices of Galois Geometries

7.01. Linear Codes Defined by Incidence Matrices of Galois Geometries

We define the incidence matrix $A = (a_{ij})$ of points and hyperplanes in the projective space PG(N,q), $q = p^h$, p prime, $h \geq 1$, as the matrix whose rows are indexed by the hyperplanes of PG(N,q) and whose columns are indexed by the points of PG(N,q), and with entry

$$a_{ij} = \begin{cases} 1 & \text{if point } j \text{ belongs to hyperplane } i, \\ 0 & \text{otherwise.} \end{cases}$$

The p-ary linear code of points and hyperplanes of PG(N,q), $q = p^h$, p prime, $h \geq 1$, is the $\mathbb{F}_p$-span of the rows of the incidence matrix A. We denote this code by C(N,q). We identify the support of a codeword with the corresponding set of points of PG(N,q).

The fundamental parameters n,k, and d of these linear codes C(N,q) are known [1,54]:

1. $n = q^N + q^{N-1} + \cdots + q + 1$,
2. $k = \binom{p+N-1}{N}^h + 1$,
3. $d = q^{N-1} + \cdots + q + 1$.

7.02. Small Weight Codewords

We note that the minimum distance of this code $C(N,q)$ is known. Moreover, the codewords of minimum weight also have been classified.

Theorem 7.1.1 (Assmus and Key [1]). *Every codeword of weight $d = q^{N-1} + \cdots + q + 1$ in $C(N,q)$ is, up to a scalar multiple, the incidence vector of a hyperplane.*

So the question arises: what is the second smallest weight of $C(N,q)$, and what are the codewords of this second smallest weight: can they be characterized in a geometrical way?

The difference of the incidence vectors of two hyperplanes of $\mathrm{PG}(N,q)$ defines a codeword of weight $2q^{N-1}$. So, the preceding questions can also be formulated in the following way: *is $2q^{N-1}$ the second smallest weight of $C(N,q)$; and if this is indeed the case, are all the codewords of weight $2q^{N-1}$ equal, up to a scalar multiple, to the difference of the incidence vectors of two hyperplanes?*

For the code $C(2,q)$, it has effectively been proven that the second weight is equal to $2q$.

Theorem 7.2.2 (Lavrauw *et al.* [53]). *There are no codewords of weight in the interval $[q+2, 2q-1]$ in the linear code $C(2,q)$.*

Techniques 7.3.3. *To prove that $2q$ is the second smallest weight of the code $C(2,q)$, $q = p^h$, p prime, $h \geq 1$, one can proceed in the following way.*

The minimum weight of $C(2,q) \cap C(2,q)^{\perp}$ is $2q$ [1]. So, only codewords in $C(2,q) \setminus C(2,q)^{\perp}$ having weight in the interval $[q+2, 2q-1]$ need to be eliminated.

A possible codeword $c \in C(2,q) \setminus C(2,q)^{\perp}$ of weight $w(c) \in [q+2, 2q-1]$ satisfies $c.\ell = \alpha \neq 0$, for some constant α valid for all lines ℓ of $\mathrm{PG}(2,q)$. Hence, supp(c) defines a blocking set of $\mathrm{PG}(2,q)$. This fact already makes the link to the article [6] on blocking sets. A detailed study of these possible codewords c shows that supp(c) must satisfy the following property: supp(c) must share $1 \pmod{p}$ points with every small linear blocking set of $\mathrm{PG}(2,q)$; see e.g. [6, Section 4] for the definition of linear blocking sets. Imposing this condition on supp(c) leads to the proof that supp(c) is equal to a line m of $\mathrm{PG}(2,q)$, but then supp(c) has weight $q+1$, which contradicts the fact that supp$(c) \in [q+2, 2q-1]$. This eliminates the existence of codewords of weight in $[q+2, 2q-1]$ in $C(2,q) \setminus C(2,q)^{\perp}$. Gathering all results gives that the second smallest weight of $C(2,q)$ is equal to $2q$.

So here, geometrical ideas again lead to results on linear codes.

For $q = p$ prime, the following results are valid.

Theorem 7.4.4 (Fack *et al.* [27]). *The only codewords c, with $0 < w(c) \leq 2p + (p-1)/2$, in the p-ary linear code $C(2,p)$, p prime, $p \geq 11$, are:*

- *codewords with weight $p+1$: the scalar multiples of the incidence vectors of the lines of $\mathrm{PG}(2,p)$,*
- *codewords with weight $2p$: $\alpha(c_1 - c_2)$, c_1 and c_2 the incidence vectors of two distinct lines of $\mathrm{PG}(2,p)$, $\alpha \in \mathbb{F}_p \setminus \{0\}$,*
- *codewords with weight $2p+1$: $\alpha c_1 + \beta c_2$, $\alpha, \beta \in \mathbb{F}_p \setminus \{0\}$, $\beta \neq -\alpha$, with c_1 and c_2 the incidence vectors of two distinct lines of $\mathrm{PG}(2,p)$.*

The general results on the second smallest weight of the codes $C(N,q)$ are as follows.

Theorem 7.5.5 (Lavrauw *et al.* [53]). *There are no codewords of weight in the interval* $[v_N+1, 2q^{N-1}-1]$ *in the code* $C(N,q)$.

8.0 A Geometrical Result Obtained via Linear Codes

To conclude this article, we also wish to give a geometrical result obtained via coding-theoretical arguments.

Definition 8.1.1. A *strong representative system* S in $\mathrm{PG}(N,q)$ is a set of points such that every point $P \in S$ belongs to at least one tangent hyperplane $T_P(S)$ to S, i.e., every point P of S belongs to at least one hyperplane $T_P(S)$ only sharing P with S.

The $(q+1)$-arcs of $\mathrm{PG}(2,q)$ and the ovoids of $\mathrm{PG}(3,q)$ (see e.g. [25]) are particular examples of strong representative systems. Moreover, results of Bruen and Thas [11, 12] show that $q\sqrt{q}+1$ is the largest size for a strong representative system in $\mathrm{PG}(2,q)$, q square, and that q^2+1 is the largest size for a strong representative system in $\mathrm{PG}(3,q)$. For $N \geq 4$, Bruen and Thas prove that the size of every strong representative system S satisfies the bound $|S| < q^{(N-1)/2}$. But the linear codes defined by the incidence matrices of points and hyperplanes of $\mathrm{PG}(N,q)$ lead to great improvements for large dimensions N.

Techniques 8.2.2. *The idea of obtaining an upper bound on the size of strong representative systems in* $\mathrm{PG}(N,q)$ *via linear codes is as follows.*

Every point P of S has a tangent hyperplane $T_P(S)$. Enumerate the points $P_1,\ldots,P_{v_{N+1}}$ of $\mathrm{PG}(N,q)$ *such that $S=\{P_1,\ldots,P_{|S|}\}$. Select for every point P_i of S a particular tangent hyperplane $T_{P_i}(S)$, and enumerate the hyperplanes $\pi_1,\ldots,\pi_{v_{N+1}}$ such that $\pi_i = T_{P_i}(S)$, $i=1,\ldots,|S|$. Then the incidence matrix A of* $\mathrm{PG}(N,q)$ *has the following form:*

$$A = \begin{pmatrix} I_{|S|} & B \\ C & D \end{pmatrix},$$

with $I_{|S|}$ the identity matrix of rank $|S|$.

This implies that $rank(A) \geq |S|$, but $rank(A)$ is known (Subsection 7.01.), *so we find that the size of a strong representative system in* $\mathrm{PG}(N,q)$ *must satisfy*

$$\binom{p+N-1}{N}^h + 1 \geq |S|.$$

For large dimensions N, depending on the characteristic p, this upper bound on $|S|$ improves greatly on the upper bound $q^{(N-1)/2}$ found via the standard counting arguments.

Acknowledg nents

The first author was supported by the Strategic Development Fund of the New Bulgarian University under Contract 357/14.05.2009. The second author was supported by the project *Combined algorithmic and theoretical study of combinatorial structures* between the Research Foundation – Flanders (Belgium) (FWO) and the Bulgarian Academy of Sciences. This research also takes place within the project *Linear codes and cryptography* of the Research Foundation – Flanders (Belgium) (FWO) (project nr. G.0317.06).

References

[1] E. F. Assmus, Jr. and J. D. Key, *Designs and their codes*, vol. 103 of Cambridge Tracts in Mathematics, Cambridge University Press, Cambridge, 1992.

[2] S. Ball, *The polynomial method in Galois geometries*, in Current research topics in Galois geometry, Nova Sci. Publ., New York, 2011, ch. 5, pp. 103–128.

[3] B. I. Belov, V. N. Logacev, and V. P. Sandimirov, *Construction of a class of linear binary codes meeting the Varsamov-Griesmer bound*, Probl. Peredaci Inform., 10 (1974), pp. 36–44.

[4] A. Beutelspacher, *Blocking sets and partial spreads in finite projective spaces*, Geom. Dedicata, 9 (1980), pp. 425–449.

[5] J. Bierbrauer and Y. Edel, *Large caps in projective Galois spaces*, in Current research topics in Galois geometry, Nova Sci. Publ., New York, 2011, ch. 4, pp. 85–102.

[6] A. Blokhuis, P. Sziklai, and T. Szőnyi, *Blocking sets in projective spaces*, in Current research topics in Galois geometry, Nova Sci. Publ., New York, 2011, ch. 3, pp. 61–84.

[7] E. Boros and T. Szőnyi, *On the sharpness of a theorem of B. Segre*, Combinatorica, 6 (1986), pp. 261–268.

[8] R. C. Bose, *Mathematical theory of the symmetrical factorial design*, Sankhyā, 8 (1947), pp. 107–166.

[9] R. C. Bose and R. C. Burton, *A characterization of flat spaces in a finite geometry and the uniqueness of the Hamming and the MacDonald codes*, J. Combinatorial Theory, 1 (1966), pp. 96–104.

[10] A. E. Brouwer and M. van Eupen, *The correspondence between projective codes and 2-weight codes*, Des. Codes Cryptogr., 11 (1997), pp. 261–266.

[11] A. A. Bruen and J. A. Thas, *Blocking sets*, Geom. Dedicata, 6 (1977), pp. 193–203.

[12] ——, *Hyperplane coverings and blocking sets*, Math. Z., 181 (1982), pp. 407–409.

[13] A. A. Bruen, J. A. Thas, and A. Blokhuis, *On M.D.S. codes, arcs in* $\mathrm{PG}(n,q)$ *with* q *even, and a solution of three fundamental problems of B. Segre*, Invent. Math., 92 (1988), pp. 441–459.

[14] K. A. Bush, *Orthogonal arrays of index unity*, Ann. Math. Statistics, 23 (1952), pp. 426–434.

[15] L. R. A. Casse, *A solution to Beniamino Segre's "Problem* $I_{r,q}$*" for* q *even*, Atti Accad. Naz. Lincei Rend. Cl. Sci. Fis. Mat. Natur. (8), 46 (1969), pp. 13–20.

[16] L. R. A. Casse and D. G. Glynn, *The solution to Beniamino Segre's problem* $I_{r,q}$, $r = 3$, $q = 2^h$, Geom. Dedicata, 13 (1982), pp. 157–163.

[17] ——, *On the uniqueness of* $(q+1)_4$*-arcs of* PG$(4,q)$, $q = 2^h$, $h \geq 3$, Discrete Math., 48 (1984), pp. 173–186.

[18] A. A. Davydov and P. Östergård, *New linear codes with covering radius* 2 *and odd basis*, Des. Codes Cryptogr., 16 (1999), pp. 29–39.

[19] ——, *New quaternary linear codes with covering radius 2*, Finite Fields Appl., 6 (2000), pp. 164–174.

[20] ——, *On saturating sets in small projective geometries*, European J. Combin., 21 (2000), pp. 563–570.

[21] ——, *Linear codes with covering radius* $R = 2, 3$ *and codimension* tR, IEEE Trans. Inform. Theory, 47 (2001), pp. 416–421.

[22] ——, *Recursive constructions of complete caps*, J. Statist. Plann. Inference, 95 (2001), pp. 167–173. Special issue on design combinatorics: in honor of S. S. Shrikhande.

[23] J. De Beule, K. Metsch, and L. Storme, *Characterization results on arbitrary non-weighted minihypers and on linear codes meeting the Griesmer bound*, Des. Codes Cryptogr., 49 (2008), pp. 187–197.

[24] ——, *Characterization results on weighted minihypers and on linear codes meeting the Griesmer bound*, Adv. Math. Commun., 2 (2008), pp. 261–272.

[25] F. De Clerck and N. Durante, *Constructions and characterizations of classical sets in* PG(n,q), in Current research topics in Galois geometry, Nova Sci. Publ., New York, 2011, ch. 1, pp. 1–32.

[26] S. Dodunekov and J. Simonis, *Codes and projective multisets*, Electron. J. Combin., 5 (1998), pp. Research Paper 37, 23 pp. (electronic).

[27] V. Fack, Sz. L. Fancsali, L. Storme, G. Van de Voorde, and J. Winne, *Small weight codewords in the codes arising from Desarguesian projective planes*, Des. Codes Cryptogr., 46 (2008), pp. 25–43.

[28] J. C. Fisher, J. W. P. Hirschfeld, and J. A. Thas, *Complete arcs in planes of square order*, in Combinatorics '84 (Bari, 1984), vol. 123 of North-Holland Math. Stud., North-Holland, Amsterdam, 1986, pp. 243–250.

[29] E. M. Gabidulin, A. A. Davydov, and L. M. Tombak, *Linear codes with covering radius* 2 *and other new covering codes*, IEEE Trans. Inform. Theory, 37 (1991), pp. 219–224.

[30] D. G. Glynn, *The nonclassical* 10*-arc of* PG$(4,9)$, Discrete Math., 59 (1986), pp. 43–51.

[31] P. Govaerts and L. Storme, *On a particular class of minihypers and its applications. II. Improvements for q square*, J. Combin. Theory Ser. A, 97 (2002), pp. 369–393.

[32] J. H. Griesmer, *A bound for error-correcting codes*, IBM J. Res. Develop., 4 (1960), pp. 532–542.

[33] B. R. Gulati and E. G. Kounias, *On bounds useful in the theory of symmetrical factorial designs*, J. Roy. Statist. Soc. Ser. B, 32 (1970), pp. 123–133.

[34] N. Hamada, *Characterization, respectively nonexistence of certain q-ary linear codes attaining the Griesmer bound*, Bull. Osaka Women's Univ., 24 (1985), pp. 1–47.

[35] ——, *A characterization of some* $[n,k,d;q]$*-codes meeting the Griesmer bound using a minihyper in a finite projective geometry*, Discrete Math., 116 (1993), pp. 229–268.

[36] N. Hamada and T. Helleseth, *A characterization of some q-ary codes* ($q > (h-1)^2$, $h \geq 3$) *meeting the Griesmer bound*, Math. Japon., 38 (1993), pp. 925–939.

[37] ——, *Codes and minihypers*, in Proceedings of the Third Euro Workshop on Optimal Codes and Related Topics, Sunny Beach, Bulgaria, 2001, pp. 79–84.

[38] N. Hamada and T. Maekawa, *A characterization of some q-ary linear codes* ($q > (h-1)^2$, $h \geq 3$) *meeting the Griesmer bound. II*, Math. Japon., 46 (1997), pp. 241–252.

[39] U. Heim, *On t-blocking sets in projective spaces*. manuscript, 1994.

[40] R. Hill, *An extension theorem for linear codes*, Des. Codes Cryptogr., 17 (1999), pp. 151–157.

[41] R. Hill and P. Lizak, *Extensions of linear codes*, in Proc. Intern. Symposium on Inform. Theory, Whistler, BC, Canada, 1995.

[42] R. Hill and H. Ward, *A geometric approach to classifying Griesmer codes*, Des. Codes Cryptogr., 44 (2007), pp. 169–196.

[43] J. W. P. Hirschfeld, *Projective geometries over finite fields*, Oxford Mathematical Monographs, The Clarendon Press Oxford University Press, New York, second ed., 1998.

[44] J. W. P. Hirschfeld and G. Korchmáros, *On the embedding of an arc into a conic in a finite plane*, Finite Fields Appl., 2 (1996), pp. 274–292.

[45] ——, *Arcs and curves over a finite field*, Finite Fields Appl., 5 (1999), pp. 393–408.

[46] J. W. P. Hirschfeld, G. Korchmáros, and F. Torres, *Algebraic curves over a finite field*, Princeton Series in Applied Mathematics, Princeton University Press, Princeton, NJ, 2008.

[47] J. W. P. Hirschfeld and L. Storme, *The packing problem in statistics, coding theory and finite projective spaces: update 2001*, in Finite geometries, vol. 3 of Dev. Math., Kluwer Acad. Publ., Dordrecht, 2001, pp. 201–246.

[48] T. Honold and I. Landjev, *Codes over rings and ring geometries*, in Current research topics in Galois geometry, Nova Sci. Publ., New York, 2011, ch. 7, pp. 159–184.

[49] B. C. Kestenband, *A family of complete arcs in finite projective planes*, Colloq. Math., 57 (1989), pp. 59–67.

[50] I. Landjev and A. Rousseva, *An extension theorem for arcs and linear codes*, Probl. Peredaci Inform., 42 (2006), pp. 65–76.

[51] I. Landjev and L. Storme, *A weighted version of a result of Hamada on minihypers and on linear codes meeting the Griesmer bound*, Des. Codes Cryptogr., 45 (2007), pp. 123–138.

[52] ——, *A study of* $(x(q+1), x; 2, q)$*-minihypers*, Des. Codes Cryptogr., 54 (2010), pp. 135–147.

[53] M. Lavrauw, L. Storme, P. Sziklai, and G. Van de Voorde, *An empty interval in the spectrum of small weight codewords in the code from points and k-spaces of* $\mathrm{PG}(n,q)$, J. Combin. Theory Ser. A, 116 (2009), pp. 996–1001.

[54] F. J. MacWilliams and N. J. A. Sloane, *The theory of error-correcting codes*, North-Holland Publishing Co., Amsterdam, 1977. North-Holland Mathematical Library, Vol. 16.

[55] W.-A. Jackson, K. M. Martin, and M. B. Paterson, *Applications of Galois Geometry to Cryptology*, in Current research topics in Galois geometry, Nova Sci. Publ., New York, 2011, ch. 9, pp. 213–241.

[56] T. Maruta, *On the extendability of linear codes*, Finite Fields Appl., 7 (2001), pp. 350–354.

[57] ——, *Extendability of linear codes over* $\mathrm{GF}(q)$ *with minimum distance d,* $\gcd(d,q) = 1$, Discrete Math., 266 (2003), pp. 377–385. The 18th British Combinatorial Conference (Brighton, 2001).

[58] ——, *A new extension theorem for linear codes*, Finite Fields Appl., 10 (2004), pp. 674–685.

[59] F. Pambianco and L. Storme, *Small complete caps in spaces of even characteristic*, J. Combin. Theory Ser. A, 75 (1996), pp. 70–84.

[60] B. Segre, *Curve razionali normali e k-archi negli spazi finiti*, Ann. Mat. Pura Appl. (4), 39 (1955), pp. 357–379.

[61] ——, *Ovals in a finite projective plane*, Canad. J. Math., 7 (1955), pp. 414–416.

[62] ——, *Sui k-archi nei piani finiti di caratteristica due*, Rev. Math. Pures Appl., 2 (1957), pp. 289–300.

[63] ——, *Introduction to Galois geometries*, Atti Accad. Naz. Lincei Mem. Cl. Sci. Fis. Mat. Natur. Sez. I (8), 8 (1967), pp. 133–236.

[64] G. Solomon and J. J. Stiffler, *Algebraically punctured cyclic codes*, Information and Control, 8 (1965), pp. 170–179.

[65] L. Storme and J. A. Thas, *Complete k-arcs in* $\mathrm{PG}(n,q), q$ *even*, Discrete Math., 106/107 (1992), pp. 455–469. A collection of contributions in honour of Jack van Lint.

[66] ——, *M.D.S. codes and arcs in* $\mathrm{PG}(n,q)$ *with q even: an improvement of the bounds of Bruen, Thas, and Blokhuis*, J. Combin. Theory Ser. A, 62 (1993), pp. 139–154.

[67] ——, *k-arcs and dual k-arcs*, Discrete Math., 125 (1994), pp. 357–370. 13th British Combinatorial Conference (Guildford, 1991).

[68] J. A. Thas, *Connection between the Grassmannian* $G_{k-1;n}$ *and the set of the k-arcs of the Galois space* $S_{n,q}$, Rend. Mat. (6), 2 (1969), pp. 121–134.

[69] J. F. Voloch, *On the completeness of certain plane arcs. II*, European J. Combin., 11 (1990), pp. 491–496.

[70] H. N. Ward, *Divisibility of codes meeting the Griesmer bound*, J. Combin. Theory Ser. A, 83 (1998), pp. 79–93.

[71] V. K. Wei, *Generalized Hamming weights for linear codes*, IEEE Trans. Inform. Theory, 37 (1991), pp. 1412–1418.

In: Current Research Topics in Galois Geometry
Editors: J. De Beule and L. Storme
ISBN:978-1-61209-523-3

Chapter 9

APPLICATIONS OF GALOIS GEOMETRY TO CRYPTOLOGY

Wen-Ai Jackson[1,*], Keith M. Martin[2,†] and Maura B. Paterson[3,‡]
[1] School of Mathematical Sciences, the University of Adelaide, SA 5005, Australia
[2] Information Security Group, Royal Holloway, University of London, Egham, Surrey TW20 0EX, U.K.
[3] Department of Economics, Mathematics and Statistics, Birkbeck, University of London, Malet Street, London WC1E 7HX, U.K.

Abstract

Cryptology is the study of mathematical techniques for implementing core information security services such as confidentiality and authentication. Galois geometry has played an important role in developing the theory of cryptology in a number of different areas. We begin this article by commenting on reasons why Galois geometry arises in cryptology and the relevance of its application. We then discuss five separate applications. Secret sharing schemes are primitives for distributing partial information about a secret in such a way that only authorised coalitions of shareholders can recover the secret from the partial information. Authentication codes model the unconditionally secure provision of an authentication service. Key predistribution schemes are techniques for advance arrangement of cryptographic keys within a network so that users have access to the keys that they need to secure their future communications. Cryptographic security is sometimes based on the difficulty of solving multivariate equations systems, which can be interpreted geometrically. Finally, the Advanced Encryption Standard is arguably the most important current symmetric encryption algorithm and has some application of Galois geometry in its core.

Key Words: Cryptology, Cryptography, Secret sharing schemes, Authentication codes, Key predistribution, Multivariate cryptography, Algebraic cryptanalysis, Advanced Encryption Standard.

AMS Subject Classification: 11T71, 94A60, 94B27

*E-mail address: wen.jackson@adelaide.edu.au
†E-mail address: Keith.Martin@rhul.ac.uk
‡E-mail address: m.paterson@bbk.ac.uk

1. Introduction

We begin this article with a brief introduction to cryptology. We then consider why Galois geometry arises in cryptology and what impact its application has to real systems. The remaining five sections are each devoted to a different application of Galois geometry to cryptology. Section 2. discusses secret sharing schemes. Section 3. deals with authentication codes. Section 4. examines key predistribution schemes. Section 5. considers multivariate equations systems. Finally, Section 6. looks at geometric aspects of the Advanced Encryption Standard.

1.1. Cryptography

The science of *cryptology*, often referred to as *cryptography*, is the study of mathematical techniques, algorithms and protocols for implementing the core security services that are required to support electronic information protection. These security services include *confidentiality* (restricting access to the contents of communicated data), *data integrity* (protecting data from manipulation), *data origin authentication* (correctly attributing the originator of some data) and *non-repudiation* (providing evidence of the occurrence of a data exchange that cannot later be denied). Cryptographic *primitives* (the basic techniques) are widely used to protect banking transactions (for example ATM communication), mobile telecommunications, secure web access (using the SSL protocol), secure email, password storage on computer operating systems and so on.

Most cryptographic primitives critically rely on the use of *keys*, which are generally numbers selected from a large space by some random process. As the majority of cryptographic algorithms are publicly described, the security of a cryptographic primitive typically relies on the protection of the relevant keys. The nature of these keys provides a classification of cryptographic primitives into *symmetric*, where the secret keys employed by the sender and the receiver of the data are identical, and *public-key*, where only one of the keys needs to be secret, the other can be publicly known.

Cryptographic research focusses not only on the design of cryptographic primitives, but also on the study of subversion of cryptographic primitives, known as *cryptanalysis*, and on the design of supporting infrastructures for cryptography, one aspect of which is *key management*. The applications that we examine in this article cover all three research areas. We also demonstrate applications in two cryptographic security models. In *unconditional security*, the security of a cryptographic primitive is independent of the resources available to an attacker. This is a strong model, the cost of attainment of which, measured in terms of efficiency, is often too high for real applications. Hence, most implemented cryptography relies on *computational security*, where the security of a cryptographic primitive is based on the perceived computational hardness of a mathematical problem.

There are many introductory texts that provide a basic primer in cryptography. For a comprehensive coverage of modern cryptography, we highly recommend [74]. Surveys of combinatorial applications to cryptography can be found in [7, 16, 57].

1.2. Galois Geometry in Cryptography

Galois geometry is not a natural source of influence in the design of cryptographic primitives such as encryption algorithms, since the inherent structure of a geometry is precisely what is *not desired* for any process that is effectively randomising data. We will demonstrate three different (but closely related) ways in which Galois geometry plays a role in cryptography:

1. *As an encapsulation of linearity over finite fields.* The most important example of this influence is secret sharing schemes, which are cryptographic primitives that require rich internal structure. In contrast to encryption, linearity in secret sharing schemes is a desirable property. It thus makes sense to base secret sharing schemes on Galois geometry. Since secret sharing schemes often require slightly asymmetric properties due to the nature of their access structures (see Section 2.), Galois geometry also provides a convenient vehicle within which to conceptualise potential solutions.

2. *As a source of interesting combinatorial designs.* Some areas of cryptography are appropriately modeled by a combinatorial design. Examples of this include unconditionally secure authentication codes and certain families of key predistribution schemes. Thus although these models are not inherently geometric, Galois geometry has the potential to contribute interesting constructions.

3. *As a means of interpreting sets of multivariate polynomial equations.* One of the hard problems on which cryptographic primitives are sometimes based is the difficulty of solving large sets of multivariate equations. Geometric interpretations of such equation systems can give insight into their behaviour. Thus Galois geometry can be useful for both cryptographic design (we will show an example in Section 5.) and, more significantly, cryptanalysis (see Sections 5. and 6.).

It should be noted that Galois geometry is not used directly in the types of cryptographic application that are generally deployed in the "real world" to process data. Galois geometry, rather, contributes to our understanding of key management (see Sections 2. and 4.), to development of cryptographic models offering the ideal notion of unconditional security (see Sections 2. and 3.), and as a tool for studying complex equation systems (see Sections 5. and 6.).

2. Secret Sharing Schemes

A *secret sharing scheme* (SSS) is a protocol by which a *secret* piece of information can be protected among a finite set $\mathcal{P}$ of *players* in such a way that only certain predetermined subsets of players can jointly compute the secret. Secret sharing schemes were introduced independently by Blakley [8] and Shamir [69]. Since then, the area has developed in many directions.

Secret sharing schemes are fundamental cryptographic primitives that underpin many cryptographic mechanisms proposed for distributed environments. In the real world, secret sharing schemes are commonly used to protect cryptographic master keys.

The main relationship between secret sharing and Galois geometry arises through the study of *linear* secret sharing schemes. In this section, we will briefly introduce secret sharing schemes from a geometric perspective and discuss research issues where Galois geometry plays a role (see [41] for a more detailed review).

2.1. Model for Secret Sharing

The simplest unconditionally secure model for secret sharing involves an honest *dealer* who securely communicates shares of a *secret* s to a group of honest players. The *access structure* is the collection Γ of subsets of $\mathcal{P}$ which can jointly compute (through a combiner function) the secret from their shares. An *authorised* (resp. *unauthorised*) set is a subset of $\mathcal{P}$ which is in (resp. not in) the access structure. It is reasonable to assume that if A is an authorised set, then so is any set containing A.

A *perfect* SSS is one where any unauthorised set obtains no information about the secret if their shares are input to the combiner function. The *information rate* ρ measures the efficiency of the secret by comparing the players' share sizes to that of the secret. Let $\mathrm{H}(s)$ denote the size (formally, the entropy, see [38] for more details and how this relates to combinatorial measures) of the secret s, and $\mathrm{H}(p)$ the size of player p's share. For perfect schemes $\mathrm{H}(p) \geq \mathrm{H}(s)$ and so the information rate:

$$\rho = \max_{p \in \mathcal{P}} \mathrm{H}(s)/\mathrm{H}(p)$$

is at most 1. Perfect SSSs where $\rho = 1$ are said to be *ideal*.

A SSS consists of two phases. In the *sharing phase*, a dealer sends shares to each player and in the *reconstruction phase*, the players send their shares as inputs to the combiner function. If all authorised inputs are valid and correct, the output of the combiner function will be the secret.

2.2. Linear Secret Sharing Schemes

The most important class of SSSs are *linear SSSs* (LSSSs). These have the property that any linear combination of shares of different secrets results in shares for the same linear combination of the secrets. LSSSs are efficient to implement and have many interesting properties. They have been defined from many different (and equivalent) perspectives, such as Galois geometry [8, 73], vector spaces [11], linear codes [6, 77], and monotone span programs [46]. We will, naturally, adopt the Galois geometry approach in the following discussion.

Let $\Sigma = \mathrm{PG}(d,q)$ be the projective space of dimension d over the finite field $\mathbb{F}_q$, where a point P is represented by its homogeneous coordinates $(x_0, x_1, \ldots, x_d)$ ($x_i \in \mathbb{F}_q$). Let $\langle S_1, S_2 \rangle$ denote the subspace generated by the subspaces S_1 and S_2, and let $[\Sigma]$ denote the collection of all subspaces of Σ.

For an access structure Γ, consider the following assignment $\sigma \colon \mathcal{P} \cup \{s\} \to [\Sigma]$ such that:

1. if $A \in \Gamma$, then $s^\sigma \subseteq A^\sigma$,
2. if $A \notin \Gamma$, then $s^\sigma \cap A^\sigma = \emptyset$,

where $A^\sigma = \langle x^\sigma \| x \in A \rangle$. For each $x \in s \cup \mathcal{P}$, let $[x]$ be any matrix whose row space is equal to x^σ (and more generally for $A \subseteq \mathcal{P}$). Let H be the collection of the q^{d+1} distinct $(d+1)$-tuples over $\mathbb{F}_q$.

The mapping σ gives rise to a LSSS for Γ (referred to as a *geometric scheme* in [41]) in the following manner. Firstly, each matrix $[x]$, for $x \in \mathcal{P} \cup \{s\}$, is public knowledge. The dealer randomly and secretly selects $h \in H$ and distributes $[p]h^t$ as a share for each player p of the secret $[s]h^t$. If $A \in \Gamma$, then $s^\sigma \subseteq A^\sigma$ and so the row space of $[A]$ contains the row space of $[s]$. Thus from $[A]h^t$ and the publicly known $[A]$, $[s]h^t$ can be calculated. Using a similar argument, if $A \notin \Gamma$ then $A^\sigma \cap s^\sigma = \emptyset$, and so knowing $[A]h^t$ gives no information about $[s]h^t$. Thus this is a perfect SSS. The linearity follows immediately.

The most famous LSSS is *Shamir's scheme* [69] for realizing the (t,n) *threshold access structure* $T(t,n) = \{A \subseteq \mathcal{P} \| |A| \geq t\}$. This has an interpretation as a geometric scheme σ where the points s^σ and p^σ ($p \in \mathcal{P}$) are distinct points of a normal rational curve $\{(0,0,\ldots,0,1)\} \cup \{(1,\alpha,\alpha^2,\ldots,\alpha^{t-1}) \| \alpha \in \mathbb{F}_q\}$ in $\mathrm{PG}(t-1,q)$. This curve has the property that every set of t points are independent and so generate the whole space. Consequently, for a subset A of t points, $A^\sigma \supseteq s^\sigma$, and for a subset B of $t-1$ or fewer points, $B^\sigma \cap s^\sigma = \emptyset$.

2.3. Ideal Secret Sharing Schemes

Ideal LSSSs are those where every player has the same size of share as the secret, as in Shamir's scheme. In classical secret sharing, a fundamental research question is *which access structures give rise to ideal SSSs*?

This question has proved to be surprisingly difficult to answer. In [12], it was shown that the access structure of an ideal SSS induces a matroid T on $\mathcal{P} \cup \{s\}$, whose circuits through s are exactly the sets $A \cup \{s\}$ for each *minimal* set A of Γ (by minimal we mean that for all $p \in A$, $A \backslash \{p\} \notin \Gamma$). Furthermore, an ideal geometric scheme in $\mathrm{PG}(d,q)$, where s^σ is a point, exists for Γ if and only if the matroid T is representable over $\mathbb{F}_q$.

Each access structure Γ is associated with the *dual* access structure Γ^* whose authorised sets are the sets B of players, where B has a non-empty intersection with every (minimal) authorised set in Γ. In the ideal case, the concept of the dual matroid corresponds to that of the dual access structure. Furthermore, duality in the geometric sense corresponds to duality in the access structure and matroid sense. Indeed, from any geometric scheme we can easily construct a geometric scheme for the dual access structure with share size no larger than the original [35].

The study of access structures that give rise to ideal SSSs remains an area of innovative research. A related question, of closer relevance to Galois geometry is *which access structures give rise to an ideal LSSS*? For recent results on both these questions, see [28,55].

2.4. Efficient Linear Secret Sharing Schemes

Another fundamental research question is, given an access structure Γ, *what is the optimal information rate for any SSS (LSSS) realising* Γ?

It has been shown [5, 34, 73] that LSSSs exist for every access structure. However the LSSSs constructed using these generic techniques tend not be efficient. To see this, consider the access structure Γ with maximal unauthorised sets $B_1,\ldots,B_k$. Let e_i be the all zero vector with a 1 in the ith position ($1 \leq i \leq k$). The *cumulative scheme* σ for Γ is $\sigma\colon \mathcal{P}\cup\{s\} \to [\mathrm{PG}(k-1,q)]$, where $s^\sigma = (1,\ldots,1)$ and $x^\sigma = \langle e_i \,\|\, x \notin B_i\rangle$. The significance of the cumulative scheme is that not only can such a LSSS be constructed for any access structure Γ, but it is also the "worst" possible scheme, since every "sensible" geometric scheme with a point secret is contained within it [37]. Thus the cumulative scheme provides the solution space within which to search for more efficient geometric schemes, although all known search algorithms are exponential [6,37,76].

Research has focussed on studying this problem for specific families of access structures. For example, access structures on five participants [36] and *multipartite access structures* [28]. Generic upper bounds on the information rate for LSSSs are also known. However given an arbitrary access structure Γ, the design of an efficient LSSS for Γ often requires an ad hoc approach.

The Galois geometric interpretation of LSSSs can be a useful conceptual tool for constructing LSSSs. To illustrate this, consider a simple example. Let σ for Γ in $\pi = \mathrm{PG}(d,q)$ and τ for Γ' in $\alpha = \mathrm{PG}(d',q)$ be geometric schemes on the same player set $\mathcal{P}$ with point secrets. Now embed σ and τ in $\mathrm{PG}(d+d',q)$ in such a way that $\pi\cap\alpha = s^\tau = s^\alpha$. Then we can define $\phi\colon \mathcal{P}\cup\{s\} \to [\mathrm{PG}(d+d',q)]$ by $s^\phi = s^\sigma (= s^\tau)$ and for $x \in \mathcal{P}$ let $x^\phi = \langle x^\sigma, x^\tau\rangle$. It is easy to show that ϕ is a scheme for $\Gamma+\Gamma' = \{A \,\|\, A \in \Gamma\cup\Gamma'\}$. Similarly, to construct a geometric scheme for $\Gamma\Gamma' = \{A\cup B \,\|\, A\in\Gamma,\ B\in\Gamma'\}$ in $\mathrm{PG}(d+d'+1,q)$, we can embed σ and τ in such a way that $\pi\cap\alpha = \emptyset$. Define the new scheme by $\phi\colon \mathcal{P}\cup\{s\} \to [\mathrm{PG}(d+d'+1,q)]$ for $\Gamma\Gamma'$ by $s^\phi = s^\tau + s^\sigma$ (adding them as vectors) and for $x \in \mathcal{P}$ let $x^\phi = \langle x^\sigma, x^\tau\rangle$. Several other constructions for obtaining new geometric schemes from existing ones can be found in [41].

2.5. Specific Families of Access Structures

There are many different access structures for which a SSS could be defined. However relatively few of these access structures are of any real interest to potential applications. Attention in the literature tends to focus on SSSs for families of access structures that "make sense" from an application perspective. Such families include *threshold* access structures [69], *compartmented* access structures [72], *multilevel* access structures [11], and *multipartite* [28].

As well as seeking SSSs (LSSSs) with good information rates, another question is *how to maximise the number of possible players for particular families of access structures*? For example, with respect to threshold schemes, Shamir schemes are limited by the number of possible points on a normal rational curve. While this is not likely to be an issue in applications where the number of secrets is large (for example the secret is a cryptographic key), there are applications of SSSs where it is desirable to share a smaller secret. Galois geometry is a natural place to look for specific constructions that permit many players. For example, in [33] it was shown how to construct a *three-level* LSSS using a twisted cubic of $\mathrm{PG}(3,q)$ that allows more players than the generic multilevel LSSS discussed in [11].

2.6. Secret Sharing Schemes with Extended Capabilities

A great deal of research on SSSs concerns schemes with *extended capabilities*, by which we mean any additional properties that are not specified in the traditional secret sharing model discussed so far.

In the traditional SSS model, we assume that the dealer is honest and that players do not cheat, hence the adversary is any unauthorised group of players. Various alternative secret sharing models can be defined in terms of the capability of an *adversary* who tries to corrupt the scheme. If the dealer is honest, but the players may cheat, then it is necessary to construct SSSs with *cheater detection* or *cheater correction*. On the other hand, if the dealer can also cheat then a *verifiable* SSS is required. All these alternative models involve players being given additional information to their share that helps them to confront malicious behaviour. Galois geometry has played a role in some of these constructions. For a review of adversary models for secret sharing schemes, see [58].

The traditional SSS model is also static and does not consider the problem of how to efficiently change the scheme parameters over time, for example some players may become corrupted and have to leave the scheme. Various models for coping with such situations without the need for a complete share refresh have been considered. Again, Galois geometry has been used to construct optimal SSSs in several of these models, for example [4]. For a review of updating SSSs, see [56].

In the remainder of this section, we discuss two interesting extended capabilities.

2.6.1. Multiplicative Linear Secret Sharing Schemes

The study of *multi-party computation* involves devising protocols which allow players connected by a complete network to securely compute functions on the players' inputs. This is important for deploying cryptography in distributed environments. Recall that in a LSSS, the secret is a *linear* function on the inputs. A stronger requirement would be for an *arithmetic* function (that is, involving addition and multiplication) on the inputs. To this end, we define a *multiplicative* LSSS (or MLSSS) to be a LSSS σ with the additional property that for two secret values $s^{\sigma}h, s^{\sigma}h'$, the product $(s^{\sigma}h)(s^{\sigma}h')$ can be obtained by a (fixed) linear combination of the products of the shares of the players. Note that there is no specification as to how many players are needed in this process, hence the entire player set may be required.

Let the *adversary structure* of a SSS be the sets not in the access structure. The adversary structure is said to be Q^2 if no two unauthorised sets cover the player set. In [22], it was shown that an access structure Γ admits a multiplicative secret sharing scheme if and only if it satisfies Q^2, which happens if and only if $\Gamma^* \subseteq \Gamma$.

In [22], a combination of geometric schemes was used to show that for any LSSS with an access structure satisfying Q^2, there is an efficient procedure to obtain a MLSSS with share size at most twice that of the LSSS. However it remains unknown how to construct MLSSSs with optimal information rates. In particular, it is unknown exactly which access structures admit ideal MLSSSs (see [23] for recent results).

2.6.2. Multisecret Sharing Schemes

The problem of *multisecret sharing* concerns how to share more than one secret amongst a collection $\mathcal{P}$ of n players. In the case of m secrets, the access structure is a collection $\Gamma = \{\Gamma_1, \ldots, \Gamma_m\}$, where each Γ_i is a monotone set of subsets of $\mathcal{P}$.

We consider the case of a (w,t,k)-multithreshold scheme, where each k-set K of $\mathcal{P}$ is associated with a secret s_K, at least t participants of K are required to determine s_K, and no w-set W of $\mathcal{P}$ can obtain s_K unless $|W \cap K| \geq t$. Hence, the collection Γ_K of authorised sets of s_K is $\Gamma_K = \{A \subseteq \mathcal{P} \parallel |A \cap K| \geq t\}$ and the collection Δ_K of unauthorised sets of s_K is $\Delta_K = \{A \subseteq \mathcal{P} \parallel |A| \leq w\} \setminus \Gamma_K$. A multisecret sharing scheme for the set $\mathcal{S}$ of $m = \binom{n}{t}$ secrets s_K is called a (w,t,k)-*multithreshold scheme* (MTS) and can be generated using the assignment $\sigma \colon \mathcal{P} \cup \{\mathcal{S}\} \to [\Sigma]$ such that:

1. if $A \in \Gamma_K$, then $s_K^\sigma \subseteq A^\sigma$,
2. if $A \in \Delta_K$, then $s_K^\sigma \cap A^\sigma = \emptyset$,

for all $A \subseteq \mathcal{P}$ and each k-set K of $\mathcal{P}$. The resulting MTS is linear.

In [39], it was shown that for most meaningful choices of w, each user in a (w,t,k)-MTS needs to be given a secret value that is at least $\binom{w+k-2t+1}{k-t}$ times larger than the size of any secret. This bound is a generalisation of the bound on user storage for a KPS that was proved in [9] (see Section 4.). It is thus of particular interest to find MTSs that meet this bound.

In [9], an optimal $(w,1,k)$-MTS is constructed. Optimal (w,t,n)-MTSs correspond to optimal *ramp schemes*, which are a generalisation of threshold schemes. An optimal (w,k,k)-MTS is also easily constructed [39]. However, the task of constructing optimal MTSs for $1 < t < k$ appears to be difficult and only two constructions are known. In [40], a family of optimal $(n-k+1,2,k)$-MTSs were constructed, and in [3], a family of optimal $(w,2,3)$-MTSs. Both these constructions were based on Galois geometry.

3. Authentication Codes

Data origin authentication is one of the most important cryptographic services in the commercial world, since it is of extreme importance to know both the origin and correctness of received data. Although most implemented authentication mechanisms are based on more efficient computationally secure mechanisms, it is important to understand authentication in the unconditionally secure model. The study of *authentication codes* is in many ways the analogue of the work conducted for confidentiality (secrecy) by Claude Shannon in the 1940s [70].

Galois geometry has made an important contribution to the theory of authentication codes from the very outset, since the seminal paper by Gilbert, MacWilliams, and Sloane [32] used projective planes to construct an important family of authentication codes. Since then a substantial body of research has investigated the close relationship between authentication codes and combinatorial designs. The area also benefits from a monograph by Dingyi Pei [68], which provides a comprehensive overview of the relevant research. We thus restrict ourselves in this section to a brief introduction and an indication of the role of Galois geometry.

3.1. A-codes

The standard model for an authentication code (often referred to as an *A-code*) involves a *transmitter* who communicates a sequence of distinct *source states* from a set $\mathcal{S}$ to a *receiver* by encoding them using one from a set $\mathcal{E}$ of *encoding rules*. Each encoding rule is an injective mapping from $\mathcal{S}$ into a set $\mathcal{M}$ of *messages*. The receiver recovers the source states from the received messages by determining their (unique) pre-images under the agreed encoding rule. The receiver accepts a message as authentic if it lies in the image of the agreed encoding rule.

While the transmitter and the receiver trust one another, an *opponent* observes the resulting sequence of messages and attempts to determine another message which will be accepted by the receiver as authentic, thereby deceiving (*spoofing*) the receiver.

3.2. A^2-codes

An alternative model has been proposed for the case where the transmitter and receiver do not trust one another. In this case they need to operate through an *arbiter*. Authentication codes in this model are normally referred to as *authentication codes with arbitration*, or *A^2-codes*. In this case, we have a set of *encoding rules* for use between the transmitter and the arbiter, and a set of *decoding rules* for use between the arbiter and the receiver. The receiver secretly agrees a decoding rule f with the arbiter. The arbiter then forms an encoding rule e, which has the property that all messages of e are valid under f, and secretly gives e to the transmitter. Since the transmitter does not know f and the receiver does not know e, the arbiter can be used to resolve any disputes that later arise between the transmitter and receiver. An opponent in an A^2-code operates in a similar way to an opponent in an A-code.

3.3. Research Approaches

The main goal of an authentication code is to keep the opponent's probability of succeeding in an attack as low as possible, while also minimising the number of encoding (decoding) rules required, since the latter is a measure of the efficiency of the code.

Two approaches have been taken to studying authentication codes in either of the two models. In one approach, combinatorial bounds on the probability that the opponent succeeds in deceiving the receiver may be given in terms of the numbers of source states, messages and intercepted messages. In turn, bounds on the number of encoding rules required to achieve these bounds on the probability of deception may be derived and combinatorial characterisations of authentication systems with a minimum number of encoding rules given.

In the other approach, bounds on the probability of deception are given in terms of the information about the encoding rule contained in the intercepted messages. This provides direct bounds on the number of encoding rules, and combinatorial characterisations of authentication systems attaining these bounds may also be given. The combinatorial characterisations arising from the two approaches are distinct. The information theoretic approach imposes less restrictive conditions on the probability of deception and schemes with fewer encoding rules arise from the characterisation.

3.4. Geometric Constructions

In [67], it was shown that optimal A-codes are equivalent to a special class of designs known as *strongly partially balanced t-designs*. Thus, constructing good A-codes reduces to the problem of finding such designs. If the A-codes have the additional property that they are *Cartesian*, which means that each message represents at most one source state (and hence the A-code offers no secrecy), then one class of optimal A-codes corresponds to orthogonal arrays of index 1. Galois geometry can be used to construct orthogonal arrays of index 1, with a notable construction arising from projective planes, as pointed out in [32]. Other examples of optimal Cartesian A-codes based on Galois geometry include constructions from symplectic spaces [78] and unitary spaces [30]. Cartesian A-codes were also constructed from generalised quadrangles in [25]. A family of non-Cartesian optimal A-codes is constructed in [68] from normal rational curves.

Optimal A^2-codes are also closely related to special families of designs referred to as *restricted partially balanced designs* (see [68] for details). Likewise, these designs can arise from Galois geometry, with one of the first examples being found in [45].

While authentication codes do not, by default, offer secrecy (and in the case of Cartesian codes they explicitly do not), they can also be designed to offer levels of secrecy. Several of the geometric constructions referred to above can be adapted to offer secrecy (see [68] for details).

Hence the main contribution of Galois geometry to authentication theory is as a source of interesting constructions. It should be noted that while Galois geometry could almost certainly be used to construct many new types of authentication codes, the only constructions strictly of interest will be those that offer new parameter sets for optimal (or close to optimal) authentication codes.

4. Key Predistribution Schemes

The management of cryptographic keys in any information system is one of the most challenging aspects of implementing cryptography. One of the most important key management processes is *key establishment*, which governs the placement of cryptographic keys in a network. This is especially relevant in applications of symmetric cryptography, where it is necessary to ensure that all parties who are authorised to access (or verify) a cryptographically protected piece of information have the appropriate key.

Symmetric key establishment almost always involves a trusted third party, which we will term a *key management authority* (KMA), at some stage in the process. In some environments this KMA is online and available at time of use. However, in many other environments it is not possible for a KMA to form part of a live network and to assist in online key establishment. In this case, the KMA can only be involved in initialisation processes that take place prior to deployment of the network. At this stage, the KMA must equip each node in the network with the necessary cryptographic keys for facilitating security services after the nodes are deployed in the network. Key establishment schemes of this type are usually referred to as *key predistribution schemes* (KPSs) because the keys are distributed in advance and cannot be generated "on the fly".

A major current trend in computing technologies is a shift from centralised, relatively stable, wired networks consisting of powerful devices, to distributed, dynamic, wireless networks consisting of lightweight devices. An example of this type of network is a *wireless sensor network* (WSN). Such networks have several important characteristics that include the need to conduct basic network services using the network nodes themselves (rather than via a centralised infrastructure) and the need for highly efficient network protocols due to the power and energy constraints of the nodes. These characteristics lend themselves to the use of key predistribution of symmetric keys. Features of such environments that add to the challenges of designing an appropriate KPS include the following:

- *Highly constrained nodes.* The nodes are very small battery-powered devices and are highly constrained with respect to memory storage and power.
- *Lack of central control.* After deployment, all network functionality must be achieved through co-operation between the nodes.
- *Hop-based communication.* In networks where battery-operated nodes are using radio communication, the constrained nature of the nodes means that in most cases the communication range of a node will be much smaller than the network diameter. Thus nodes communicate by *hopping*, meaning that a node passes data to a node within range, who then passes it onto a node within its range, etc.
- *Nodes vulnerable to compromise.* The constrained nature of the nodes means that strong security protection such as tamper-resistance is usually not viable. Thus it is normally assumed that nodes can be fairly easily captured and that any sensitive information (such as keys) that is stored on them is likely to be exposed.

A KMA thus needs to load keys onto nodes prior to deployment using a KPS to determine which keys are allocated to which nodes. After deployment, two nodes will be able to use a cryptographic service on a network link (such as encryption or a message authentication code) if they:

1. are in radio communication range of one another; and
2. share at least one key.

If either of these conditions is not met, then the nodes will have to seek a path of network links connecting them such that these conditions are met on each of the intermediate hops. Key establishment in such networks can thus be regarded as consisting of the following three stages:

1. *Key predistribution.* The KMA chooses a KPS defined on the n nodes $\mathcal{U} = \{U_1, \ldots, U_n\}$ in the network. Following [51], this KPS can be modelled by a set system $(\mathcal{I}, \mathcal{B})$ (sometimes referred to as a *key ring*), where $\mathcal{I} = \{x_i \parallel 1 \leq i \leq v\}$ is a set of v *key identifiers* and $\mathcal{B} = \{B_j \parallel 1 \leq j \leq n\}$ is a set of n *node allocations*. For each key identifier x_i, the KMA randomly selects a key K_i. The KMA then associates each node U_j in the network with a node allocation B_j and issues U_j with the keys $L_j = \{K_i \parallel x_i \in B_j\}$. Note that the association of U_j with B_j does not need to be secret, however the instantiation of B_j by L_j must be.

2. *Shared key discovery.* If two nodes within communication range of one another wish to deploy a cryptographic service, they first need to determine if they have any keys in common. The default method is to broadcast their node allocations to one another, but more efficient techniques can sometimes be found. If they have key identifiers in common, then a session key can be generated from the common keys associated with these identifiers by means of a suitable key derivation function.

3. *Path-key establishment.* If two nodes fail to identify common keys during shared key discovery, then they need to find a secure path between one another that employs intermediate nodes which do have common keys. Obviously, the shorter this secure path the better.

4.1. Requirements

The main challenge in designing a KPS that is suitable for this type of environment is that a balance must be sought between competing, and to an extent contradictory, requirements:

- *Storage.* Nodes are memory constrained and thus the number of keys stored on each node should be kept as low as possible.

- *Connectivity.* Each node should store enough keys that secure paths through the network can be established when needed. Measures of *global connectivity* assess the connectivity of the entire network. If the node allocations for any two nodes have non-empty intersection, then we will refer to the network as having *full connectivity.* Measures of *local connectivity*, which assess the ability of nodes to form secure paths with nodes in their close neighbourhood are probably most appropriate. One such, from [53], is the probability that U_i and U_j have at least one key in common (i.e. $B_i \cap B_j \neq \emptyset$).

- *Resilience.* Keys should be distributed in such a way that the damage caused by exposure of the keys stored on a node is controlled. One suggested measure of resilience is $\mathrm{fail}(s)$ [50], which is the probability that a link between two non-compromised nodes U_i and U_j is affected after s other nodes $\mathcal{S}$ are compromised at random, where a link is *affected* if $B_i \cap B_j \neq \emptyset$ and $B_i \cap B_j \subseteq \cup_{U_k \in \mathcal{S}} B_k$.

- *Efficiency.* Several processes involved in key establishment for a WSN, including *computation*, *shared key discovery* and *path-key establishment*, have the potential to involve a large amount of processing power. Since WSN nodes have limited battery power, it is thus desirable to make these processes as economical as possible.

- *Network size.* Since many applications of WSNs involve large numbers of nodes, it is important that a KPS can support a large number of nodes.

The main challenge in designing KPSs is that several of these requirements tend to compete with one another. For example, increasing the maximum number of nodes that can be supported often involves increasing the storage at each node. Also, many KPSs trade off measures of connectivity against resilience. The need for such trade-offs is illustrated by the limitations of the following trivial schemes:

Single key KPS. This KPS consists of a single key that is stored by each node in the network. It provides optimal connectivity and storage, but has very poor resilience since all communication links are affected by a single node capture.

Complete pairwise key KPS. In this KPS, a unique key is assigned to each pair of nodes. This scheme has full connectivity and optimal resilience, since compromise of one node does not affect any pair of non-compromised nodes. However, this KPS requires each node to store $n-1$ keys, which is infeasible if n is large (which will be the case in many WSNs).

Thus we make the following observations concerning the building of KPSs for WSNs:

1. *Full connectivity is not necessary.* Full connectivity is a nice feature, but unnecessary in a KPS for a WSN.

2. *Deterministic schemes have some advantages.* The obvious advantage of deterministic KPSs is that we can generally make definitive statements about their properties, which aids analysis. Also, it may be possible to exploit the structure of deterministic schemes to give very efficient shared key discovery.

3. *Flexibility is attractive.* It is useful to be able to vary the resilience, connectivity, and storage to suit requirements.

There have been a large number of proposals for KPSs for constrained networks. There are also several surveys [14, 59, 81], each of which takes a slightly different approach. We will now list some examples of schemes that have been based on geometric structures.

4.2. KPSs Based on Geometry

Projective planes. Çamtepe and Yener proposed a KPS based on a finite projective plane [13]. In this scheme, the set $\mathcal{I}$ of key identifiers is given by the set of points of a projective plane Π of order q, and the set $\mathcal{B}$ of node allocations is given by the set of lines of Π. Not only do such KPSs have full connectivity, but amongst other advantages they have efficient shared key discovery [64] (at least in the case where the projective plane is cyclic). However the significant "catch" with using a projective plane is the restriction on the number of nodes relative to the size of the node allocation. This means that facilitating a very large number of nodes comes at the unattractive cost of relatively large key storage for each node (in this case each node allocation contains q identifiers, where q is approximately the square root of the maximum number of nodes).

Generalised quadrangles. In order to provide KPSs that require less storage given the number of nodes in the network than their projective plane schemes, Çamtepe and Yener also considered schemes based on the classical generalised quadrangles $\mathcal{Q}(4,q)$, $\mathcal{Q}(5,q)$, and $H(4,q^2)$, with the key identifiers being given by the points and the node allocations by the lines of the GQ [13]. KPSs of this type do not have full connectivity since some pairs of lines do not meet.

Common intersection designs. The idea behind the use of GQ(s,t)'s as key rings was generalised in [50]:

Definition 4..1. Let $(\mathcal{I},\mathcal{B})$ be a (v,b,r,k)-configuration. We say that $(\mathcal{I},\mathcal{B})$ is a *(v,b,r,k,μ)-common intersection design* (CID) if for any distinct pair of blocks $B_i,B_j \in \mathcal{B}$, we have: $|\{B_k \in \mathcal{B} \mid\mid B_i \cap B_k \neq \emptyset \text{ and } B_j \cap B_k \neq \emptyset\}| \geq \mu$.

Thus any key ring based on a CID provides the guarantee that if two nodes do not share a key, there will be at least μ nodes who could act as intermediaries in a secure two-hop path between the original nodes. From a local connectivity perspective, it is desirable for μ to be as large as possible since this increases the chance that one of these intermediary nodes is within communication range. Several upper bounds on μ were established in [52] and optimal CIDs were constructed using group-divisible designs, strongly-regular graphs, and generalised quadrangles.

Transversal designs. One useful class of CIDs is provided by transversal designs. In [50], Lee and Stinson propose a KPS based on transversal designs, which can be regarded as a variation on the projective plane KPS that permits an additional trade-off between the connectivity and the storage requirements. This scheme can be described as follows:

- Let q be a prime power, and let k be an integer between 1 and q.
- Let P be a point of PG$(2,q)$ and let $l_1,l_2,\ldots,l_k$ be lines that pass through P.
- The key identifiers are given by the points of $l_i \setminus \{P\}$ for $i = 1,2,\ldots,k$.
- The node allocations are given by the lines of PG$(2,q)$ not passing through P.

The resulting KPSs, termed *linear schemes* in [53], have several interesting properties. The values of k and n can be varied to produce key rings with a range of compromises between the storage k, maximum number of nodes n^2, local connectivity $\frac{k}{n+1}$, and resilience. Also, the local connectivity and resilience can be computed using formulae that were derived in [53]. Furthermore, as was the case with KPSs built from projective planes, they have a very efficient shared-key discovery phase.

Generalised transversal designs. Lee and Stinson also proposed schemes they refer to as *quadratic*, based on structures they call generalised transversal designs [53]. These schemes allow larger numbers of nodes than the linear schemes, and can be described as follows:

- Let q be a prime power, and let k be an integer between 1 and q.
- Let P be a point of PG$(2,q)$ and let $l_1,l_2,\ldots,l_k$ be lines that pass through P.
- Associate a key with each point of $l_i \setminus \{P\}$ for $i = 1,2,\ldots,k$.
- Let $l \notin \{l_1,l_2,\ldots,l_k\}$ be a line through P, and let $\mathcal{C}_{P,l}$ be the set of all non-singular conics of PG$(2,q)$ that contain the point P and have l as a tangent.
- The key identifiers are given by the points of $l_i \setminus \{P\}$ for $i = 1,2,\ldots,k$.
- The node allocations are given by the set of lines of PG$(2,q)$ not passing through P, together with the set of conics in $\mathcal{C}_{P,l}$.

The performance of these quadratic schemes was analysed in [53] and was shown to offer some interesting tradeoffs. For example, they offered better resilience than linear schemes for low levels of compromised nodes, while providing similar levels of local connectivity.

The main challenge in designing KPSs for WSNs from a geometric perspective is to find new deterministic KPSs that offer different tradeoffs between the important parameters to those schemes already discussed.

5. Multivariate Equation Systems

The problem of finding a common solution to a large number of multivariate polynomial equations over a finite field arises in cryptology in both a constructive context (*e.g.* in the design of public-key encryption schemes) and a destructive context (*e.g.* in algebraic cryptanalysis). Such equation systems have a natural geometric interpretation, although this is not widely appreciated in the cryptographic community. In this section, we outline the areas of cryptography in which multivariate equation systems play a role, and we indicate ways in which geometric ideas give insight into the behaviour of these equation systems.

Finding solutions to a multivariate equation system is believed to be computationally infeasible in general; for $d = 2$, the problem of finding common solutions in the case where the polynomials f_i are chosen uniformly at random is called the *multivariate quadratic (MQ) problem* and it is known to be NP-complete [31].

5.1. Multivariate Cryptography

There have been several attempts to exploit the hardness of the MQ problem in the design of public-key cryptosystems. This is partially inspired by the fact that some other hard problems used in cryptography (such as the problem of factorising the product of two large primes) could be solved by a quantum computer [71], whereas there are no known quantum algorithms for solving the MQ problem efficiently. As an illustration, we will describe a multivariate signature scheme, although systems of multivariate equations have also been used to construct other cryptographic primitives such as public-key encryption schemes (see [80] for a survey of various proposed schemes).

5.1.1. Digital Signatures

A digital signature can be thought of as a means of binding the identity of a signer to the message that is being signed, much as a traditional handwritten signature on a document links that document to the signer. A digital signature scheme consists of a *key generation algorithm*, a *signing algorithm*, and a *verification algorithm*. The key generation algorithm is used to produce a pair (s_u, v_u) of keys for each user u; the *signing key* s_u is known only to u, whereas the *verification key* v_u is made public. In order to sign a message M, the user u runs the signing algorithm with M and s_u as inputs; the output $X_{s_u}(M)$ of this algorithm is a digital signature. The verification algorithm takes as inputs a digital signature S, a message m, and a verification key v_u; it returns true if X is a valid output of the signature algorithm on inputs M and s_u, and false otherwise. A commonly accepted security requirement for a digital signature scheme is that it should be computationally infeasible for an adversary who doesn't know s_u to produce a message/signature pair (M,X) such that the verification algorithm returns true on inputs M, X, and v_u.

5.1.2. The Oil and Vinegar Signature Scheme

In order to make use of a hard problem in cryptography, it is necessary to be able to generate instances of that problem with a *trapdoor*, which is some extra information that enables solutions to be efficiently computed, while ensuring that the problem remains intractable for adversaries who do not have access to the trapdoor information. In the case of the MQ problem, this is usually done by generating polynomials of a particular form that makes it easy to find solutions, then changing coordinates in an attempt to disguise the fact that the polynomials have that special form. The trouble with this approach is that while it is believed to be hard to solve a random instance of the MQ problem, there is the risk that the instances generated in such a specialised manner may prove to be weak in some sense. If the transformations applied to the initial systems of equations do not disguise their special structure sufficiently, it may be possible to break the scheme. Indeed, attacks have been found against most of the multivariate schemes that have been proposed to date (*e.g.* [10, 27, 66]).

To illustrate this, we will consider the case of the *Oil and Vinegar signature scheme* proposed by Patarin [65], which was subsequently broken by Kipnis and Shamir [47]. This scheme is based on a system of k equations in $2k$ variables over $\mathbb{F}_q$ of the form

$$\begin{aligned} G_1(x_1,x_2,\ldots,x_{2k}) &= m_1, \\ G_2(x_1,x_2,\ldots,x_{2k}) &= m_2, \\ &\vdots \\ G_k(x_1,x_2,\ldots,x_{2k}) &= m_k, \end{aligned} \tag{1}$$

where the G_e are homogeneous quadratic polynomials[1]. In order to be able to find solutions to this system, Patarin proposed constructing the G_e in the following manner:

1. For $e = 1,2,\ldots,k$, let F_e be a randomly chosen $2k \times 2k$ matrix over $\mathbb{F}_q$ for which the first k entries in the top k rows are all zero. Then F_e defines a quadratic form $F_e(Y)$, given in the usual way by $Y^t F_e Y$, where $Y = (y_1,y_2,\ldots,y_{2k})$.

2. Let A be a randomly chosen nonsingular $2k \times 2k$ matrix over $\mathbb{F}_q$.

3. For $e = 1,2,\ldots,k$, set $G_e = A^t F_e A$.

The structure of the F_e implies that at most one of the variables $y_1,y_2,\ldots,y_k$ (referred to as the *oil variables*) occurs in each monomial of $F_e(Y)$. Consider the following system

[1] For clarity we are considering a slightly modified version of the scheme here, as described in [47]. The attack can still be extended to the original version, however; see [47] for details.

of equations:

$$
\begin{aligned}
F_1(y_1, y_2, \ldots, y_{2k}) &= m_1, \\
F_2(y_1, y_2, \ldots, y_{2k}) &= m_2, \\
&\vdots \\
F_k(y_1, y_2, \ldots, y_{2k}) &= m_k.
\end{aligned} \tag{2}
$$

If we randomly assign values to the variables $y_{k+1}, y_{k+2}, \ldots, y_{2k}$ (referred to as *vinegar variables*), this becomes a system of k linear equations in the k oil variables. If this system is singular, we reselect different values for the vinegar variables, otherwise it has a unique solution that can be found efficiently. A solution Y for this system can be translated into a solution X for the system (1) by setting $X = A^{-1}Y$, since

$$
\begin{aligned}
X^t G_e X &= Y^t A^{-t} A^t F_e A A^{-1} Y \\
&= Y^t F_e Y = m_e.
\end{aligned}
$$

Thus, knowledge of the transformation A permits efficient solution of the system (1), whereas it is hoped that without A it should be infeasible to find such a solution. In order to use this trapdoor as a signature scheme, the signing key is taken to be A, and the verification key is the set of forms G_e. To sign a message M, a user who knows A computes a solution to the system $G(X) = M$ as described above. This solution is the user's signature on M; anybody can verify the signature by checking that it is indeed a valid solution to this system of equations. We will now see, however, that geometric considerations can be used to break this scheme.

5.1.3. Kipnis and Shamir's Cryptanalysis of the Oil and Vinegar Signature Scheme

The separation of the coordinates $y_1, y_2, \ldots, y_{2k}$ into 'oil' and 'vinegar' variables enables solutions to be found for a large system of quadratic equations. It was hoped that the transformation A would suffice to 'mix' the oil and vinegar variables, and thus disguise the special structure of these equations. However, Kipnis and Shamir showed in [47] that given only the system (1), it is possible to determine a change of variables that will effectively separate the variables again, which amounts to a key recovery attack.

Consider the set of quadrics in $\mathrm{PG}(2k-1,q)$ described by the equations $Y^t F_e Y = 0$. Kipnis and Shamir note that the $(k-1)$-dimensional space spanned by the points $P_1 = (1,0,0,\ldots,0)$, $P_2 = (0,1,0,\ldots,0)$, $\ldots$, P_k (referred to as the *oil subspace*) is contained in each of these quadrics. Furthermore, if we consider the polarities arising from the quadrics described by any of the F_e that are non-singular, we observe that the oil subspace is also self polar with respect to each of these quadrics. The collineation induced by the matrix A transforms these quadrics into those given by the equations $X^t G_e X = 0$; therefore, there exists some $(k-1)$-dimensional space contained on each of these quadrics that is self-polar with respect to each of them. If we can determine this space, then we can construct a collineation transforming it into the oil subspace, which will thus permit us to separate the oil and vinegar variables. For non-singular G_i and G_j, the map from $\mathrm{PG}(2k-1,q)$ to itself obtained by composing the polarity with respect to G_j followed by the polarity with respect to G_i is the collineation induced by the matrix $G_i^{-1}G_j$; by the previous argument, it fixes the image under A of the oil subspace. Kipnis and Shamir argue that this subspace can be determined efficiently by computing all the matrices $G_i^{-1}G_j$ and finding their common eigenspace. Once this subspace is determined, any collineation that maps this subspace onto the oil subspace will transform the system of quadratic equations into one for which solutions can be easily computed as described above. Such a collineation thus acts as an alternative signing key corresponding to the verification key given by the forms G_e, and thus permits the efficient forgery of signatures.

This example illustrates the fact that linear changes of coordinates transformations can preserve the underlying geometry of equation systems, even if they make the equations look random from an algebraic point of view. Great care must therefore be taken when using such techniques to generate trapdoors; the inherent difficulty of this is reflected in the fact that the majority of multivariate schemes proposed to date have subsequently been broken.

5.2. Algebraic Cryptanalysis

We have seen that the problem of solving multivariate polynomial equations has been used explicitly in the construction of cryptographic primitives. However, it also arises when considering the security of certain systems that are not based directly on the problem.

A *symmetric cipher* consists of an encryption function E that takes a *plaintext* message P, together with a *key* K and returns a *ciphertext* $C = E(K,M)$, together with an encryption function D that takes a ciphertext and a key, and returns the corresponding plaintext, so that $D(E(K,M),K) = M$. One basic security property that is required is that an adversary who does not know K should not be able to recover the plaintext given only the ciphertext. Additionally, an adversary with knowledge of M and $C(M,K)$ for one or more values of M should not be able to determine the value of the key K that was used (this is referred to as a *known plaintext attack*).

If we treat each of the bits of the plaintext, ciphertext, and key as a variable, then the encryption function can be expressed as a system of polynomial equations over $\mathbb{F}_2$. An adversary with knowledge of the ciphertext corresponding to a single plaintext message could potentially recover the key that was used, provided that it was able to solve the system of equations; this is known as an *algebraic attack*. In general we might expect that the degrees of the polynomials involved will be so high that analysis of the system is impractical. However, there are some notable exceptions. In [19], it was shown that the plaintext, ciphertext, and key bits (as well as certain bits of the internal state) of the Rijndael block cipher, which forms the basis for the Advanced Encryption Standard (see Section 6.), could be related by a system of 9600 quadratic equations in 1600 variables over $\mathbb{F}_2$. Furthermore, in [61], it was noted that this cryptosystem could be described alternatively in terms of a system of 9600 very sparse quadratic equations in 1600 variables over $\mathbb{F}_{2^8}$ (see Section 6.2.3.). This provoked a lot of interest at the time. However, although algebraic attacks have been shown to be effective against certain stream ciphers ([2,18]), they have yet to provide a convincing break of a block cipher such as AES.

A system of equations such as those required to describe AES is so large that we cannot expect to solve it in practice. However, from an academic point of view, a cipher is considered to be broken if an attack is found whose complexity is less than that of performing a *brute force search* (*i.e.* trying each possible key in turn), regardless of whether it is actually feasible in practice to carry out the attack. Thus, in order to determine whether a cipher such as AES can be broken by an algebraic attack, we require an understanding of the complexity of algorithms for solving systems of multivariate equations over finite fields, especially in the case of fields of even characteristic. Unfortunately, despite much attention being paid to this question, this is a subject in which we have few definitive answers.

As the MQ problem is known to be NP-complete, it is unrealistic to expect an efficient technique for solving general instances of the problem. However, as noted in the previous section, it may be the case that the structure of the system of equations arising from a particular cryptosystem could somehow cause it to be more amenable to solution. For instance, Gröbner basis algorithms for equation solving are known to be doubly exponential in the worst case (although their behaviour tends to be much better than this on average [20, 21]). However, the systems of equations arising from the so-called HFE multivariate cryptosystems have proved to be easier to solve using Gröbner basis techniques than would be predicted from their size alone [29].

The cryptographic literature contains several proposals of techniques for potentially solving the MQ problem. Perhaps the most simple technique considered is that known as *linearisation*. To linearise a system of multivariate equations, each monomial is treated as a variable in its own right, so that a system of linear equations is obtained. If a unique solution exists, it can be translated into a solution for the original system. From a geometric point of view, this approach corresponds to using the Veronese mapping between quadrics in $\mathrm{PG}(n,q)$ and hyperplanes of $\mathrm{PG}(\frac{1}{2}n(n+3),q)$. However, even if the quadrics have a unique point of intersection, the hyperplanes may intersect in a space of dimension ≥ 1, thus giving spurious points that do not correspond to solutions of the original system. In fact, the point required is the intersection of the hyperplanes with the appropriate Veronese variety.

In [48], Kipnis and Shamir proposed a technique they call *relinearisation*, in which the set of quadratic equations describing this variety is restricted to the appropriate subspace to obtain a new system of equations which is then linearised in turn, the process being repeated until a solution is found. Unfortunately, the behaviour of this technique is not well understood, due to the difficulty in determining the extent of the linear dependencies between the equations that are obtained.

Courtois *et al.* proposed a technique they called *extended linearisation*, or XL for short [17], with the intention of generalising the basic relinearisation method. It involves embedding a system of equations into a larger system of higher degree, in an attempt to find univariate polynomials (or bivariate, in the homogeneous case) whose factors may give information about potential solutions. Subsequent research has related this technique to certain Gröbner basis algorithms [75], and indicated that the original estimates for its complexity were somewhat optimistic [26]. Murphy and Paterson have demonstrated that the complexity of the XL technique can be substantially affected by a linear change of coordinates, and have shown that by considering the underlying geometry, an appropriate choice of coordinates can be found [60].

This suggests that a geometric approach has a role to play in better understanding the behaviour of multivariate equation systems and techniques for finding solutions. Solving such a system of equations amounts to finding the intersection of a system of hypersurfaces. As the essential properties of such a system, and its intersection, are invariant under collineations, we might expect the true difficulty of finding solutions to be determined by techniques that are themselves invariant.

The question of solving generic systems of equations over the complex numbers is certainly not a new problem (for example, see [54]). However, the equation systems that are of greatest cryptographic interest are defined over finite fields of small characteristic, are expected to have zero-dimensional solutions, and are highly *overdefined* (*i.e.* the number of equations greatly outnumbers the number of variables) and thus far from generic. Further progress in this area is required in order to shed insight onto such vexed questions as the security of AES in the face of algebraic attacks.

6. The Advanced Encryption Standard

A *block cipher* is a symmetric cipher that takes a block of plaintext of a specific length together with a key, and returns a ciphertext which is a function of both the key and the plaintext. Towards the end of the last century, the US government's National Institute of Standards and Technology held an open competition for the design of a new block cipher to become the *Advanced Encryption Standard* (AES), in order to replace the old Data Encryption Standard (DES) whose short key length makes it vulnerable to exhaustive key search. The winner (the Rijndael block cipher proposed by Belgian cryptographers Joan Daemen and Vincent Rijmen [24]) was announced in 2000, and approved as an official standard in the following year [62].

The AES has stood up well in the face of considerable cryptologic scrutiny. A report on its security by the Symmetric Techniques Virtual Lab of the ECRYPT European Network of Excellence in Cryptology states "the conclusion of this report is that, five years after publication, there are still no discernible cryptographic weaknesses in the AES" [1]. However, several researchers have noticed that the algebraic nature of some of the components of the Rijndael encryption function cause them to exhibit geometric properties that are not typically found in a block cipher. In Subsection 6.1., we briefly discuss the design of Rijndael, without giving full technical details of its specification (these can be found in [15, 24], for example). In Subsection 6.2., we describe some of the ways in which this design has been shown to give rise to unexpected geometric structure.

6.1. The Design of AES

The AES takes an input block of 16 bytes of data, together with a key of either 128, 192, or 256 bytes, which is expanded using the *key schedule* to give ten 16-byte *round keys*. Each byte consists of 8 bits, and can thus be interpreted as an element of the field $\mathbb{F} = \mathbb{F}_{2^8}$ expressed as an extension of $\mathbb{F}_2$ by the roots of the polynomial $m(x) = x^8 + x^4 + x^3 + x + 1$. The encryption function of the AES involves 10 rounds[2], and is based on a structure known as a *substitution-permutation network*. In each round, the input bytes are replaced by other

[2]The first nine rounds are identical, but one operation is omitted from the final round so that the decryption process can be related more directly to the encryption process, facilitating simpler implementation.

bytes according to a look-up table known as an *S-box* (the S stands for substitution), an $\mathbb{F}$-linear transformation is applied (it acts as a permutation), and the 16 bytes of the round key are combined with the resulting bytes through addition in $\mathbb{F}$.

One of the main criteria motivating the design of the AES round function is the need to resist two powerful attacks known as *linear cryptanalysis* and *differential cryptanalysis*. Linear cryptanalysis is based on finding affine approximations that hold with high probability across multiple rounds of the cipher, whereas differential cryptanalysis involves studying the statistical properties of the differences in the internal state of the cipher that arise when a pair of messages with a given difference are encrypted. If pairs of messages with a given difference are encrypted using the same key then after the first round, and successive rounds, a distribution on the differences between the resulting states is induced. A sequence of differences between the states over a sequence of rounds is known as a *differential path*; if a path that holds with (relatively) high probability is found, then it may be possible to exploit it in a key recovery attack. The various components of the AES round function have been selected with a view to reducing the occurrence of linear or differential paths within the cipher.

6.1.1. The AES S-box

The substitution operation is the only non-linear component of the AES encryption function. It is of particular interest to cryptanalysts, for although it is usually implemented in the form of a look-up table, it is in fact defined entirely in algebraic terms. It is based around the operation that interprets each byte $\mathbf{b}$ as an element of $\mathbb{F}$ and replaces it by the element $\mathbf{b}^{2^8-2}$ (for elements of $\mathbb{F}^*$ this is simply inversion in $\mathbb{F}$; 0 is mapped to 0). This map was chosen as it is highly non-linear, and it provides good protection against differential cryptanalysis since the output differences that arise from a given difference in inputs are distributed relatively evenly (for details, see [63]). In order to overcome the algebraic simplicity of this operation, it is then combined with a $\mathbb{F}_2$-linear map, and a constant is added.

6.1.2. Diffusion in AES

Claude Shannon observed that for a cipher to defeat statistical analysis, it is necessary for it to provide good *diffusion*: each bit of the ciphertext should be influenced by the majority of the bits of the ciphertext [70]. In the case of AES, the diffusion is provided by $\mathbb{F}$-linear operations that were designed around the parity-check matrix for an MDS code (see e.g. [49]), in an effort to resist linear and differential attacks (for details, see [24]).

6.2. Geometric Properties of AES

The fact that the various design elements of AES can all be interpreted in terms of operations over $\mathbb{F}$ leads to it being more directly algebraic in nature than many contemporary block ciphers. Various authors have observed the corresponding occurrence of geometric structure in AES. While this has yet to lead to an actual attack, it has certainly aroused interested within the cryptanalytic community. We will now give a brief description of some of these observations.

6.2.1. The Group Generated by AES

The round function of an iterated block cipher with a particular key can be considered to be an element of a group that acts on the set of possible internal states of the cipher, and it has been argued in the literature that it is desirable for the group generated by a cipher's round function using all possible keys to be either the alternating or symmetric group in order to resist certain attacks (although this alone is no guarantee of strength). In this vein, it has been demonstrated that the round functions of AES generate the alternating group $A_{2^{128}}$ [79]. However, Jackson and Murphy point out that the inversion operation used in the AES S-box can be regarded as performing inversion on the projective line, except for its behaviour at the points (0,1) and (1,0). Jackson and Murphy show that the set of "AES-like transformations" of the projective line of the form

$$\mathrm{PG}(1,\mathbb{F}) \to \mathrm{PG}(1,\mathbb{F}) : (1,x) \mapsto (x, kx+1), \quad k \in \mathbb{F},$$

generate $\mathrm{PGL}(2,\mathbb{F})$, rather than the alternating group, and suggest this may in fact give a better indication of the vulnerability of AES to attacks such as those described in [43, 44] (more details can be found in [15, 42]).

6.2.2. The AES Difference Table

One tool that is used for studying the properties of a function $f\colon \mathbb{F} \to \mathbb{F}$ is the so-called *difference table*. This is a $2^8 \times 2^8$ table whose entry in position (i,j) represents the number of $x \in \mathbb{F}$ for which $f(x+i)+f(x) = j$. It thus gives an illustration of how differences in pairs of outputs vary with respect to the difference in the corresponding inputs to the function. Jackson and Murphy showed that the table obtained by considering the inversion function from the AES S-box (with the omission of the row/column corresponding to the input/output difference of 0) can be used to define an incidence structure isomorphic to that given by the points and hyperplanes of $\mathrm{PG}(7,2)$ in a natural way [42].

6.2.3. The BES Representation of AES

Murphy and Robshaw showed that it is possible to embed AES in a larger cipher that they call the Big Encryption System (BES) [61], which operates on 128-byte blocks and has a round function that consists entirely of an $\mathbb{F}$-linear mapping, inversion in $\mathbb{F}$, and the addition of the key (unlike the round function of AES, which also includes a $\mathbb{F}_2$-linear transformation as part of the S-box). Let $\phi\colon \mathbb{F} \to \mathbb{F}^8$ be given by $a \mapsto (a^{2^0}, a^{2^1}, a^{2^2}, a^{2^3}, a^{2^4}, a^{2^5}, a^{2^6}, a^{2^7})$. Then ϕ can be used to map a 16-byte element of the AES state space to a 128-byte element of the BES state space by applying it to each byte of the space in turn. The BES is defined so that when it is applied to elements that are derived from the AES state space in this manner then it induces the action of AES on such elements. It can thus be regarded as providing a description of the AES round function that is particularly simple from an algebraic point of view; in particular, it can be used to derive a sparse system of quadratic equations over $\mathbb{F}$ that describes the behaviour of AES.

7. Concluding Remarks

We have shown that Galois geometry has played an influential role in several different areas of cryptology. While the applications to secret sharing and authentication codes are well-known, the development of new technologies such as WSNs has led to new applications in key predistribution, and the more recent applications to cryptanalysis are perhaps the most surprising. There is no reason to suppose that Galois geometry cannot continue to be a source of constructions and modeling tools for understanding some diverse aspects of cryptology. It is also hoped that some of the problems generated by these applications are of independent interest to Galois geometry research.

Acknowledg ments

This work of the third author was conducted at Royal Holloway and supported by EPSRC grant EP/D053285/1.

References

[1] D.STVL.2AESsecurityreport.http://www.ecrypt.eu.org/ecrypt1 /documents/D.STVL.2-1.0.pdf,2006.

[2] E. Barkan, E. Biham, and N. Keller, *Instant ciphertext-only cryptanalysis of GSM encrypted communication*, in Advances in cryptology—CRYPTO 2003, D. Boneh, ed., vol. 2729 of Lecture Notes in Comput. Sci., Springer, Berlin, 2003, pp. 600–616.

[3] S. G. Barwick and W.-A. Jackson, *An optimal multisecret threshold scheme construction*, Des. Codes Cryptogr., 37 (2005), pp. 367–389.

[4] S. G. Barwick, W.-A. Jackson, and K. M. Martin, *Updating the parameters of a threshold scheme by minimal broadcast*, IEEE Trans. Inform. Theory, 51 (2005), pp. 620–633.

[5] J. Benaloh and J. Leichter, *Generalized secret sharing and monotone functions*, in Advances in cryptology—CRYPTO '88 (Santa Barbara, CA, 1988), S. Goldwasser, ed., vol. 403 of Lecture Notes in Comput. Sci., Springer, Berlin, 1990, pp. 27–35.

[6] M. Bertilsson and I. Ingemarsson, *A construction of practical secret sharing schemes using linear block codes*, in Advances in cryptology - AUSCRYPT '92 (Gold Coast, Queensland, Australia, 1992), J. Seberry, ed., vol. 718 of Lecture Notes in Comput. Sci., Springer, 1993, pp. 67–79.

[7] S. R. Blackburn, *Combinatorial schemes for protecting digital content*, in Surveys in combinatorics, 2003 (Bangor), vol. 307 of London Math. Soc. Lecture Note Ser., Cambridge Univ. Press, Cambridge, 2003, pp. 43–78.

[8] G. R. Blakley, *Safeguarding cryptographic keys*, in Proceedings of AFIPS 1979 National Computer Conference, vol. 48, 1979, pp. 313–317.

[9] C. Blundo, A. De Santis, A. Herzberg, S. Kutten, U. Vaccaro, and M. Yung, *Perfectly-secure key distribution for dynamic conferences*, in Advances in cryptology - CRYPTO '92 (Santa Barbara, CA, 1992), E. F. Brickell, ed., vol. 740 of Lecture Notes in Comput. Sci., Springer, Berlin, 1993, pp. 471–486.

[10] A. Braeken, C. Wolf, and B. Preneel, *A study of the security of unbalanced oil and vinegar signature schemes*, in Topics in cryptology—CT-RSA 2005, A. Menezes, ed., vol. 3376 of Lecture Notes in Comput. Sci., Springer, Berlin, 2005, pp. 29–43.

[11] E. F. Brickell, *Some ideal secret sharing schemes*, J. Combin. Math. Combin. Comput., 6 (1989), pp. 105–113.

[12] E. F. Brickell and D. M. Davenport, *On the classification of ideal secret sharing schemes*, J. Cryptology, 4 (1991), pp. 123–134.

[13] S. A. Çamtepe and B. Yener, *Combinatorial design of key distribution mechanisms for wireless sensor networks*, in Computer Security – ESORICS 2004, P. Samarati, P. Ryan, D. Gollmann, and R. Molva, eds., vol. 3193 of Lecture Notes in Comput. Sci., Springer, 2004, pp. 293–308.

[14] ——, *Key distribution mechanisms for wireless sensor networks: a survey*. Rensselaer Polytechnic Institute, Computer Science Department, Technical Report TR-05-07, March 2005.

[15] C. Cid, S. Murphy, and M. Robshaw, *Algebraic aspects of the advanced encryption standard*, Springer, New York, 2006.

[16] C. J. Colbourn, J. H. Dinitz, and D. R. Stinson, *Applications of combinatorial designs to communications, cryptography, and networking*, in Surveys in combinatorics, 1999 (Canterbury), vol. 267 of London Math. Soc. Lecture Note Ser., Cambridge Univ. Press, Cambridge, 1999, pp. 37–100.

[17] N. Courtois, A. Klimov, J. Patarin, and A. Shamir, *Efficient algorithms for solving overdefined systems of multivariate polynomial equations*, in Advances in cryptology—EUROCRYPT 2000 (Bruges), B. Preneel, ed., vol. 1807 of Lecture Notes in Comput. Sci., Springer, Berlin, 2000, pp. 392–407.

[18] N. Courtois and W. Meier, *Algebraic attacks on stream ciphers with linear feedback*, in Advances in cryptology—EUROCRYPT 2003, E. Biham, ed., vol. 2656 of Lecture Notes in Comput. Sci., Springer, Berlin, 2003, pp. 345–359.

[19] N. Courtois and J. Pieprzyk, *Cryptanalysis of block ciphers with overdefined systems of equations*, in Advances in cryptology—ASIACRYPT 2002, Y. Zheng, ed., vol. 2501 of Lecture Notes in Comput. Sci., Springer, Berlin, 2002, pp. 267–287.

[20] D. A. Cox, J. Little, and D. O'Shea, *Using algebraic geometry*, vol. 185 of Graduate Texts in Mathematics, Springer, New York, second ed., 2005.

[21] ——, *Ideals, varieties, and algorithms*, Undergraduate Texts in Mathematics, Springer, New York, third ed., 2007. An introduction to computational algebraic geometry and commutative algebra.

[22] R. Cramer, I. Damgård, and U. Maurer, *General secure multi-party computation from any linear secret-sharing scheme*, in Advances in cryptology—EUROCRYPT 2000 (Bruges), B. Preneel, ed., vol. 1807 of Lecture Notes in Comput. Sci., Springer, Berlin, 2000, pp. 316–334.

[23] R. Cramer, V. Daza, I. Gracia, J. Jiménez-Urroz, G. Leander, J. Martí-Farré, and C. Padró, *On codes, matroids, and secure multiparty computation from linear secret-sharing schemes*, IEEE Trans. Inform. Theory, 54 (2008), pp. 2644–2657.

[24] J. Daemen and V. Rijmen, *The design of Rijndael*, Information Security and Cryptography, Springer-Verlag, Berlin, 2002. AES—the advanced encryption standard.

[25] M. De Soete, *Some constructions for authentication-secrecy codes*, in Advances in cryptology—EUROCRYPT '88 (Davos, 1988), C. G. Günther, ed., vol. 330 of Lecture Notes in Comput. Sci., Springer, Berlin, 1988, pp. 57–75.

[26] C. Diem, *The XL-algorithm and a conjecture from commutative algebra*, in Advances in cryptology—ASIACRYPT 2004, P. J. Lee, ed., vol. 3329 of Lecture Notes in Comput. Sci., Springer, Berlin, 2004, pp. 323–337.

[27] V. Dubois, P.-A. Fouque, A. Shamir, and J. Stern, *Practical cryptanalysis of SFLASH*, in Advances in cryptology—CRYPTO 2007, A. Menezes, ed., vol. 4622 of Lecture Notes in Comput. Sci., Springer, Berlin, 2007, pp. 1–12.

[28] O. Farràs, J. Martí-Farré, and C. Padró, *Ideal multipartite secret sharing schemes*, in Advances in cryptology—EUROCRYPT 2007, M. Naor, ed., vol. 4515 of Lecture Notes in Comput. Sci., Springer, Berlin, 2007, pp. 448–465.

[29] J.-C. Faugère and A. Joux, *Algebraic cryptanalysis of hidden field equation (HFE) cryptosystems using Gröbner bases*, in Advances in cryptology—CRYPTO 2003, D. Boneh, ed., vol. 2729 of Lecture Notes in Comput. Sci., Springer, Berlin, 2003, pp. 44–60.

[30] R.-Q. Feng and Z.-X. Wan, *A construction of Cartesian authentication codes from geometry of classical groups*, J. Combin. Inform. System Sci., 20 (1995), pp. 197–210.

[31] M. R. Garey and D. S. Johnson, *Computers and intractability*, W. H. Freeman and Co., San Francisco, Calif., 1979. A guide to the theory of NP-completeness.

[32] E. N. Gilbert, F. J. MacWilliams, and N. J. A. Sloane, *Codes which detect deception*, Bell System Tech. J., 53 (1974), pp. 405–424.

[33] M. Giulietti and R. Vincenti, *Three-level secret sharing schemes from the twisted cubic*, Discrete Math., 310 (2010), pp. 3236–3240.

[34] M. Ito, A. Saito, and T. Nishizeki, *Secret sharing scheme realizing general access structure*, in GLOBECOM'87, Proceedings of the IEEE Global Telecom. Conf., Springer, 1988, pp. 612–613.

[35] W.-A. Jackson and K. M. Martin, *Geometric secret sharing schemes and their duals*, Des. Codes Cryptogr., 4 (1994), pp. 83–95.

[36] ——, *Perfect secret sharing schemes on five participants*, Des. Codes Cryptogr., 9 (1996), pp. 267–286.

[37] ——, *An algorithm for efficient geometric secret sharing schemes*, Util. Math., 54 (1998), pp. 127–150.

[38] ——, *Combinatorial models for perfect secret sharing schemes*, J. Combin. Math. Combin. Comput., 28 (1998), pp. 249–265.

[39] W.-A. Jackson, K. M. Martin, and C. M. O'Keefe, *Multisecret threshold schemes*, in Advances in cryptology—CRYPTO '93 (Santa Barbara, CA, 1993), D. R. Stinson, ed., vol. 773 of Lecture Notes in Comput. Sci., Springer, Berlin, 1994, pp. 126–135.

[40] ——, *A construction for multisecret threshold schemes*, Des. Codes Cryptogr., 9 (1996), pp. 287–303.

[41] ——, *Geometrical contributions to secret sharing theory*, J. Geom., 79 (2004), pp. 102–133.

[42] W.-A. Jackson and S. Murphy, *Projective aspects of the AES inversion*, Des. Codes Cryptogr., 43 (2007), pp. 167–179.

[43] T. Jakobsen and L. R. Knudsen, *The interpolation attack on block ciphers*, in Fast Software Encryption, 4th International Workshop, FSE '97 (Haifa), E. Biham, ed., vol. 1267 of Lecture Notes in Comput. Sci., Springer, 1997, pp. 28–40.

[44] ——, *Attacks on block ciphers of low algebraic degree*, J. Cryptology, 14 (2001), pp. 197–210.

[45] T. Johansson, *Lower bounds on the probability of deception in authentication with arbitration*, IEEE Trans. Inform. Theory, 40 (1994), pp. 1573–1585.

[46] M. Karchmer and A. Wigderson, *On span programs*, in Proc. of the 8th IEEE Structure in Complexity Theory, IEEE Computer Society Press, 1993, pp. 102–111.

[47] A. Kipnis and A. Shamir, *Cryptanalysis of the oil and vinegar signature scheme*, in Advances in cryptology—CRYPTO '98 (Santa Barbara, CA, 1998), H. Krawczyk, ed., vol. 1462 of Lecture Notes in Comput. Sci., Springer, Berlin, 1998, pp. 257–266.

[48] ——, *Cryptanalysis of the HFE public key cryptosystem by relinearization*, in Advances in cryptology—CRYPTO '99 (Santa Barbara, CA), M. J. Wiener, ed., vol. 1666 of Lecture Notes in Comput. Sci., Springer, Berlin, 1999, pp. 19–30.

[49] I. Landjev and L. Storme, *Galois geometries and coding theory*, in Current research topics in Galois geometry, Nova Sci. Publ., New York, 2011, ch. 8, pp. 185–212.

[50] J. Lee and D. R. Stinson, *A combinatorial approach to key predistribution for distributed sensor networks*, in IEEE Wireless Communications and Networking Conference 2005, vol. 2, March 2005, pp. 1200–1205.

[51] ——, *Deterministic key predistribution schemes for distributed sensor networks*, in Selected areas in cryptography, H. Handschuh and M. A. Hasan, eds., vol. 3357 of Lecture Notes in Comput. Sci., Springer, Berlin, 2005, pp. 294–307.

[52] ——, *Common intersection designs*, J. Combin. Des., 14 (2006), pp. 251–269.

[53] ——, *On the construction of practical key predistribution schemes for distributed sensor networks using combinatorial designs*, ACM Trans. Inf. Syst. Secur., 11 (2008), pp. 1–35.

[54] F. S. Macaulay, *The algebraic theory of modular systems*, Cambridge Mathematical Library, Cambridge University Press, Cambridge, 1994. Revised reprint of the 1916 original, With an introduction by Paul Roberts.

[55] J. Martí-Farré and C. Padró, *On secret sharing schemes, matroids and polymatroids*, in Theory of cryptography, S. P. Vadhan, ed., vol. 4392 of Lecture Notes in Comput. Sci., Springer, Berlin, 2007, pp. 273–290.

[56] K. M. Martin, *Dynamic access policies for unconditionally secure secret sharing schemes*, in Proceedings of IEEE Information Theory Workshop on Theory and Practice in Information-Theoretic Security, 2005., Oct. 2005, pp. 61–66.

[57] ——, *The combinatorics of cryptographic key establishment*, in Surveys in combinatorics 2007, vol. 346 of London Math. Soc. Lecture Note Ser., Cambridge Univ. Press, Cambridge, 2007, pp. 223–273.

[58] ——, *Challenging the adversary model in secret sharing schemes*, in Contact Forum Coding Theory and Cryptography II, Proceedings of the Royal Flemish Academy of Belgium for Science and the Arts, Royal Flemish Academy of Belgium for Science and the Arts, 2008, pp. 45–63.

[59] K. M. Martin and M. B. Paterson, *An application-oriented framework for wireless sensor network key establishment*, Electron. Notes Theor. Comput. Sci., 192 (2008), pp. 31–41.

[60] S. Murphy and M. B. Paterson, *A geometric view of cryptographic equation solving*, J. Math. Cryptol., 2 (2008), pp. 63–107.

[61] S. Murphy and M. J. B. Robshaw, *Essential algebraic structure within the AES*, in Advances in cryptology—CRYPTO 2002, M. Yung, ed., vol. 2442 of Lecture Notes in Comput. Sci., Springer, Berlin, 2002, pp. 1–16.

[62] National Institute of Standards and Technology, *The advanced encryption standard.* Federal Information Processing Standards Publication (FIPS) 197, 2001.

[63] K. Nyberg, *Differentially uniform mappings for cryptography*, in Advances in cryptology—EUROCRYPT '93 (Lofthus, 1993), T. Helleseth, ed., vol. 765 of Lecture Notes in Comput. Sci., Springer, Berlin, 1994, pp. 55–64.

[64] E. C. Park and I. F. Blake, *Reducing communication overhead of key distribution schemes for wireless sensor networks*, in Proceedings of 16th International Conference on Computer Communications and Networks, 2007. ICCCN 2007., Aug. 2007, pp. 1345–1350.

[65] J. Patarin, *The oil and vinegar algorithm for signatures.* Presented at the Dagstuhl Workshop on Cryptography, September 1997.

[66] J. Patarin, *Cryptanalysis of the Matsumoto and Imai public key scheme of Eurocrypt'88*, Des. Codes Cryptogr., 20 (2000), pp. 175–209.

[67] D. Y. Pei, *Information-theoretic bounds for authentication codes and block designs*, J. Cryptology, 8 (1995), pp. 177–188.

[68] ——, *Authentication codes and combinatorial designs*, Discrete Mathematics and its Applications (Boca Raton), Chapman & Hall/CRC, Boca Raton, FL, 2006.

[69] A. Shamir, *How to share a secret*, Comm. ACM, 22 (1979), pp. 612–613.

[70] C. E. Shannon, *Communication theory of secrecy systems*, Bell System Tech. J., 28 (1949), pp. 656–715.

[71] P. W. Shor, *Algorithms for quantum computation: discrete logarithms and factoring*, in Proceedings of the 35th Annual Symposium on Foundations of Computer Science, 1994, Nov 1994, pp. 124–134.

[72] G. J. Simmons, *How to (really) share a secret*, in Advances in cryptology—CRYPTO '88 (Santa Barbara, CA, 1988), S. Goldwasser, ed., vol. 403 of Lecture Notes in Comput. Sci., Springer, Berlin, 1990, pp. 390–448.

[73] G. J. Simmons, W.-A. Jackson, and K. M. Martin, *The geometry of shared secret schemes*, Bull. Inst. Combin. Appl., 1 (1991), pp. 71–87.

[74] D. R. Stinson, *Cryptography*, Discrete Mathematics and its Applications (Boca Raton), Chapman & Hall/CRC, Boca Raton, FL, third ed., 2006. Theory and practice.

[75] M. Sugita, M. Kawazoe, and H. Imai, *Relation between the XL Algorithm and Gröbner Basis Algorithms*, IEICE Transactions, 89-A (2006), pp. 11–18.

[76] M. van Dijk, *A linear construction of perfect secret sharing schemes*, in Advances in cryptology—EUROCRYPT '94 (Perugia), A. De Santis, ed., vol. 950 of Lecture Notes in Comput. Sci., Springer, Berlin, 1995, pp. 23–34.

[77] ——, *On the information rate of perfect secret sharing schemes*, Des. Codes Cryptogr., 6 (1995), pp. 143–169.

[78] Z. X. Wan, B. Smeets, and P. Vanroose, *On the construction of Cartesian authentication codes over symplectic spaces*, IEEE Trans. Inform. Theory, 40 (1994), pp. 920–929.

[79] R. Wernsdorf, *The Round Functions of RIJNDAEL Generate the Alternating Group*, in Fast software encryption. 9th international workshop, FSE 2002, Leuven, J. Daemen and V. Rijmen, eds., vol. 2365 of Lecture Notes in Comput. Sci., Springer, 2002, pp. 143–148.

[80] C. Wolf, *Multivariate Quadratic Polynomials in Public Key Cryptography*, PhD thesis, Katholieke Universiteit Leuven, 2005. `http://eprint.iacr.org/2005/393`.

[81] Y. Xiao, V. K. Rayi, B. Sun, X. Du, F. Hu, and M. Galloway, *A survey of key management schemes in wireless sensor networks*, Comput. Commun., 30 (2007), pp. 2314–2341.

In: Current Research Topics in Galois Geometry
Editors: J. De Beule and L. Storme
ISBN: 978-1-61209-523-3

Chapter 10

GALOIS GEOMETRIES AND LOW-DENSITY PARITY-CHECK CODES

***Marcus Greferath*[1,*], *Cornelia Rößing*[1,†] *and Leo Storme*[2,‡]**
[1] School of Mathematical Sciences, University College Dublin,
Belfield, Dublin 4, Ireland
[2] Ghent University, Department of Mathematics, Krijgslaan 281-S22,
9000 Ghent, Belgium

Abstract

Low-Density Parity-Check (LDPC) codes form an important class of error correcting codes for today's communication systems. These codes allow for a highly efficient decoding scheme that is known as message passing decoding. Although so-called random LDPC codes perform close to a theoretical limit derived in coding theory, there is a demand for systematic code design in order to keep the encoding process efficient. This article revisits a collection of geometric constructions with an emphasis on those which are based on triangle free geometries.

Key Words: Low-density parity-check codes, sparse matrices, partial linear spaces, triangle free geometries, $(0,1)$-geometries, inversive spaces.

AMS Subject Classification: 05B20, 05B25, 51E20, 94B05, 94B35

Introduction

Reliable communications are in great demand at present. Common applications desire higher bandwidth communications in devices consuming less and less energy. It is, therefore, of high importance to use transmission systems that are as effective as possible. The science of finding efficient schemes by which information can be coded for reliable transmission through a noisy channel is called *coding theory*. The basic idea behind coding and

*E-mail address: marcus.greferath@ucd.ie
†E-mail address: roessing@maths.ucd.ie
‡E-mail address: ls@cage.ugent.be

error correction is to add redundant data with each transmitted word of information so that, even if errors occur, sufficient protection exists to reliably recover the original message.

During the recent 15 years, a class of codes that exhibit performance near a theoretical limit (aka Shannon limit for noisy channels) has been developed. This is the class of *low-density parity-check codes*, denoted by LDPC codes in the sequel. Originally they were discovered by R. Gallager [10] in the sixties of the previous century. Due to a lack of computational resources, they were not fully appreciated until their rediscovery by D. MacKay [18] in the nineties. LDPC codes can yield high performance on the binary symmetric channel as well as on the additive white Gaussian noise channel, and have been shown to outperform other code classes in many applications. The algorithm used for decoding is called *message passing*, and one of its versions is known as the *sum-product algorithm*. This algorithm uses a graphical representation of the code, called the *Tanner Graph*.

Decoding schemes based on message passing (in sparse graphs) are highly efficient, and it is desirable to have the encoding process of these codes most efficient as well. To achieve this goal, scholars seek for systematic constructions of LDPC codes. Many good constructions are known nowadays, but so far only a very few of them outperform random constructions of LDPC codes. For this reason, there is further demand for a systematic construction of LDPC codes with excellent performance.

Constructions

LDPC codes have been systematically constructed in various ways. Regarding graph based construction, we refer to Margulis [19], Rosenthal and Vontobel [28], and Lafferty and Rockmore [14]. For most of these approaches, it turned out that Ramanujan graphs, which are optimal relative to a certain expansion property, are of particular value for the construction.

Regarding algebraic approaches, the reader is referred to Bond, Hui and Schmidt [5], and later Greferath, O'Sullivan, and Smarandache [23]. Here, linear congruences are used to relate the row and column numbers of the nonzero entries of a sparse parity-check matrix.

During the short history of LDPC codes, it turned out soon that geometric approaches can be used in the construction of LDPC codes. An important construction was proposed by Kou, Lin, and Fossorier [13] and makes use of the general concept of incidence structures. In the paper mentioned, the underlying incidence structures were either affine or projective spaces over the finite field $\mathbb{F}_{2^s}$.

Vontobel and Tanner [33] discovered a way to use finite generalized polygons (FGPs) to construct Tanner graphs and LDPC codes. This associated graph has the property that its girth is exactly twice its diameter. This is the largest possible girth.

The approach in [33] can be viewed within a similar framework, namely that the occuring points are points of a projective space and the lines form a subset of the lines in that space determined by a bilinear form. Comparable work has been done recently in [12,21,26] by exploiting quadratic forms in such spaces.

Later in this paper, we will study a construction of LDPC codes involving incidence structures that are known as *circle geometries*. Among other postulates, any three distinct points define exactly one circle of this geometry. There is also a notion of *tangent* circles, and a maximal set of circles, mutually tangent in one and the same point, is referred to as a *pencil*. Under suitable assumptions, a derived incidence structure consisting of these pencils as points, and the original circles as lines can be shown to be a triangle free partial linear space—a structure that is also known as a $(0,1)$-geometry. Our performance simulations of these LDPC codes show that they are of high quality.

Structure of This Article

Our goal is to discuss and promote geometry based constructions for LDPC codes. To keep things self-contained and easy to read, we have decided in favour of the following structure of presentation. In the first section, we will discuss basic definitions and facts from the theory of LDPC codes. In the second section, we will explain in more detail how message passing decoders work, and how a performance analysis of an LDPC code is generally done. Section 3 will explain how the quality of a given LDPC code is usually assessed. We will see how a usually vast number of Monte Carlo simulations finally yields what is called a waterfall diagram that contains relevant information regarding the performance of the code in question. Section 4 will briefly explain how incidence structures are natural sources of parity-check matrices for LDPC codes. After that we will see in Section 5 how incidence matrices of linear spaces, and particularly affine and projective spaces, yield classes of high-quality LDPC codes. In the 6th and last section, we will focus on incidence matrices of partial linear spaces. We will study codes that are based on generalised quadrangles and other triangle-free structures. We will particularly discuss recent approaches that involve triangle free incidence structures induced by finite (circle) geometries.

1. Low-Density Parity-Check Codes

Low-density parity-check codes (or LDPC codes for short) are linear block codes that possess a sparse check matrix. To get familiar with this notion, let us briefly revisit the theory of linear block codes.

Throughout the entire article let $\mathbb{F}_2$ denote the finite field with 2 elements. A (binary) *linear block code* of length n is a subspace C of the vector space $\mathbb{F}_2^n$. If C is of dimension m then we will refer to C as an $[n,m]$-code. The $[n,m]$-code C can also be thought of as the row space of a binary $m \times n$ matrix M. This matrix will be called a *generator matrix* of C.

For any message vector $v \in \mathbb{F}_2^m$ the vector vM is a codeword of C; that means encoding of messages can be done comparably efficient via matrix multiplication. There are applications however where matrix multiplication is considered too complex, and for this reason coding schemes that lead to lower encoding complexity are desirable (cf. [27]).

A *parity-check matrix* for C is a binary $k \times n$ matrix H such that C is the null space of H^T. This means that a word $c \in \mathbb{F}_2^n$ is a codeword if and only if $cH^T = 0$. So, C is the set of solutions of k equations: each codeword satisfies parity-check equations with regard to the rows of H. LDPC codes are those codes which possess a parity-check matrix H that is sparse, i.e. that has only a few nonzero entries. More precisely, we require that the number of $1's$ is small compared to the number of $0's$ in H (cf. the statement after Definition 1.2).

A further feature that a check matrix may or may not have is what is called *regularity*.

Definition 1.1. Let H be a $k \times n$ parity-check matrix of a binary code C. We call H and also C, by abuse of notation, (γ, ρ)-*regular* if:

(i) Every row of H has exactly ρ nonzero entries.

(ii) Every column of H has exactly γ nonzero entries.

Definition 1.2. The *density* δ of a regular $k \times n$ check matrix H of an LDPC code with row weight ρ and column weight γ is defined as

$$\delta := \frac{\rho}{n} = \frac{\gamma}{k}.$$

Practical applications are interested in densities $\delta \approx 6/n$, where $256 \leq n \leq 8192$.

The reader might have noticed that parity-check matrices for LDPC codes do not necessarily need to be of full rank. In fact, it turns out that for real world applications a moderate set of extra rows in the check matrix of an LDPC code can contribute to an improved performance of the implemented decoder. As a matter of fact, most of the parity-check matrices based on the geometric structures discussed later are highly redundant in the sense that they have a considerable number of dependent rows.

It is clear that our regular parity-check matrices are nothing but incidence matrices of *combinatorial designs*. We will return to this point later and first discuss a graph theoretical notion that is connected with these matrices.

Definition 1.3. The *Tanner graph* of a $k \times n$ parity-check matrix H is a bipartite graph on the vertex set $S \cup T$ where S is a set of k check vertices and T is a set of n bit vertices. An edge is drawn between check vertex $s \in S$ and bit vertex $t \in T$ if and only if the entry H_{st} of the parity-check matrix H is nonzero.

Consider the following example that illustrates the current setup.

Example 1.4. Consider the code of length $n = 7$ and dimension $m = 4$ with parity-check matrix

$$H = \begin{bmatrix} 1 & 1 & 1 & 0 & 0 & 0 & 0 \\ 0 & 0 & 1 & 1 & 0 & 1 & 0 \\ 0 & 0 & 1 & 0 & 1 & 0 & 1 \end{bmatrix}.$$

Its corresponding Tanner graph G is represented by the diagram in Figure 1.

Here, let $S := \{A, B, C\}$ and $T := \{1, \ldots, 7\}$. The boxed vertices (i.e. the elements of S) label the rows of H, and the circled vertices (i.e. the elements of T) label the columns of H. For every 1 that occurs in row s and column t of H there is an edge between check vertex s and bit vertex t in the associated graph G.

In this example, G happens to be a tree, i.e. it does not contain any cycles. It is known that LDPC codes with Tanner graphs that are trees are not overly useful in theory and applications. On the other hand, the correctness of message passing decoders used for LDPC codes can only be proved if the underlying Tanner graph is a tree. It turns out, however, that if the size of the smallest cycle in that graph is not too small then iterative message passing decoding still works reasonably well. This size, known as the *girth* of the graph, must obviously be even, as the Tanner graph is bipartite.

We conclude that for the sake of good decoding performance one of the main goals in the design of LDPC codes is to find check matrices such that the associated Tanner graph is of large enough girth, typically at least 6.

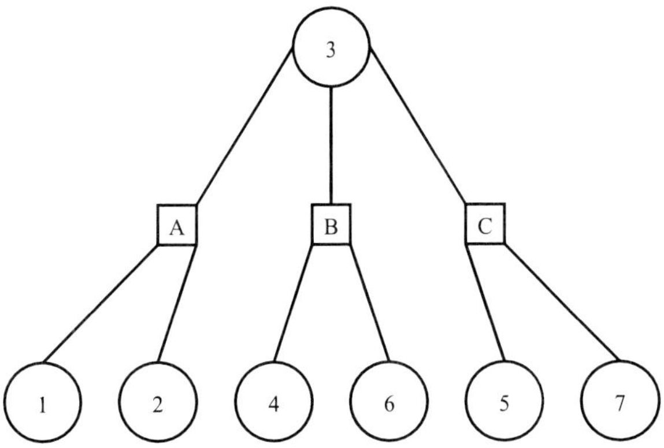

Figure 1. Tanner graph G of the parity-check matrix H.

A further important parameter measuring the quality of a code C is what is called the *minimum distance* of C. This parameter is essential for describing the error-correction capabilities of C.

Definition 1.5. The *Hamming distance* d_H of two binary words $x, y \in \mathbb{F}_2^n$ is defined as the number of positions in which these words differ:

$$d_H(x,y) := |\{1 \leq i \leq n \mid x_i \neq y_i\}|.$$

The *Hamming weight* w_H of a word x is the number of positions with a nonzero entry, which means $w_H(x) := d_H(x, 0)$ where 0 denotes the all zero vector. The *minimum distance* $d(C)$ of a code is

$$d(C) := \min\{d_H(x,y) \mid x, y \in C,\ x \neq y\}.$$

We will now try to find out more about the minimum distance of LDPC codes in terms of their check matrices.

Lemma 1.6. *The number $Z(\ell)$ of* 1*'s in a linear combination of ℓ columns of a check matrix of girth at least* 6 *with constant column weight γ satisfies*

$$Z(\ell) \geq \ell\gamma - \ell(\ell - 1).$$

Proof. We will proceed by induction and first observe that $Z(1) = \gamma$ in accordance with the claim. Assume that $Z(\ell) \geq \ell\gamma - \ell(\ell-1)$ for some $\ell \geq 1$, and assume that $\ell + 1$ distinct columns are given. Then by the assumption on the girth of the underlying matrix, the $(\ell+1)$st column shares at most one 1-entry with each of the preceding ℓ columns, and hence the number of 1's in the sum is given by $Z(\ell+1) \geq Z(\ell) + \gamma - 2\ell \geq \ell\gamma + \gamma - 2\ell - \ell(\ell-1) = (\ell+1)\gamma - (\ell+1)\ell$ which finishes the proof. □

Proposition 1.7. *For a parity-check matrix of girth at least* 6 *with column weight γ, a non-empty set of linearly dependent columns has at least $\gamma + 1$ elements.*

Proof. Assume there are ℓ linearly dependent columns in the matrix where ℓ is assumed to be minimal. Then $0 \geq Z(\ell) \geq \ell\gamma - \ell(\ell-1)$ according to the preceding proposition. Solving this for ℓ yields $\ell \geq \gamma + 1$. □

According statements can clearly be made about the rows of a regular parity-check matrix. The preceding statement has an immediate consequence for the minimum distance of the LDPC code in question.

Theorem 1.8. *The minimum distance of an LDPC code of girth at least* 6 *with column weight* γ *is at least* $\gamma + 1$.

Proof. It is well known that the minimum distance of a linear code is d, if every choice of $d-1$ columns of a parity-check matrix for this code is linearly independent but there are d linearly dependent columns. With Proposition 1.7 we conclude that the minimum distance of an LDPC code under the above assumptions is at least $\gamma + 1$. □

2. Decoding of LDPC Codes

It is known in the literature (cf. [2]) that the general decoding problem for block codes is computationally hard. Linearity of a code *can* reduce this complexity but does not do so necessarily. Low-density parity-check codes are decoded by what are called *iterative message passing decoders*. These decoders exploit the structure of the underlying Tanner graph of the given LDPC code and are efficient due to the fact that the given parity-check matrix of the code is sparse.

In order to get started we need to mention a traditional distinction between hard decision and soft decision decoding. Under hard decision, each received information bit is interpreted as either 0 or 1 at the receiving end of the channel. A word of length n consisting of symbols of the given alphabet is passed to a decoder that needs to correct possible errors and outputs a codeword that most likely has been transmitted.

Soft decision associates a certain probability distribution on $\mathbb{F}_2$ to each signal received at the end of the channel. As we are dealing with the binary case here only, this information could be represented by a single real number $p \in [0,1]$ measuring the probability that the received signal comes from a transmitted 1. For the sake of generality, we will however handle them as distributions on $\mathbb{F}_2$, i.e. as pairs of non-negative real numbers (p_0, p_1) with $p_0 + p_1 = 1$. Words of length n consisting of such pairs are then passed to the decoder, which in turn transforms these to codewords using a message passing algorithm on the Tanner graph.

We will illustrate this inspired by an example given by Wiberg [34].

2.1. The Sum-product Algorithm

Assume that at the receiving end of the communication channel, a vector r of soft information about a binary word is given in the form

$$r = \big[[.9,.1],[.6,.4],[.1,.9],[.1,.9],[.8,.2],[.8,.2],[.6,.4]\big].$$

Under hard decision, r yields the binary word $[0,0,1,1,0,0,0]$ and it is easily checked that

$$[0,0,1,1,0,0,0]\begin{bmatrix} 1 & 1 & 1 & 0 & 0 & 0 & 0 \\ 0 & 0 & 1 & 1 & 0 & 1 & 0 \\ 0 & 0 & 1 & 0 & 1 & 0 & 1 \end{bmatrix}^T = [1,0,1] \neq [0,0,0],$$

which shows that $[0,0,1,1,0,0,0]$ is not a codeword, and particularly cannot be the transmitted word.

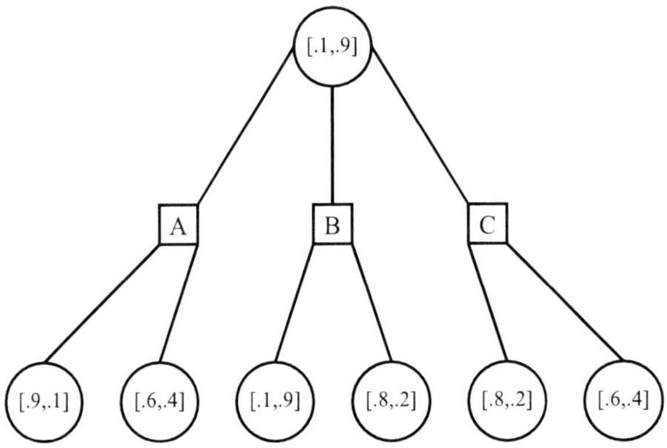

Figure 2. Sum-product algorithm – initialisation.

Figure 2 shows the initialisation and the first step of the sum-product algorithm applied to r: it stores the initial distributions r_t in the bit vertices $t \in T$, and passes these distributions as messages $p_{t,s}$ from bit vertex $t \in T$ to check vertex $s \in S$ along the edges, as is depicted in Figure 3.

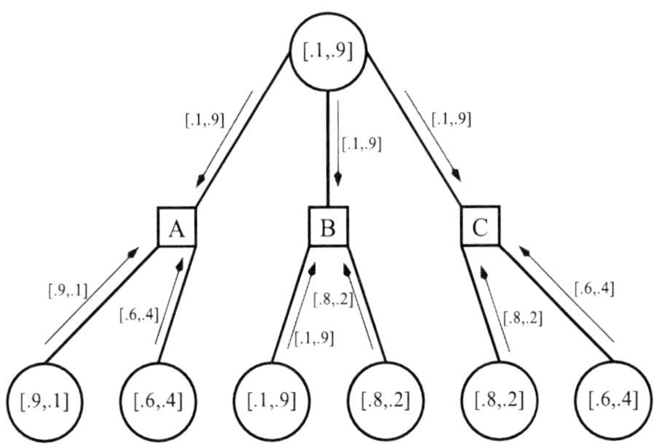

Figure 3. Sum-product algorithm – passing bit messages to the checks.

Next, as depicted in Figure 4, for each check vertex the algorithm computes convolutions of the distributions that reach the check, and passes the results back to the bit vertices.

More precisely: let $T(s)$ be the neighbourhood of check vertex $s \in S$, and denote by $p_{t,s}$ the distribution that reaches s from bit vertex $t \in T(s)$. Let $q_{s,t}$ denote the distribution that is passed back from s to t. Then

$$q_{s,t} := \bigotimes_{\substack{u \in T(s) \\ u \neq t}} p_{u,s} \quad \text{for all } t \in T(s).$$

The symbol $\otimes$ denotes the (additive) convolution of two distributions, which means

$$a \otimes b(0) := a(0)\,b(0) + a(1)\,b(1) \quad \text{and} \quad a \otimes b(1) := a(0)\,b(1) + a(1)\,b(0).$$

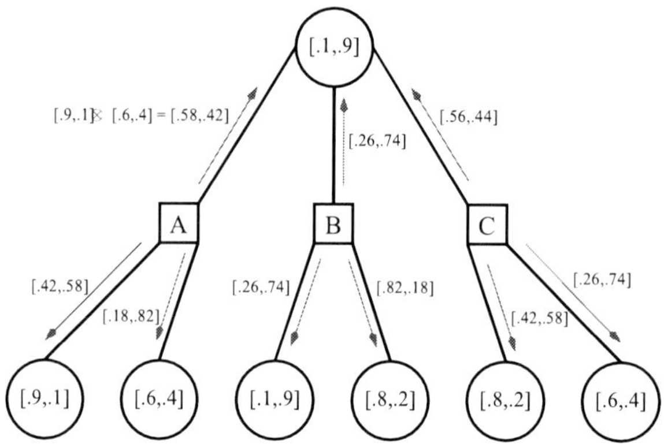

Figure 4. Sum-product algorithm – convolution step.

After convolution and passing the messages $q_{s,t}$ back to the bit vertices, the latter compute Hadamard products (componentwise multiplication) of the distributions passed to them along with the initial distribution. These products are then normalized, to make them distributions again, and then passed back to the check vertices. A (normalized) copy of the Hadamard product of all incoming distributions with the received one is kept as an update of the bit vertex distribution. All this can be seen in Figure 5.

More precisely: let $S(t)$ be the neighbourhood of bit vertex $t \in T$, and denote by $q_{s,t}$ the distribution that reaches t from check vertex $s \in S(t)$. As above, let $p_{t,s}$ denote the distribution that is passed back from t to s. Then for all $s \in S(t)$, we set

$$p_{t,s} := \frac{1}{N} r_t \cdot \prod_{\substack{v \in S(t) \\ v \neq s}} q_{v,t} \qquad \text{with} \quad N := r_t(0) \prod_{\substack{v \in S(t) \\ v \neq s}} q_{v,t}(0) + r_t(1) \prod_{\substack{v \in S(t) \\ v \neq s}} q_{v,t}(1).$$

Here $\prod$ (resp. $\cdot$) denotes the pointwise Hadamard product of two vectors, which means

$$a \cdot b(0) := a(0)\,b(0) \quad \text{and} \quad a \cdot b(1) := a(1)\,b(1).$$

At the same time, a set of updated distributions $\hat{r}$ is stored, and has value

$$\hat{r}_t := \frac{1}{N} r_t \cdot \prod_{v \in S(t)} q_{v,t} \qquad \text{with} \quad N := r_t(0) \prod_{v \in S(t)} q_{v,t}(0) + r_t(1) \prod_{v \in S(t)} q_{v,t}(1),$$

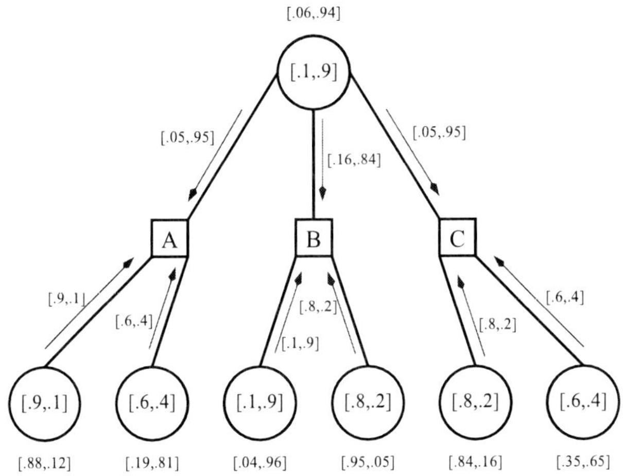

Figure 5. Sum-product algorithm – update and begin of second iteration.

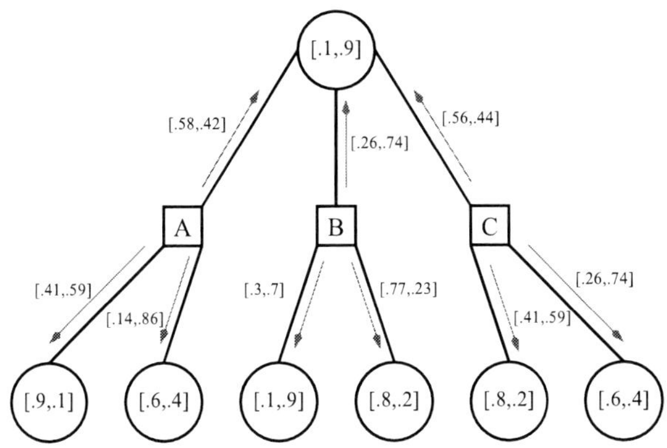

Figure 6. Sum-product algorithm – check-to-bit communication in second iteration.

for all $t \in T$.

This procedure is repeated a number of times. After each iteration, a hard decision on vector $\hat{r}$ of updated probability distributions is done. Once this yields a codeword, the algorithm terminates.

In Figure 7 we can see this: at the end of the second iteration we obtain from the vector

$$\hat{r} = \big[[.86,.14],[.2,.8],[.06,.94],[.05,.95],[.74,.26],[.93,.07],[.3,.7]\big]$$

of updated distributions the binary word $[0,1,1,1,0,0,1]$, and it is easily checked that

$$[0,1,1,1,0,0,1]\begin{bmatrix} 1 & 1 & 1 & 0 & 0 & 0 & 0 \\ 0 & 0 & 1 & 1 & 0 & 1 & 0 \\ 0 & 0 & 1 & 0 & 1 & 0 & 1 \end{bmatrix}^T = [0,0,0],$$

showing that our resulting word is a codeword. This is the appropriate time to terminate the algorithm.

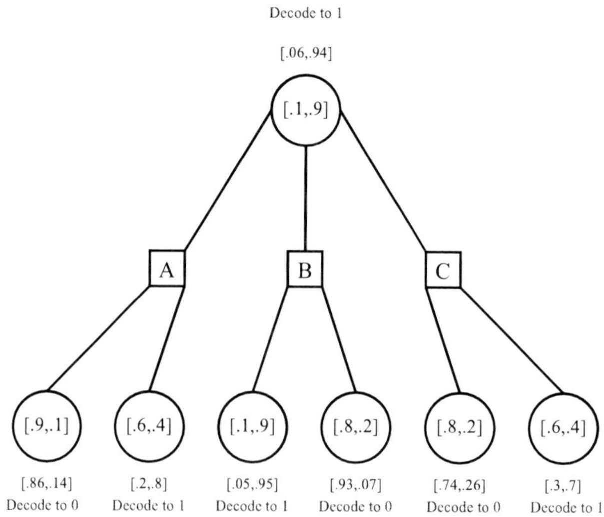

Figure 7. Sum-product algorithm – terminating step.

3. Assessing the Quality of an LDPC Code

Given an arbitrary LDPC code, there is generally no easy theoretical way to predict its performance. Despite the fact that there are some parameters giving an indication how it will perform, the only way to decide the quality of a code is by doing an extensive simulation of its behavior in a communication channel.

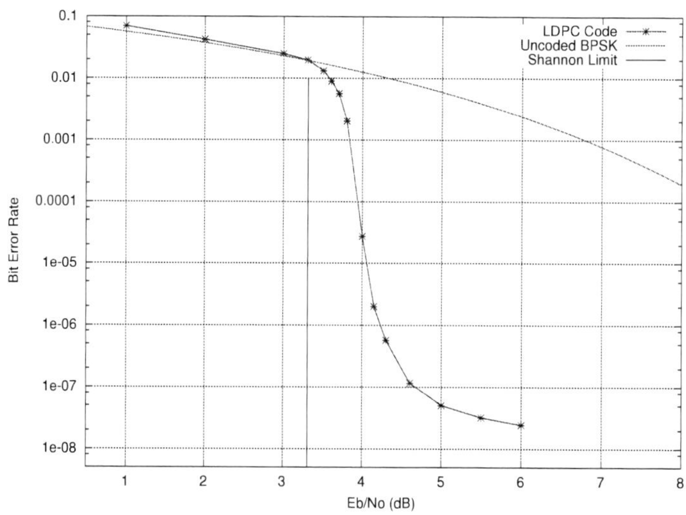

Figure 8. A waterfall diagram.

The performance for LDPC codes is generally expressed in the form of a *waterfall diagram*. This is a graphical diagram representing the *bit-error-ratio* (BER) as a function of the *signal-to-noise ratio* (SNR). Here, the bit-error-ratio is the fraction of erroneous bits in a stream of output bits, and the signal-to-noise ratio is a quantity measuring the quality of the channel. Obtaining this diagram requires simulation, namely the decoding of (putatively) received words using the code under inspection, the channel model and the decoder at the receiver.

To test a code and a decoder at a specific noise level, the simulation procedure is as follows: a message vector is randomly generated, the message is encoded using an encoder (e.g. multiplication by a generator matrix, or hopefully even more efficiently) to produce a codeword, which is modulated to create the desired signal. Noise is generated randomly according to the channel's signal-to-noise ratio and added to the modulated codeword to simulate the channel. The resulting vector of distributions is considered as received word, and then decoded using the sum-product algorithm discussed before. This procedure is repeated a number of times assuming we use the same code and decoder, but obviously varying the noise level in the channel. The results are recorded and represented in the performance diagram.

For a given code, this process of coding, transmitting and decoding must usually be repeated a massive number of times, particularly if the code is tested at a high SNR. An idea of how many times the process is repeated can be understood if we consider that for a good code the performance graph must be drawn for values of BER that range from 1 to 10^{-7}, and for higher rate to 10^{-8} and sometimes even lower down to 10^{-12} and less.

Note that values of BER around 10^{-8} mean that the decoder makes an error (decides for a message different from the transmitted one) on average every one hundred thousand received messages, where we assume a block length of $n = 1000$. Moreover the variance associated with every point of the graph depends on how many errors have been found for that particular value of SNR, so to have a realistic and precise representation of the performance of a code, billions of simulations must be carried out.

In a waterfall diagram three curves are set in relation to each other. One is a vertical line that is usually referred to as the *Shannon limit*. It marks the point on the x-axis right of which the benefit of coding should be expected. A further curve of very moderate slope marks the behaviour of a communication system in which no coding is performed at all. This curve essentially maps the signal-to-noise ratio (SNR) of the channel in a one-to-one manner to the bit-error-ratio (BER) in the received word. It might be considered as a gauging curve. The third curve, the actual waterfall, is measuring the performance of the given code on an *additive white Gaussian noise* AWGN channel. The steeper this curve, and the closer it is to the vertical Shannon limit, the better.

Most waterfall curves exhibit what is called an *error floor*. This is a region in which the plotted curve flattens out. Its slope approaches that of a horizontal line, and hence, an improvement of the channel quality does not yield any further improvement in bit error ratio that results from using the code. Regarding possible applications, it is considered to be ineffective to use a given code on a channel with SNR where the code has the error floor, or even worse, to use a code that exhibits an error floor at too high BER. Developers therefore strive for the construction of codes exhibiting error floors only at a very low bit error ratio.

4. Finite Incidence Structures and LDPC Codes

We mentioned earlier that parity-check matrices for LDPC codes can be constructed geometrically based on incidences between points and blocks of finite incidence structures. This can, as we will see, be done in more than one way. Regarding notations, we will essentially follow the presentation in [16].

Definition 4.1. Let $\mathbb{G} := (P, B)$ be a finite incidence structure consisting of a set P of n points and a set B of k blocks.

- **(a)** We define the *type-I matrix* of $\mathbb{G}$ as a binary $k \times n$ matrix $H^{(1)}$, where entry $H^{(1)}_{ij} = 1$ if point $j \in P$ is incident with block $i \in B$ and where $H^{(1)}_{ij} = 0$ otherwise.
- **(b)** The *type-II matrix* of $\mathbb{G}$ is defined as the binary $n \times k$ matrix $H^{(2)}$ where $H^{(2)}_{ij} = 1$ if point $i \in P$ is incident with block $j \in B$ and where $H^{(2)}_{ij} = 0$ otherwise.

Obviously, $H^{(2)}$ is the transpose of $H^{(1)}$, and hence, their densities in the sense of Definition 1.2 are the same. The null space of $H^{(1)}$ is called the *type-I LDPC code* of $\mathbb{G}$. It is a code of length n. Accordingly, the null space of $H^{(2)}$ is called the *type-II LDPC code* of $\mathbb{G}$, and it is a code of length k. Note, that except coming from the same incidence structure, the two codes checked by $H^{(1)}$ and $H^{(2)}$ usually have nothing in common.

We learned earlier that the girth of the associated Tanner graph of $H^{(1)}$ or $H^{(2)}$ should not be 4 in order to make the sum-product algorithm perform well. In the language of geometry, this means that the incidence structure in question should not contain 2-gons, i.e. a set of two distinct blocks that meet in more than one point. Incidence structures of this type are known as *partial linear spaces*, and we will assume this property for the remainder of this article. Partial linear spaces where every pair of distinct points is connected by exactly one line are called *linear spaces*.

5. LDPC Codes from Linear Spaces

Code constructions based on finite linear spaces were described by Kou, Lin and Fossorier [13]. That article deals with projective and affine spaces over a Galois field of characteristic 2, but the approach can be discussed for any characteristic as clarified in [16].

5.1. LDPC Codes Derived from Affine Spaces

For the finite q-element field $\mathbb{F}_q$, consider the m-dimensional affine space $\mathrm{AG}(m,q)$ which contains q^m points and $\frac{q^{m-1}(q^m-1)}{q-1}$ lines. Each line contains q points, and each point is contained in $\frac{q^m-1}{q-1}$ lines. Forming the matrices $H^{(1)}$ and $H^{(2)}$ in the fashion described above, we see that these are of density $\delta = \frac{1}{q^{m-1}}$.

The LDPC code checked by $H^{(1)}$ is a code of length $n = q^m$. According to Theorem 1.8, its minimum distance is lower bounded by $d_{\min} \geq \frac{q^m-1}{q-1} + 1$.

The type-II LDPC code checked by $H^{(2)}$ is of length $n = \frac{q^{m-1}(q^m-1)}{q-1}$. Again by Theorem 1.8, its minimum distance is lower bounded by $d_{\min} \geq q+1$.

5.2. LDPC Codes Derived from Projective Spaces

It is not surprising that the methods that we have just applied to affine spaces can also be applied to projective spaces. We will briefly discuss the properties of the resulting codes.

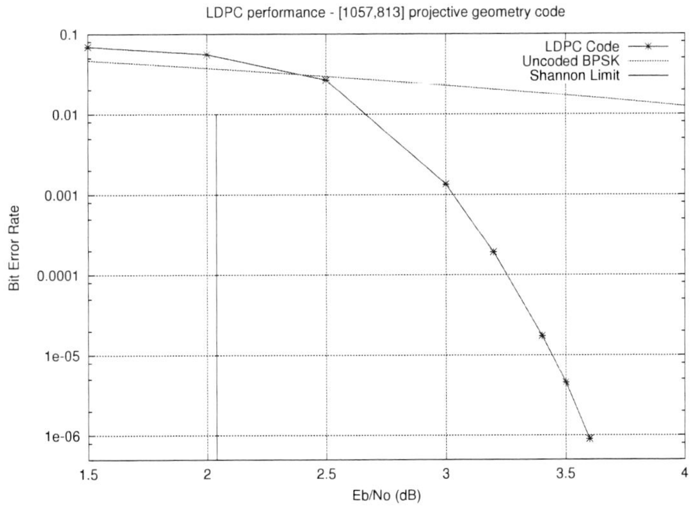

Figure 9. Projective geometry based LDPC code.

This time we consider the m-dimensional projective space $\mathrm{PG}(m,q)$ which contains $\frac{q^{m+1}-1}{q-1}$ points and $\frac{(q^{m+1}-1)(q^m-1)}{(q^2-1)(q-1)}$ lines. Each line contains $q+1$ points, and each point is contained in $\frac{q^m-1}{q-1}$ lines. Forming the check matrices $H^{(1)}$ and $H^{(2)}$, we see that these are of density $\delta = \frac{q^2-1}{q^{m+1}-1}$.

The LDPC code checked by $H^{(1)}$ is a code of length $n = \frac{q^{m+1}-1}{q-1}$. By Theorem 1.8, its minimum distance is lower bounded by $d_{\min} \geq \frac{q^m-1}{q-1} + 1$. This code was known and studied as projective geometry code long before the interest in LDPC coding arose, namely in the framework of what are called *majority-logic* decodable codes (cf. [4,20]). It can be seen that this code is equivalent to a cyclic code and hence the encoding procedure is of accordingly low complexity.

The type-II LDPC code checked by $H^{(2)}$ is of length $n = \frac{(q^{m+1}-1)(q^m-1)}{(q^2-1)(q-1)}$. Again by Theorem 1.8, its minimum distance is lower bounded by $d_{\min} \geq q+2$. This code can be put into quasi-cyclic form which again decreases the encoding complexity.

5.3. Variations and Concluding Remarks

Finite geometry has been used for constructions of LDPC codes in ways alternative to those described above. For projective or affine geometries of dimension m over $\mathbb{F}_q$ and varying $\mu \leq m-1$, Tang et al. [29] considered an induced incidence structure whose set of points is given by the set of μ-dimensional subspaces, and whose blocks are the $(\mu+1)$-dimensional subspaces of the given geometry.

The associated Tanner graphs of all the finite-geometry based LDPC code constructions discussed so far have a comparably modest girth (namely 6). Nevertheless, on the Gaussian channel under the sum-product algorithm they show a performance close enough to the Shannon limit, and may hence considered to be very good LDPC codes.

The reader may be reminded of the fact that for an understanding of the strength or weakness of an LDPC code with respect to message-passing iterative decoding, classical notions like minimum Hamming distance, or a simple reference to the girth of the underlying graph are usually not sufficient. There are further structural features of the graph like *stopping sets*, *trapping sets*, and *absorbing sets* of vertices, that are relevant. Also, a quite extensive theory of what are called *pseudo codewords* has been developed to advance the understanding of the performance of LDPC codes. For details the reader is referred to the literature (cf. [22,31,32]).

6. LDPC Codes from Partial Linear Spaces

All of the incidence structures considered in the following will belong to the class of partial linear spaces. As seen earlier, these should form the next more general class of incidence structure inducing LDPC codes of interest.

Definition 6.1. A *partial linear space* of order $(s,t) \in \mathbb{N}^2$ is an incidence structure $\mathbb{S} = (P,L)$ consisting of a set P of *points* and a set $L \subseteq 2^P$ of *lines* satisfying the following axioms:

(i) Every line of $\mathbb{S}$ contains exactly $s+1$ points and every point of $\mathbb{S}$ is contained in exactly $t+1$ lines.

(ii) Two distinct points of $\mathbb{S}$ are connected by at most one line.

Partial linear spaces have enjoyed intensive investigation during the recent 20 years. For a comprehensive treatment, the reader is referred to [6].

6.1. LDPC Codes Derived from Generalized Quadrangles

A first interesting class of partial linear spaces is that of the generalized quadrangles.

Definition 6.2. A *generalized quadrangle* is a partial linear space $\mathrm{GQ} = (P,L)$such that the following property holds: if $p \in P$ is a point that is not incident with the line $\ell \in L$ then there exists a unique point q on ℓ that is connected by a line with p.

Remark 6.3. Let $\mathrm{GQ} = (P,L)$ be a generalized quadrangle.

(a) The dual incidence structure $\mathrm{GQ}^{\perp}(L,P)$, equipped with the inverse incidence, is a generalized quadrangle. If GQ is of order (s,t) then $\mathrm{GQ}^{\perp}$ is of order (t,s). but even if $s=t$ this does not imply that these two quadrangles are isomorphic.

(b) There are $n = (st+1)(s+1)$ points and $k = (st+1)(t+1)$ lines.

(c) It can be shown that $s+t$ is a divisor of $(st+1)(s+1)(t+1)$, and that $s \leq t^2$ if $t \neq 1$, and accordingly that $t \leq s^2$ provided $s \neq 1$.

(d) The orders of the nontrivial (i.e. $s \neq 1 \neq t$) generalized quadrangles that have been found so far are (q,q), (q,q^2), (q^2,q^3) and $(q-1,q+1)$, along with their reverse pairs, where q is an arbitrary prime power.

Examples 6.4. The classical examples of the generalized quadrangles with order (s,t), $s \neq 1 \neq t$, are:

(i) The generalized quadrangle $W(q)$ of order (q,q) consisting of all the points of $\mathrm{PG}(3,q)$ and of all the totally isotropic lines under a symplectic polarity η of $\mathrm{PG}(3,q)$;

(ii) The generalized quadrangle $Q(4,q)$ of order (q,q) consisting of all the points and the lines of a non-singular parabolic quadric $Q(4,q)$ of $\mathrm{PG}(4,q)$;

(iii) The generalized quadrangle $Q^-(5,q)$ of order (q,q^2) consisting of all the points and the lines of a non-singular elliptic quadric $Q^-(5,q)$ of $\mathrm{PG}(5,q)$;

(iv) The generalized quadrangles $H(3,q^2)$ and $H(4,q^2)$ of respective orders (q^2,q) and (q^2,q^3) consisting of all the points and the lines of the non-singular Hermitian varieties in $\mathrm{PG}(3,q^2)$ and $\mathrm{PG}(4,q^2)$.

Remark 6.5. Regarding isomorphisms and dualities between these classical generalized quadrangles, the following results hold:

(a) $W(q)$, q even, and $Q(4,q)$, q even, are isomorphic.

(b) $W(q)$ and $Q(4,q)$ are self-dual if and only if q is even.

(c) $W(q)$, q odd, and $Q(4,q)$, q odd, are dual generalized quadrangles.

(d) $Q^-(5,q)$ and $H(3,q^2)$ are dual to each other.

A standard reference for generalized quadrangles is [24], where also examples of order $(s,t) = (q-1,q+1)$ are described.[1]

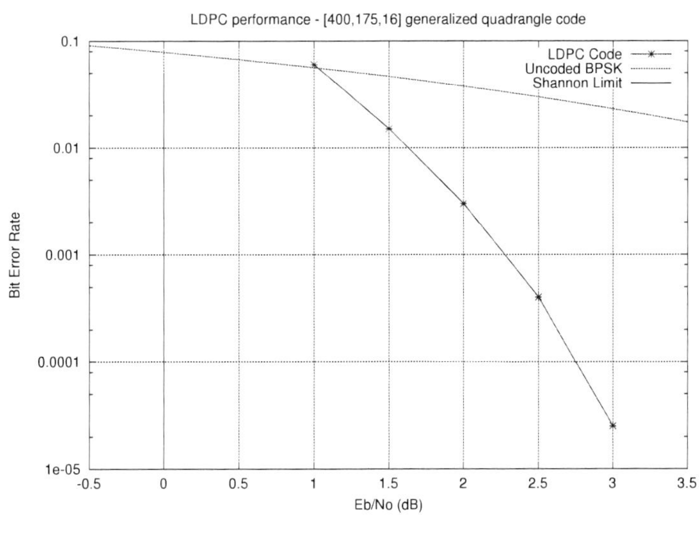

Figure 10. Performance of an LDPC code derived from a GQ of order $(7,7)$.

In 2001, Vontobel and Tanner [33] introduced codes derived from generalized quadrangles. Their examples contain order (q,q) generalized quadrangles that can be constructed

[1] see also Subsection 6.2.

based on symplectic polarities in $PG(3,q)$. For details regarding this construction the reader is referred to [6]. Figure 10 shows the performance of their $[400,175,16]$-code that is based on a generalized quadrangle of order $(7,7)$.

Remark 6.6. A further immediate generalization of these ideas (see [17] for details) is the use of generalized polygons which were introduced by J. Tits [30]. The girth of the incidence graph of a generalized n-gon is twice its diameter, namely $2n$. As mentioned earlier, this is an indicator for potentially better performance of these codes under iterative decoding algorithms.

Further Results

Further investigations in this direction have been done by Kim, Mellinger, and Storme in [12], and by Pepe, Storme, and Van de Voorde in [26].

We first mention the main results on the minimum distances of the LDPC codes defined by the classical generalized quadrangles of [12]. We recall that, by Definition 4.1, the type-I-matrix is the binary incidence matrix whose rows correspond to the lines and whose columns correspond to the points of the generalized quadrangle, while the type-II-matrix is the binary incidence matrix whose rows correspond to the points and whose columns correspond to the lines of the generalized quadrangle.

To make the table as accessible as possible to the reader, we present for the classical generalized quadrangles both the minimum distance or the lower bound on the minimum distance of the LDPC code defined by the type-I-matrix and by the type-II-matrix, and in case of equality, the description of the codewords of the smallest weight.

Table 1. LDPC codes from classical generalized quadrangles.

	GQ	d(type-I-matrix)	d(type-II-matrix)
(1)	$W(q), q=2^h$	$2(q+1)$	$2(q+1)$
(2)	$W(q)$, q odd	$2(q+1)$	$\gtrapprox \frac{(q+1)\sqrt{q}}{2}$
(3)	$Q(4,q)$, q odd	$\gtrapprox \frac{(q+1)\sqrt{q}}{2}$	$2(q+1)$
(4)	$Q^{-}(5,q)$	$\geq (q+1)(q^2-q+2)$	$2(q+1)$
(5)	$H(3,q^2)$	$2(q+1)$	$\geq (q+1)(q^2-q+2)$
(6)	$H(4,q^2)$	$\geq (q^2+1)(q^3-q^2+2)$	$\geq q\sqrt{(q^2+1)(q-1)}+q^2+2$

In case of equality for the type-I-matrix, the minimum distance of the LDPC code is obtained for the following codewords:

- for $W(q)$, q even or odd, $\text{supp}(c) = L \cup L^{\perp}$ for a non-isotropic line L,
- for $H(3,q^2)$, $\text{supp}(c) = \ell \cup \ell^{\perp}$ for a Baer subline ℓ of $H(3,q^2)$ belonging to a $(q+1)$-secant line of $PG(3,q^2)$ to $H(3,q^2)$.

In case of equality for the type-II-matrix, the minimum distance of the LDPC code is obtained for the following codewords:

- for $W(q)$, q even, $\mathrm{supp}(c) = \mathcal{R} \cup \mathcal{R}^{\mathrm{op}}$ for a regulus $\mathcal{R}$ such that both $\mathcal{R}$ and its opposite regulus $\mathcal{R}^{\mathrm{op}}$ consist completely of totally isotropic lines,
- for $Q(4,q)$ and $Q^-(5,q)$, $\mathrm{supp}(c) = \mathcal{R} \cup \mathcal{R}^{\mathrm{op}}$ for a regulus $\mathcal{R}$ such that both $\mathcal{R}$ and its opposite regulus $\mathcal{R}^{\mathrm{op}}$ are completely contained in $Q(4,q)$ or $Q^-(5,q)$.

The lower bounds on d(type-I-matrix) for $Q^-(5,q)$ and $H(4,q^2)$ most likely are not sharp. To give an idea of how these bounds relate to the smallest known weights of the codewords of these LDPC codes, we note that in [25], codewords in the corresponding LDPC codes have been constructed of respective weights $2(q^3-q^2+q)$ and $2(q^5-q^3+q^2)$; so this lower bound differs at most a factor 2 from the exact minimum distance.

6.2. LDPC Codes from Triangle-Free Geometries

What follows is motivated by the observation that the defining property of a generalized quadrangle can be weakened in order to obtain larger classes of partial linear spaces and their derived LDPC codes.

Definition 6.7. A partial linear space $\mathbb{S} = (P, L)$ is called an (α, β)-*geometry* if whenever (p, ℓ) is a non-incident point-line pair there are either α or β points on ℓ which are collinear with p.

According to this definition, generalized quadrangles are examples of $(1,1)$-geometries, and generalized polygons form a particular class of examples of $(0,1)$-geometries.

Important in this context is the fact that a $(0,1)$-geometry is a triangle-free structure. This implies that the girth of the according incidence graph is lower bounded by 8.

A large class of LDPC codes based on $(0,1)$-geometries was presented in the authors' previous work [9], where a certain class of 3-designs was used in order to construct the $(0,1)$-geometries in question. We have revisited that paper, and present its results in an improved version here.

We first recall the notions of internal structure of an incidence structure $\mathbb{I}$ with respect to a point p and of an inversive space which is a type of circle geometry.

Definition 6.8. Let $\mathbb{I} = (P, B)$ be an incidence structure and let $p \in P$ be one of its points. We define a new incidence structure $\mathbb{I}_p := (P_p, B_p)$, where

$$\begin{aligned} P_p &:= P \setminus \{p\}, \\ B_p &:= \{c \setminus \{p\} \mid c \in B \text{ and } p \in c\}. \end{aligned}$$

Then $\mathbb{I}_p$ is called the *internal structure* of $\mathbb{I}$ with respect to p.

Definition 6.9. An *inversive space* is an incidence structure $\mathbb{M} = (P, C)$ consisting of a set P of *points*, and a set C of *circles* such that the following conditions are satisfied:

(i) Any three distinct points are contained in exactly one circle.

(ii) For every point $p \in P$, the internal structure $\mathbb{M}_p$ is an affine space.

We say that two circles c and d of $\mathbb{M}$ are *touching* in p, if they both contain p and if $c \setminus \{p\}$ and $d \setminus \{p\}$ are parallel lines of $\mathbb{M}_p$.

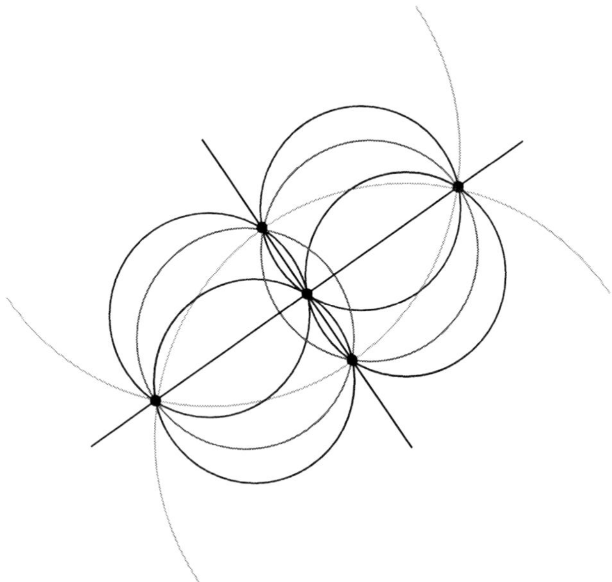

Figure 11. Illustration of the smallest inversive space of order 2 and dimension 2.

The following statements can be derived using elementary counting principles.

Remark 6.10. Let $\mathbb{M} = (P, C)$ be an inversive space. There exist positive integers q and r such that the following properties hold.

(a) All circles of $\mathbb{M}$ contain $q+1$ points.

(b) $\mathbb{M}$ contains exactly q^r+1 points. Each point is incident with exactly $q^{r-1}\frac{q^r-1}{q-1}$ circles, and for this reason $\mathbb{M}$ contains $q^{r-1}\frac{q^{2r}-1}{q^2-1}$ circles.

(c) $\mathbb{M}$ forms a 3-design with parameters $(q^r+1, q+1, 1)$.

In the preceding remark, we say q is the *order* of $\mathbb{M}$, and r is its *dimension*. For dimension 2, the above definition reduces to the definition of an *inversive plane*, also referred to as a *Möbius plane* in the literature. In this incidence structure two circles are touching if and only if they are equal or share a unique common point.

An algebraic construction of a large class of inversive spaces in terms of suitable field extensions is known. To get prepared let us agree that $\mathbb{F}x := \{\lambda x \mid \lambda \in \mathbb{F}\}$ denotes the one-dimensional subspace given by all $\mathbb{F}$-multiples of a given vector x. For elements $A = \mathbb{F}(a_1, a_2)$, $B = \mathbb{F}(b_1, b_2)$, $C = \mathbb{F}(c_1, c_2)$, and $D = \mathbb{F}(d_1, d_2)$ on the projective line $\mathrm{PG}(1, \mathbb{F})$ we recall the definition of the *cross-ratio* as:

$$\begin{bmatrix} A & B \\ D & C \end{bmatrix} := \frac{\det\begin{bmatrix} a_1 & a_2 \\ c_1 & c_2 \end{bmatrix}\det\begin{bmatrix} b_1 & b_2 \\ d_1 & d_2 \end{bmatrix}}{\det\begin{bmatrix} a_1 & a_2 \\ d_1 & d_2 \end{bmatrix}\det\begin{bmatrix} b_1 & b_2 \\ c_1 & c_2 \end{bmatrix}}.$$

This ratio takes values in $\mathbb{F} \cup \{\infty\}$.

Example 6.11. Let $\mathbb{L}:\mathbb{K}$ be a field extension of degree $r \geq 2$. We define an incidence structure $\mathbb{M}(\mathbb{L}:\mathbb{K}) = (P,C)$ such that P is the set of points of $\mathrm{PG}(1,\mathbb{L})$, and such that the 4 points $A,B,C,D \in P$ are called *concircular* if

$$\begin{bmatrix} A & B \\ D & C \end{bmatrix} \in \mathbb{K} \cup \{\infty\}.$$

Then $\mathbb{M}(\mathbb{L}:\mathbb{K})$ is an inversive space of order $|\mathbb{K}|$ and dimension r.

For $r = 2$, this reduces to the known construction for Miquelian inversive planes. In order to obtain $(0,1)$-geometries from inversive spaces, we need to consider what are called pencils.

Remark 6.12. It can easily be seen that the group $\mathrm{PGL}(2,\mathbb{L})$ is acting sharply 3-transitive on the point set P. When proving statements about (a set of) circles in $\mathbb{M}(\mathbb{L}:\mathbb{K})$ we can therefore always assume that one of these circles is defined by the points $A = \mathbb{L}(1,0)$, $B = \mathbb{L}(1,1)$, and $C = \mathbb{L}(0,1)$.

The *touching* relation in $\mathbb{M}(\mathbb{L}:\mathbb{K})$ can be nicely expressed in terms of the cross-ratio. For details see [1, p. 114].

Lemma 6.13. *Let c and d be circles in $\mathbb{M}(\mathbb{L}:\mathbb{K})$ that both contain the point P. Then c and d touch in P if and only if for all $A,A' \in c \setminus \{P\}$ and $B,B' \in d \setminus \{P\}$ there holds*

$$\begin{bmatrix} P & A \\ B & A' \end{bmatrix} - \begin{bmatrix} P & A \\ B' & A' \end{bmatrix} \in \mathbb{K}.$$

It is easy to see that if c is defined by the points $P = \mathbb{L}(0,1)$, $A = \mathbb{L}(1,0)$ and $A' = \mathbb{L}(1,1)$ and d by the points P, $B = \mathbb{L}(1,s)$ and $B' = \mathbb{L}(1,t)$, then c and d touch in P if and only if $s - t \in \mathbb{K}$. We will refer to this soon.

Definition 6.14. Let $\mathbb{M} = (P,C)$ be an inversive space of order q and dimension r, and let (p,c) be an incident point-circle pair. The set $\pi(p,c)$ of all circles touching c in p is called a *pencil* of $\mathbb{M}$. The point p is called the *carrier* of $\pi(p,c)$, and every pencil is uniquely determined by its carrier and any one of its circles.

If (p,c) is an incident point-circle pair in the inversive space $\mathbb{M}$ then the lines resulting from $\pi(p,c)$ in $\mathbb{M}_p$ form a full parallel class in $\mathbb{M}_p$. The number of circles in $\pi(p,c)$ is therefore given by q^{r-1}.

Remark 6.15. Let $\mathbb{M}$ be an inversive space of order q and dimension r.

(a) Every point of $\mathbb{M}$ is a carrier of $\frac{q^r-1}{q-1}$ different pencils.

(b) There are $\frac{q^r-1}{q-1}(q^r+1)$ different pencils in $\mathbb{M}$.

(c) Every circle of $\mathbb{M}$ is a member of $q+1$ pencils.

(d) Two distinct pencils in $\mathbb{M}$ have at most one circle in common.

We need one further element of preparation. For the proof of the following statement the authors are indebted to A. Blokhuis.

Theorem 6.16. *Let $\mathbb{M} = \mathbb{M}(\mathbb{L} : \mathbb{K})$ be an inversive space of dimension r and order q. If q is even, or q and r are odd, and π is a pencil in $\mathbb{M}$ and c a circle that does not belong to π, then there exists at most one circle $d \in \pi$ that touches c.*

Proof. Without loss of generality we may assume that we have two circles c, d of a pencil with carrier $P = \mathbb{L}(0,1)$, and that both touch a circle e that does not belong to this pencil. By Remark 6.12, we may further assume that c contains the points $A = \mathbb{L}(1,0)$ and $A' = \mathbb{L}(1,1)$, that d contains the points $B = \mathbb{L}(1,t)$ and $B' = \mathbb{L}(1,t+1)$ where $t \notin \mathbb{K}$. Finally we may assume that e contains the points A, B and $U = \mathbb{L}(1,u)$. Evaluating the corresponding cross-ratios we conclude from c touching e in A that $tu/(t-u) \in \mathbb{K}$. Likewise, we obtain from d touching e in B that $t(u-t)/u \in \mathbb{K}$, but then the product of these numbers which is given by $-t^2$ must be contained in $\mathbb{K}$. If q is even, this means $t \in \mathbb{K}$, a contradiction. If q and r are odd, then $t \in \mathbb{L}$ and $t^2 \in \mathbb{K}$ again implies $t \in \mathbb{K}$, which is a contradiction. Altogether the claim follows. □

We will construct a new incidence structure consisting of pencils and circles of the inversive space mentioned in the preceding lemma for q even. In doing so, we will take advantage of a characteristic property of these inversive spaces of even order (see Lemma 6.16): there are no three circles touching pairwise in three different points. This will ensure that the girth of the Tanner graph (cf. Definition 1.3) for these codes is at least 8 which has a positive effect on the performance of the code.

Corollary 6.17. *Let $\mathbb{M}(\mathbb{L} : \mathbb{K}) = (P, C)$ be the inversive space of dimension r and order q defined earlier. Let Π denote the set of all pencils in $\mathbb{M}$. If q is even, or q and r are odd, then the incidence structure $\mathbb{S}(\mathbb{M}) := (\Pi, C)$ where $\pi \in \Pi$ is incident with $c \in C$ if and only if $c \in \pi$, is a $(0,1)$-geometry of order $(q, q^{r-1} - 1)$.*

Like in the preceding sections we form the $k \times n$-incidence matrix $H^{(1)}$ and the $n \times k$-incidence matrix $H^{(2)}$, where $k = q^{r-1}\frac{q^{2r}-1}{q^2-1}$ and $n = (q^r+1)\frac{q^r-1}{q-1}$. These are of density $\delta = \frac{q^2-1}{q^{2r}-1}$. The quotient

$$\frac{k}{n} = \frac{q^{r-1}}{q+1} \leq 1$$

if and only if $r \leq 2$. For this reason, typically only the type-II LDPC code checked by $H^{(2)}$ will be of interest. These are of minimum distance at least $q+2$.

Example 6.18. Let $\mathbb{M}(\mathbb{F}_4 : \mathbb{F}_2)$ be the smallest Moebius space of order 2 and dimension 2. Then the induced $(0,1)$-geometry $\mathbb{S}(\mathbb{M})$ has 15 points and 10 lines. The derived parity-check matrix $H^{(1)}$ is given by

$$H^{(1)} = \begin{bmatrix} 0 & 1 & 0 & 0 & 0 & 0 & 0 & 0 & 0 & 0 & 0 & 1 & 1 & 0 & 0 \\ 0 & 0 & 0 & 1 & 0 & 0 & 0 & 1 & 0 & 0 & 0 & 0 & 1 & 0 & 0 \\ 0 & 0 & 1 & 0 & 0 & 0 & 0 & 0 & 0 & 0 & 0 & 1 & 0 & 1 & 0 \\ 0 & 0 & 0 & 0 & 0 & 0 & 0 & 1 & 1 & 0 & 1 & 0 & 0 & 0 & 0 \\ 0 & 0 & 0 & 1 & 0 & 0 & 1 & 0 & 0 & 1 & 0 & 0 & 0 & 0 & 0 \\ 0 & 1 & 0 & 0 & 1 & 0 & 0 & 0 & 0 & 0 & 0 & 0 & 0 & 0 & 1 \\ 0 & 0 & 1 & 0 & 0 & 1 & 0 & 0 & 0 & 0 & 1 & 0 & 0 & 0 & 0 \\ 0 & 0 & 0 & 0 & 0 & 1 & 1 & 0 & 0 & 0 & 0 & 0 & 0 & 0 & 1 \\ 1 & 0 & 0 & 0 & 1 & 0 & 0 & 0 & 1 & 0 & 0 & 0 & 0 & 0 & 0 \\ 1 & 0 & 0 & 0 & 0 & 0 & 0 & 0 & 0 & 1 & 0 & 0 & 0 & 1 & 0 \end{bmatrix}.$$

This matrix checks a binary $[15,6,5]$-code as the rank of $H^{(1)}$ is 9 rather than 10.

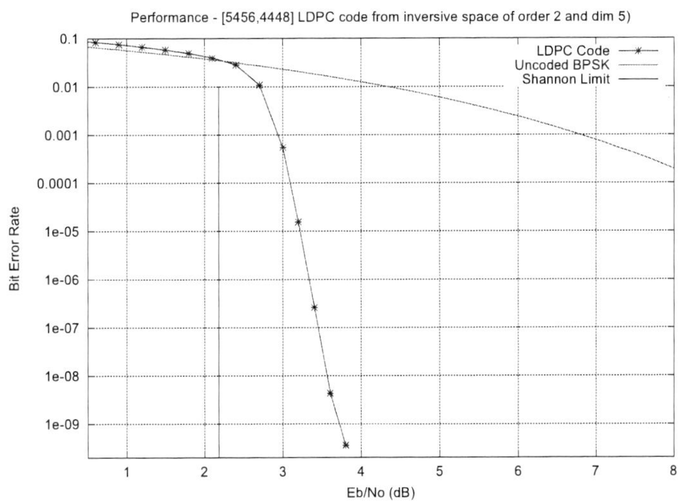

Figure 12. Performance of an LDPC code derived from $\mathbb{M}(\mathbb{F}_{2^5} : \mathbb{F}_2)$.

We conclude this section with two waterfall diagrams of type-II LDPC codes induced by inversive spaces of dimension 5 and 6 over $\mathbb{F}_2$. The resulting codes have parameters $[5456, 4448]$ and $[43680, 39603]$ respectively, and exhibit error floors only at low BER as can be seen in Figures 12 and 13. For this reason their minimum distances are expected to be better than predicted by Theorem 1.8. The resulting codes are of rate 0.82 and 0.9, respectively.

Further Constructions and Concluding Remarks

In [21], Mellinger investigated LDPC codes from triangle-free line sets. A *triangle-free line set* $\mathcal{L}$ in $AG(n,q)$ or $PG(n,q)$ is a set of lines such that no three lines of $\mathcal{L}$ form a triangle. As a consequence, the associated incidence graph has no 6-cycles, so necessarily has girth at least 8.

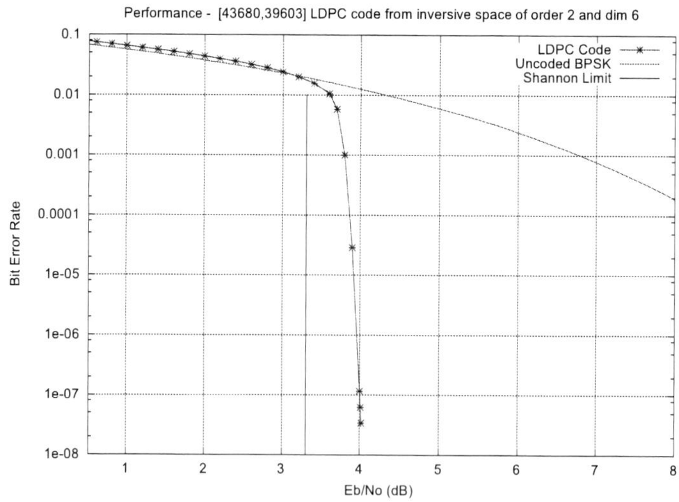

Figure 13. Performance of an LDPC code derived from $\mathbb{M}(\mathbb{F}_{2^6} : \mathbb{F}_2)$.

Mellinger first of all proves that a triangle-free line set $\mathcal{L}$ of $\mathrm{AG}(n,q)$ has at most $q^{n-1}(q^{n-2}+\cdots+q+2)$ lines.

He then considers triangle-free line sets in $\mathrm{AG}(n,q)$, constructed in the following way.

Let Σ be the n-dimensional projective space $\mathrm{PG}(n,q)$ over the finite field of order q and let H_∞ be a hyperplane of Σ such that $\Sigma' = \Sigma \setminus H_\infty$ models the affine space $\mathrm{AG}(n,q)$. Let K be a k-cap in H_∞, i.e., a set of k points, no three collinear (see e.g. [3]).

Consider now the set of lines $\mathcal{L}_K$ of Σ' whose points at infinity belong to K. Then a triangle-free line set having kq^{n-1} lines is obtained. The set of affine points of Σ' together with the set of lines $\mathcal{L}_K$ is also called the *linear representation* of the cap K in H_∞.

For q even and $n=3$, it is possible to select K equal to a hyperoval $\mathcal{O}$ of H_∞, i.e., a set of $q+2$ points, no three collinear. In this case, the set $\mathcal{L}_K$ has size $(q+2)q^2$ which is equal to the upper bound on the size of a triangle-free line set in $\mathrm{AG}(3,q)$, described above. In this case of K equal to a hyperoval $\mathcal{O}$ of H_∞, the corresponding linear representation is a generalized quadrangle of order $(s,t) = (q-1,q+1)$. In the literature on generalized quadrangles, this generalized quadrangle is denoted by $T_2^*(\mathcal{O})$.

The classical example of a hyperoval $\mathcal{O}$ is the union of a conic and its nucleus. This is called the *regular* hyperoval, and is projectively equivalent to the set of points $\{(1,t,t^2) \mid t \in \mathbb{F}_q\} \cup \{(0,1,0),(0,0,1)\}$ in $\mathrm{PG}(2,q)$, where q is even. A more general class of hyperovals is that of the *translation* hyperovals. These are the hyperovals projectively equivalent to a hyperoval $\{(1,t,t^{2^v}) \mid t \in \mathbb{F}_q\} \cup \{(0,1,0),(0,0,1)\}$, with $q = 2^h$ and with $\gcd(v,h) = 1$. But more examples of hyperovals exist. For more information on hyperovals, we refer to e.g. [8]. For the list of the known hyperovals, we refer to *Bill Cherowitzo's hyperoval page* [7].

Both Mellinger in [21] and Pepe, Storme, and Van de Voorde in [26] investigated the LDPC codes arising from the triangle-free line sets $\mathcal{L}_K$. We stress however that in [21], the description via the type-I-matrices is used, while in [26], the description via the type-II-matrices is used.

There is for every hyperoval equality for the lower bound $2q$ on the minimum distance of

Table 2. LDPC codes from linear representations of hyperovals.

	$\mathcal{O}$	d(type-I-matrix)	d(type-II-matrix)
(1)	Hyperoval	$\geq 4q$	$2q$
(2)	Translation hyperoval	$4q$	$2q$

the LDPC code arising from the type-II-matrix. Namely, take an affine plane π intersecting $\mathcal{O}$ in the two points P_1 and P_2. Then the set of the $2q$ affine planes in π through the two points P_1 and P_2 defines the support of a binary codeword of the LDPC code defined by the type-II-matrix of the linear representation of $\mathcal{O}$.

For the sharpness of the lower bound $4q$ on the minimum distance of the type-I-matrix for the linear representation of translation hyperovals, we mention the following codeword of weight $4q$ [26].

Let $q = 2^h$, $h \geq 1$. Suppose that the plane H_∞ at infinity has equation $X_2 = X_3$ and let K be the set $\{(t^{2^v}, t, 1, 1) \mid t \in \mathbb{F}_q\} \cup \{(1,0,0,0), (0,1,0,0)\}$. Let S be the set $C_1 \cup C_2 \cup C_1' \cup C_2'$, where $C_1 = \{(t^{2^v}, t, 1, 0) \mid t \in \mathbb{F}_q\}$, $C_2 = \{(t^{2^v}, t, 0, 1) \mid t \in \mathbb{F}_q\}$, $C_1' = \{(t^{2^v} + \mu, t + \mu, 1, 0) \mid t \in \mathbb{F}_q\}$ and $C_2' = \{(t^{2^v} + \mu, t + \mu, 0, 1) \mid t \in \mathbb{F}_q\}$, with $\mu \neq 0, 1$.

Let c be the vector with 1 in the coordinates corresponding to the points of $C_1 \cup C_2 \cup C_1' \cup C_2'$, and zero in the other positions. Then c is a codeword of minimum weight of the LDPC codes defined by a type-I-matrix for the translation hyperoval $\mathcal{O}$, and every such codeword of weight $4q$ arises from this construction.

We note that [26] also describes further properties of the LDPC codes defined by linear representations of geometries.

7. Open Problems

To encourage further research on LDPC codes arising from geometrical structures, we mention the following problems as specific research problems.

- Determine the exact minimum distance of the LDPC codes arising from the classical generalized quadrangles, for which the exact value is not yet known in Tables 1 and 2.

- *Partial geometries*, *semi-partial geometries*, and *(α, β)-geometries* are geometries inspired by the generalized quadrangles. For the investigation of LDPC codes arising from partial geometries or from semi-partial geometries, there has already been an incentive [11, 15]. The classes of partial, semi-partial, and more generally (α, β)-geometries contain many examples of geometries. It is of interest to investigate the properties of the linear codes they define, and to investigate which of these geometries determine the most interesting linear codes.

  This research is not only of interest for coding theory; it is also of interest for Galois geometries. Some of the most interesting results within Galois geometries have been obtained by using the links with the linear codes these geometries define.

Acknowledg nents

The authors are indebted to Marc Fossorier and Pascal O. Vontobel for many valuable suggestions that have helped to improve the quality of this article. They also wish to thank Aart Blokhuis and Geertrui Van de Voorde for the interesting discussions on inversive spaces and linear sets, and Aart Blokhuis for pointing out the proof of Theorem 6.16.

References

[1] W. Benz, *Vorlesungen über die Geometrie der Algebren*, Springer-Verlag Berlin Heidelberg New York, 1973.

[2] E. R. Berlekamp, R. J. McEliece, and H. C. A. van Tilborg, *On the inherent intractability of certain coding problems*, IEEE Trans. Inform. Theory, IT-24 (1978), pp. 384–386.

[3] J. Bierbrauer and Y. Edel, *Large caps in projective Galois spaces*, in Current research topics in Galois geometry, Nova Sci. Publ., New York, 2011, ch. 4, pp. 85–102.

[4] R. E. Blahut, *Theory and practice of error control codes*, Addison-Wesley Publishing Company Advanced Book Program, Reading, MA, 1983.

[5] J. Bond, S. Hui, and H. Schmidt, *Linear-congruence constructions of low-density parity-check codes*, in Codes, systems, and graphical models (Minneapolis, MN, 1999), 2001.

[6] F. Buekenhout, ed., *Handbook of incidence geometry*, North-Holland, Amsterdam, 1995. Buildings and foundations.

[7] B. Cherowitzo, *Bill Cherowtizo's Hyperoval Page*.
`http://www-math.cudenver.edu/~wcherowi/research/hyperoval/hypero.html`, 1999.

[8] F. De Clerck and N. Durante, *Constructions and characterizations of classical sets in* $PG(n,q)$, in Current research topics in Galois geometry, Nova Sci. Publ., New York, 2011, ch. 1, pp. 1–32.

[9] M. Flanagan, M. Greferath, and C. Rößing, *On LDPC codes from* $(0,1)$*-geometries induced by finite inversive spaces of even order*, in Proceedings of the WCC 2007, Versailles, 2007.

[10] R. Gallager, *Low density parity check codes*, MIT press, Cambridge, MA, 1963.

[11] S. J. Johnson and S. R. Weller, *Codes for iterative decoding from partial geometries*, in Proc. IEEE Int. Sym. Inform. Theory, Switzerland, June 30 - July 5, 2002.

[12] J.-L. Kim, K. E. Mellinger, and L. Storme, *Small weight codewords in LDPC codes defined by (dual) classical generalized quadrangles*, Des. Codes Cryptogr., 42 (2007), pp. 73–92.

[13] Y. Kou, S. Lin, and M. P. C. Fossorier, *Low-density parity-check codes based on finite geometries: a rediscovery and new results*, IEEE Trans. Inform. Theory, 47 (2001), pp. 2711–2736.

[14] J. Lafferty and D. Rockmore, *Codes and iterative decoding on algebraic expander graphs*, in Proceedings of the ISIT-A 2000, 2000.

[15] X. Li, C. Zhang, and J. Shen, *Regular LDPC codes from semipartial geometries*, Acta Appl. Math., 102 (2008), pp. 25–35.

[16] S. Lin and D. J. Costello Jr., *Error Control Coding: Fundamentals and Applications*, Prentice-Hall, Englewood Cliffs, NJ, 1983.

[17] Z. Liu and D. A. Pados, *LDPC codes from generalized polygons*, IEEE Trans. Inform. Theory, 51 (2005), pp. 3890–3898.

[18] D. MacKay and R. Neal, *Good codes based on very sparse matrices*, in Cryptography and Coding, 5th IMA Conference, 1995, pp. 100–111.

[19] G. Margulis, *Explicit constructions of graphs without short cycles and low density codes*, Combinatorica, 2 (1982), pp. 71–78.

[20] J. L. Massey, *Threshold decoding*, Massachusetts Institute of Technology, Research Laboratory of Electronics, Tech. Rep. 410, Cambridge, Mass., 1963.

[21] K. E. Mellinger, *LDPC codes from triangle-free line sets*, Des. Codes Cryptogr., 32 (2004), pp. 341–350.

[22] A. Orlitsky, R. Urbanke, K. Vishwanathan, and J. Zhang, *Stopping sets and the girth of Tanner graphs*, in Proc. IEEE International Symposium on Information Theory, Lausanne, Switzerland, 2002, p. 2.

[23] M. O'Sullivan, M. Greferath, and R. Smarandache, *Construction of LDPC codes from affine permutation matrices*, in Proceedings of the 40th Annual Allerton Conference on Communication, Control and Computing, 2002.

[24] S. E. Payne and J. A. Thas, *Finite generalized quadrangles*, EMS Series of Lectures in Mathematics, European Mathematical Society (EMS), Zürich, second ed., 2009.

[25] V. Pepe, L. Storme, and G. Van de Voorde, *On codewords in the dual code of classical generalised quadrangles and classical polar spaces*, Discrete Math., 310 (2010), pp. 3132–3148.

[26] ——, *Small weight codewords in the LDPC codes arising from linear representations of geometries*, J. Combin. Des., 17 (2009), pp. 1–24.

[27] T. J. Richardson and R. L. Urbanke, *Efficient encoding of low-density parity-check codes*, IEEE Trans. Inform. Theory, 47 (2001), pp. 638–656.

[28] J. Rosenthal and P. O. Vontobel, *Constructions of LDPC codes using Ramanujan graphs and ideas from Margulis*, in Proc. of the 38th Allerton Conference on Communication, Control, and Computing, 2000, pp. 248–257.

[29] H. Tang, J. Xu, Y. Kou, S. Lin, and K. Abdel-Ghaffar, *On algebraic construction of Gallager and circulant low-density parity-check codes*, IEEE Trans. Inform. Theory, 50 (2004), pp. 1269–1279.

[30] J. Tits and R. M. Weiss, *Moufang polygons*, Springer Monographs in Mathematics, Springer-Verlag, Berlin, 2002.

[31] P. O. Vontobel. `http://www.pseudocodewords.info`.

[32] P. O. Vontobel, R. Smarandache, N. Kiyavash, J. Teutsch, and D. Vukobratovic, *On the minimal pseudo-codewords of codes from finite geometries*, in Proc. IEEE Intern. Symp. on Inform. Theory 2005, Adelaide, Australia, 2005, pp. 980–984.

[33] P. O. Vontobel and R. M. Tanner, *Construction of codes based on finite generalized quadrangles for iterative decoding*, in Proc. IEEE Intern. Symp. on Inform. Theory, Washington, D.C., USA, 2001, p. 223.

[34] N. Wiberg, *Codes and Decoding on General Graphs*, PhD thesis, Linköping University, Sweden, 1996.

Index